Cereal-Based Foodstuffs: The Backbone of Mediterranean Cuisine

Fatma Boukid
Editor

Cereal-Based Foodstuffs: The Backbone of Mediterranean Cuisine

Editor
Fatma Boukid
Food Safety and Functionality Programme, Food Industry Area,
Institute of Agriculture and Food Research and Technology (IRTA)
Catalonia, Spain

ISBN 978-3-030-69227-8 ISBN 978-3-030-69228-5 (eBook)
https://doi.org/10.1007/978-3-030-69228-5

This Springer imprint is published by the registered company Springer Nature Switzerland AG
The registered company address is: Gewerbestrasse 11, 6330 Cham, Switzerland

Preface

This book provides an updated compilation of scientific literature focused on cereals-based products from the Mediterranean region, with emphasis on market trade, breeding history, processing, food quality, health aspects and safety of cereal-based products, offering a compendium of information from past to present.

Going through each chapter, the authors have made a lot of efforts to cover all aspects of **Cereals-based products from the Mediterranean Cuisine** in a precise, readable, comprehensive and critical approach to provide a full picture to the readers from farm to fork. This work is based on wide scientific references to ensure that the readers will get advantage of focused information that can be easily located.

This book will provide valuable and useful insights to market reporters, breeders, food developers, cereal scientists, food scientists, nutritionists and policymakers, and particularly to those dealing with cereals and grains sciences. This book also englobes the evolution of the Mediterranean cuisine from ancient to modern cereals, from traditional processing to modern processing, from myths to scientific evidence. This information will be extremely helpful to draw food sciences courses and to complement academic training of cereal science and technology.

I would like to thank all authors for their trust and valuable contribution for the preparation of the book. Through each chapter, each author gave valuable comprehensive review based on his expertise on the field. I would like also to thank the authors, editors and publishers who have allowed reproduction of some of the illustrations and tables included in the book.

Finally, I would also like to thank the editorial and production team at Springer Nature for their time, effort, advice and expertise. I hope that you will enjoy reading the book, and that it will serve as valuable timeless source of knowledge about cereals for the reader and as a register and a proof of the richness of the Mediterranean Cuisine.

Catalonia, Spain Fatma Boukid

Contents

Chapter 1
Cereals of the Mediterranean Region: Their Origin, Breeding History and Grain Quality Traits

Marina Mefleh

Abstract Wheat and barley are the principal founder crops in the agriculture world and they have been a part of the Mediterranean agriculture history for 8000 years. Today, wheat is the major cereal crop cultivated and consumed in the Mediterranean region. Rice was introduced to the Mediterranean agriculture in the seventh century and today plays an important socio-economic role and is a staple food for many Mediterranean countries. During the 'Green Revolution', wheat and rice were the target crops for ensuring food security of the rapid population growth. Both crops were subjected to intensive breeding with an emphasis on increasing their production and their resistance towards the biotic and abiotic stresses exerted by the rain-fed conditions of the Mediterranean region. In the last decade, improving technological and nutritional grain qualities became a big challenge for breeders in response to consumers and industrials needs. The present chapter reviews the pathways along which wheat, barley and rice were introduced to the Mediterranean region with focus on the major breeding progress that these crops went through. The introduction of modern varieties contributed into feeding the world but intensive farming system had a drastic impact on the environmental sustainability. Overall, breeding succeeded in identifying wheat grain quality traits thereby selecting improved varieties with peculiar properties. Nevertheless, the progress in rice breeding is still limited, where more research is required for selecting improved varieties.

Keywords Agriculture history · Breeding · Grain quality · Grain yield

M. Mefleh (✉)
Dipartimento di Agraria, Sezione Agronomia, Università degli Studi di Sassari, Sassari, Italy
e-mail: mmefleh@uniss.it

F. Boukid (ed.), *Cereal-Based Foodstuffs: The Backbone of Mediterranean Cuisine*, https://doi.org/10.1007/978-3-030-69228-5_1

1 Old World Agriculture

1.1 Wheat and Barley

The Neolithic founder cereal crops einkorn (*Triticum Monococcum* subsp. *Monococcum*), emmer *(Triticum Turgidum* subsp. *Dicoccum)* and barley *(Hordeum Vulgare* subsp. *Spontaneum)* were firstly domesticated 12,000 years ago from their wild relatives *Triticum Monococcum, Triticum Turgidum* subsp. *Dicoccoides and Hordeum vulgare L.* subsp. *spontaneum (C. Koch) Thell,* respectively in the Fertile Crescent region in the Middle-East spanning modern-day; Egypt, Jordan, Lebanon, Palestine, Israel, Western Syria, south turkey, Iran, Iraq and Cyprus and from there their cultivation spread to the world [1–5]. The spread of wheat and barley domestication reached the Mediterranean region nearly 8000 years ago [6] Barley and emmer were the predominant crops in Mesopotamia for Babylonians, Assyrians and Egyptians [3]. Back then, the cultivation of barley was greater than the one of emmer. Barley was the basic food in the diet of the gladiators of the Roman Empire and was considered the God food for Sumerians. Barley was principally cultivated in the Israel—Jordan area and from there its cultivation spread to the Mediterranean region.

Emmer was the first cereal to be used by Egyptians for oven bread-making [7, 8]. During the domestication process, emmer originated many tetraploid *subsp*ecies among which durum wheat (*Triticum Turgidum* subsp. *Durum*), the preferred raw material used for making traditional and staple Mediterranean dishes e.g. Pasta, couscous and bulgur. In many countries of the Mediterranean region and especially in south Italy, Lebanon, Jordan, Egypt and Syria, durum wheat is used also for making traditional bread [5, 9–11]. Bread or common wheat (*Triticum turgidum* subsp. *Aestivum*), used mainly to make bread and pastries, is the most cultivated wheat specie and exist only in a domesticated form. A natural hybrid cross of a domesticated form of Emmer and a wild diploid specie of grass *Ae. Aegilops* has led, 9000 years ago, to the birth of the hexaploid wheat specie *Triticum Aestivum* subsp. *Spelt*, the progenitor of common wheat [12]. By the first century, wheat replaced barley as a main cereal in human diet limiting barley's use to animal feeding and brewing [13]. Since theses earliest times down to the present day, wheat is a staple food in the Mediterranean region and today, its cultivation occupies 27% of the arable land [6, 14]. Wheat firstly reached Greece through Anatolia and then was shipped to south Italy and Spain, while it reached Egypt through Israel and Jordan and from there it was spread to North-Africa [2, 6, 15]. According to Moragues et al. [16] wheat reached the Iberian Peninsula either through South East Europe or North Africa.

The cultivation of emmer started to reduce at the end of the Bronze Age and at the beginning of the twentieth century it was dominated by durum and bread wheat cultivation [17]. Einkorn, emmer and spelt are still grown on a limited scale, in marginal and low fertile lands in some Mediterranean countries, though interest in these ancient wheat species has been growing of late.

1.2 Rice

The origin of rice is still a debate. Two types of domesticated rice, *Oryza (O.) sativa* and *O. glaberrima*, are cultivated worldwide and have two different origins. *O. sativa* was domesticated from the wild relative form *O. rufipogon* in China 8000 years ago and today, is cultivated all over the world. However, *O. glaberrima* was domesticated much later from the wild relative form *O. barthii* in West Africa and has been cultivated there for the last 3500 years [18, 19]. During the long-term domestication and adaptation to various ecological conditions, *O. Sativa* evolved into two major subpopulations, Japonica (short grain) and Indica (long grain) [20–22]. It is believed that the dispersal of the crop into the Mediterranean region basin occurred through Greece, by Alexander the Great, where it was considered a medicine to cure intestinal diseases at the time of Theophrastus (370–285 BC), Dioscuries (first century) and Galenos (AD 130–200). In the seventh century, it was introduced to Egypt by the Arabs or Indians and then reached the North African countries. In the eighth century, rice was brought to the Iberian Peninsula by the Arabs and at the end of the thirteenth century, rice was introduced to France. Until 1930, Rice in France was mainly grown with the purpose of building an agricultural field for other crops cultivation (mainly vines) to prevent salt marshes and it is only in the late 1500s, in the Camargue city, that rice cultivation for edible use started. According to literature, the first documented introduction of rice to Italy was by Spanish people during their visit to Naples in the fifteenth century [21, 23, 24]. Back then, rice was considered a 'luxurious' food. According to archeological evidences, rice existed in the Mesopotomia since the twelfth century BC where it was consumed as bread, porridge or cake. However, it was considered a marginal crop due to the intensive labour and high water requirements of its cultivation, which limited, its spread in the Mediterranean region until developing new technologies for water use efficiency [24].

2 From Farming Revolution to the Green Revolution

Since the dawn of agriculture, cereal crops were subjected to continuous genetic selection to improve useful traits for the humans, mainly securing a high productivity with better quality and least labour possible, and to secure crop adaptation to the environmental conditions. At first, it was an unconscious genetic selection exerted by farmers through selecting the best seeds for the next season generating the so called 'landraces' and then, a conscious genetic selection was applied by breeders using the Mendel genetic laws of inheritance and modern techniques. During the long-term domestication, new traits evolved and differentiated the cultivated varieties from their respective progenitors. The main cereal crops evolution traits were loss of spike shattering at maturity preventing seed loss at harvesting, change from hulled form to free-threshing form by loss of tough glumes, increase in seed size, reduction in number of tillers and change from long to short awns [5, 25–27].

In the mid-sixties, the green revolution focused on increasing the productivity of cereal crops to ensure food security in developing and many developed countries, where wheat and rice (Oryza sativa L.) were the protagonists. This revolution was characterized by the improvement in agriculture and management techniques coupled with a massive use of synthetic fertilizers (industrial ammonia replaced animal manure) and the introduction of a semi-dwarf gene into wheat, rice and barley resulting in a drastic reduction in the plant height [5, 28, 29]. The *Rht-1* gene identified from the Japanese wheat 'Norin' was introduced into wheat by the Nobel laureate Norman Borlaug in the International Maize and Wheat Improvement Center CIMMYT® (www.cimmyt.org) [30, 31], *sd1* gene identified from the Chinese variety 'Dijiaowujian' was introduced into rice in the International Rice Research Institute (IRRI), Asia [21]. As for barley, many semi-dwarf genes from different origins were identified, however, *sdw1* gene, originated from Japan, and *uzu1* gene, originated from the Norwegian variety 'Jotun', are the ones mostly used in barley improvement [29, 32]. The semi-dwarfing varieties released and called herein 'modern' varieties were short in size, resistant to lodging, highly responsive to nitrogen (N) fertilization supply and resistant to many pests [33]. Consequently, the green revolution resulted in a great improve in the crops harvest index. According to Lantican [30], the adoption of modern wheat varieties for cultivation in 2014 was more than 97%. Contrasting results on the performance of modern barley varieties were reported due to some undesirable agronomic traits, such as a decrease in grain size, lateness in flowering and sometimes a reduction in the quality of malt [34–36].

The Italian wheat gene pool is the most rich and diverse in the Mediterranean basin as a result of their initially richness in landraces and their subsequent use in intensive wheat breeding programs during the first half of the twentieth century [5, 37]. Thanks to the Italian geneticists and breeders and especially Nazareno Strampelli (1866–1942), Italy was the first country in the Mediterranean region—in the late nineteenth century- to start breeding work programs and consequently to release improved varieties that became the backbone of the newest varieties released in the Mediterranean region. Strampelli was the first geneticist to introduce a dwarfing gene into wheat genome. In nearly 1913, he introduced the dwarfing gene *Rht8* from the Japanese landrace 'Akakomugi', for reduced height, and *Ppd1*, for photoperiod insensitivity to bread wheat. The newly released varieties were a great success allowing Italy to be almost self-sufficient in the production of wheat. Senator Cappelli, a Mediterranean pure line durum wheat variety selected by Strampelli from the North African population Jneh el Khotifa in 1915, was a great success for its grain yield and quality and its cultivation spread to many other Mediterranean countries e.g. Spain and Turkey [37].

3 Breeding Challenges After the Green Revolution

From the beginning of breeding onwards, the Mediterranean region is involved mostly in wheat breeding activities, and especially in durum wheat. 60% of the area devoted to durum wheat cultivation is present in the Mediterranean region [1, 6].

From 'Riz au Gras' in the Camargue, France to the 'Risotto alla Milanese' in Italy, 'Paella Valenciana' in Spain, 'Tunisian Djerba Rice' in Tunisia and 'Stuffed Grape Leaves and Zucchini' in Greece, Lebanon and Syria, rice is considered a key ingredient for many traditional Mediterranean dishes. Italy is the biggest rice producer and exporter in Europe [38]. Egypt is the largest rice producer in the Middle East and North Africa region. Additionally, many Mediterranean countries other than Italy, e.g. Spain and Egypt are important rice exporters and therefore rice production impacted considerably their economy [39]. Barley is the predominant food crop in highlands, marginal and driest areas of the Mediterranean region (*e.g.* Tunisia), where it mainly used in animal feeding and brewing [40–43].

Until the advent of the green revolution, wheat and rice breeders' main task was developing high-yielding varieties resilient to biotic (pest, weeds, fungi) and abiotic (drought, salinity, high temperatures) stresses. As global food security was improved and reached in most developing countries, after the release of high-yielding modern varieties, and the standard of living was improved, consumers became more concerned about their health and more conscious about the grain quality [44]. The increase in environmental degradation caused by the intensified conventional cereal farming raised a worldwide red flag, which resulted in the re-introduction of low-input management practices for sustainable and resilient agricultural production systems [45, 46]. Additionally, modern varieties are characterized by a genetic uniformity and suitability only to conventional farming system and generally to specific favorable environments, that do not represent the diversity of local conditions. The massive adoption of these varieties has left old varieties forgotten and at risk of extinction which resulted in a huge loss in the crops genetic diversity and a decline in crops yield in marginal areas [47–50].

If the same pattern of population growth and food consumption continue, crop production must be increased by 60% by 2050 to achieve food security [51, 52]. However, a decline or stagnation in the growth trend of wheat and rice production was seen in the last decade due to the negative impacts of climate change on grain yield and quality [53–55]. The consequences of climate change- water shortage or drought, raise in temperatures and salinity, climate variability and environmental degradation- are expected to be even more pronounced in the future. An additional increase in temperature of 1.0–1.5 °C is predicted for the 2021–2050 period for the Mediterranean countries [56]. Thus, guaranteeing world food security while adopting a sustainable agriculture system has become the main strategic concern worldwide.

In response to the continuous emerging environmental and biotic threats and to the consumers' desire for functional, old and traditional foods, perceived as healthier, the main focus of research and breeding programs became to broaden the genetic diversity of wheat and rice crops in order to develop varieties highly adapted and resistant to climate variations and diseases with improved grain quality traits (without scarifying grain productivity) and to understand better the genotype per environment interaction—the behavior of crop varieties in relationship to their tolerance to environmental stresses.

3.1 Improving Grain Quality Traits for a Superior End Use Quality

3.1.1 The Case of Wheat

Wheat is the oldest crop cultivated in the Mediterranean region [1]. Today, France is the largest producer of total wheat in the Mediterranean region (35 MT in 2017/2018), followed by Turkey and Egypt (Table 1.1). Mediterranean countries are the largest importer and consumers of durum wheat end products (*e.g.* semolina, bread, pasta, couscous, and bulgur). With an average production of 4.4 million tons (MT), Italy is leader producer of durum wheat in the Mediterranean region followed by France (1.9 MT), Greece (1.07 MT) and Spain (1 MT) [27, 57–59].

Research and development in wheat crop breeding is conducted by public universities, national institutions of the Mediterranean countries, e.g. Council for Research in Agriculture and Economics - Research Centre for Cereals and Industrial Crops (CREA-CI) and the National Research Council (CNR) in Italy, Institute for Food and Agricultural Research and Technology (IRTA) in Spain, Hellenic Agricultural Organization—DEMETER Institute of Plant Breeding & Genetic Resources (HAO ELGO) in Greece, Institut National de la Recherche Agronomique (INRAE) in France, as well as international organizations such as International Center for Advanced Mediterranean Agronomic Studied (CIHEAM), International Maize and Wheat Improvement Center (CIMMYT) and Center for Agricultural Research in the Dry Areas (ICARDA), present in the region. Worldwide, CIMMYT holds the greatest wheat collection followed by USDA-ARS genebanks, the Australian Grains Genebank and ICARDA [61]. One of the main reasons behind the worldwide high interest in wheat crop is the versatility of its end products beside its nutritional composition. The quality of wheat based-products depends primarily on the grain protein content and composition. In fact, grain protein percentage contributes the most (40%) into the EU Quality index for durum wheat (European Commission Regulation No. 2237/2003, December 23, 2003), followed by gluten strength (30%). Wheat grain protein content is one of the main parameters that

Table 1.1 Average wheat production in Millions of tons per hectare of the main wheat producers in the Mediterranean region in 2017/2018 [60]

Country	Average production (millions of tons)
France	35
Turkey	20
Egypt	8.8
Spain	7.9
Morocco	7.3
Italy	6.9
Algeria	4
Tunisia	1.5
Syria	1.2
Greece	1

affect the price of the grains and flour/semolina exchanged between farmers, millers and bakers [62, 63]. A minimum grain protein percentage of 12–15 is required in the pasta industry while a minimum grain protein percentage of 10–11 is required by bakers for bread-making. Though, Furtado et al. [64] identified a new gene that is expressed in common wheat developing seeds, called the wheat bread making (*wbm*) gene associated with a superior bread-making quality.

Wheat grain proteins (8–20% of total dry weight) consist of metabolic (20–25%), albumins and globulins, and storage proteins (75–80%), gluten. This latter is classified into gliadins and glutenins [27]. Gliadins are mainly responsible of dough viscosity and extensibility while glutenins- subdivided into high and low molecular weight (HMW and LMW) glutenins subunits (GS)- are responsible of the dough elasticity and strength [65, 66]. The allelic polymorphism of gluten proteins accounts for 38% of the genetic variability of dough strength. Modern varieties have a lower proteins percentage than their respective old varieties due to grain dilution effect mediated by an increase in grain number per unit area [67]. However, the technological and textural properties of wheat end-products were not modified because the lower grain protein percentage was counterbalanced by an improvement in the gluten strength (defined as the ability of the wheat grain proteins to form a satisfactory viscoelastic matrix and specify the wheat end-use suitability) through the incorporation of improved gliadin and glutenin alleles present on the gene locus *Glu-A1*, *Glu-B1* and *Glu-B3*. As a result, modern varieties have a lower gliadin to glutenin ratio, and so the extensibility over the strength, and a higher gluten index (GI), a parameter used to assess the strength of the gluten [57, 68–73].

Some studies reported a similar quality of pasta made with emmer, old and modern durum wheats [74–76]. This could be due to the fact that besides from grain protein content and composition, the quality of pasta is highly affected by the drying and cooking conditions. In fact, under high and ultra-high temperature the quality of cooking and pasta is controlled by only protein content independently from the gluten strength [77]. The color of pasta is an important criterion in the pasta market which fueled wheat breeders to improve the grain carotenoid content of modern durum wheat through the introgression of a yellow pigment gene (Yp) [37, 70, 78]. However, De Vita et al. [57] did not find differences in semolina color between old and modern varieties.

Mefleh et al. [44] compared the productivity and grain and bread-making quality of a set old and modern durum wheat varieties grown under low input conditions and reported that old varieties could achieve a yield similar to that of modern varieties coupled with a higher grain nitrogen content and percentage. Additionally, the quality of old durum wheat bread was superior to that of modern bread.

Common bread wheat has a lower grain protein percentage and longer shelf-life than that of durum bread wheat. The difference in protein percentage is due to the higher grain number of common wheat compared to durum wheat resulting in lower grain protein percentage [79–81]. However, durum bread wheat has a low loaf volume as a result of its tenacious gluten, which triggered genecists to introduce the Glu-D genome, present in common wheat and responsible of dough elasticity and bread volume, to old and modern durum varieties. The properties of old durum wheat based-bread improved, while those of modern varieties did not change [44, 82].

Environmental conditions are important factors influencing wheat grain quality traits [55, 83]. Grain proteins percentage may increase or decrease depending on the level and duration of the heat stress [56]. For instance, high temperature and water deficit decrease the duration of gain filling period thereby increase protein content, while the rate of grain starch accumulation is not affected [84]. In a set of old and modern durum wheat varieties grown under high temperatures caused by a late sowing date, Giunta et al. [56] showed that the protein percentage of the grain and semolina of the old durum wheat varieties were higher than that of the modern ones. Domestication caused an insignificant loss of genetic diversity overtime, however, breeding for elite and high-yielding varieties caused a wheat genetic erosion. Most importantly, breeding for a stronger gluten has reduced considerably the genetic variability of gluten characteristics [49, 85]. The higher wheat allelic variability presents today belong to the HMW-GS [85–88]. Analyzing the allelic variants of gluten proteins is crucial to understand the relationship between the genetic type of gluten proteins and the quality of dough and wheat end-products and the effect of environment on the quantity of storage proteins in order to create genotypes with different superior allelic combination [87].

In the last decade, breeders are rediscovering ancient and old wheat varieties and numerous studies comparing the performance and quality of ancient, old and modern wheat varieties are carried out. This high interest in these species is due to the great sustainability of ancient and old wheats cultivation and their high genetic variability [6]. They are late in flowering, susceptible to lodging and do not benefit from high sowing and nitrogen rates and so they perfectly adapt and perform well in low fertile soil and marginal areas where the productivity of modern varieties is scarce [66]. Consequently, ancient and old wheat varieties could play an important role in improving again the biodiversity, achieving food security and could help local communities to deal better with climate change. Even though a genetic diversity in modern germplasm exists, the creation of genotypes with improved traits for yield stability and quality requires the exploration of ancient and old wheats germplasm pool, and the introduction of the favorable traits into the modern elite germplasm [89]. An interspecific hybridization between an emmer and a durum wheat was made recently creating a new set of modern emmer cultivars characterized by an improved gluten strength suitable for making a good pasta [57]. The Mediterranean region, hold an important and rich germplasm of ancient, landraces and old wheats, and national and international institutes made sure to preserve the legacy of the wheat crop [90].

Moragues et al. [16] and MacKey et al. [91] identified two clusters of durum landraces in the Mediterranean basin, one belonging to the North and East of the basin and the other one to North Africa and the Iberian Peninsula. Soriano et al. [92] found a higher diversity in the western region of the basin (West Mediterranean countries and West Balkan and Egypt) than in the eastern one (East Balkan and Turkey and East Mediterranean countries) [92]. According to Balfourier et al. [93], bread wheat belonging to the Mediterranean region can be grouped according to their geographical distribution into; western and southern Europe, the eastern Mediterranean Basin, North Africa, Turkey, the Balkans and France.

Migliorini et al. [94], compared the agronomic performance, adaptability and grain quality of a set of landraces, old and modern wheat cultivated in an organic farming system. Even though landraces and old genotypes had a lower yield than modern ones, they were more adapted in organic farming system. Additionally, the majority of consumers voted for old wheat bread. Results of human trials comparing the health benefits of ancient vs. modern wheats products revealed that the intake of ancient wheat products improves antioxidant compounds, blood glucose and lipid profile in the body [95]. However, in order to confirm the health benefits and the claims of ancient wheat products, more studies need to be performed using a vast range of genotypes (ancient and modern) due to the wide variation in composition existing within species. Additionally, it is of high importance that the genotypes be grown in the same field trial under the same conditions to exclude the confounding effect of environment on the grain composition [96].

3.1.2 The Case of Rice

Japonica rice varieties are predominantly cultivated (80%) in the Mediterranean region. Even though, they have a lower genetic diversity than Indica rice varieties [97] they are better adapted to the temperate Mediterranean climate and offer interesting and favorable agronomic traits. Among the Mediterranean countries, Egypt is the leading rice producer (4.9 MT in 2017/2018) followed by Italy, Turkey and Spain, (Table 1.2). Italy and Spain together contribute by more than 80% of the rice production in Europe [98, 99].

In the Mediterranean region, the main public research institutes and centers that play a key role in the rice field include; National agriculture research and extension systems (NARES) in Egypt, National Institute of Agricultural Research of Morocco (INRAM) in Morocco, Central Institute of Freshwater Aquaculture (CIFA)and IRTA in Spain; National Agricultural Research Foundation (NAGREF) in Greece; The Council for Research and Experimentation in Agriculture - Rice Research Unit (CRA-RIS) and Ente Nazionale Risi (ENR) in Italy and Centre de coopération internationale en recherche agronomique pour le développement (CIRAD), L'Institut de reserche pour le développement (IRD), and INRAE in France. CIRAD and IRD are also members of the Global Rice Science Partnership (GRiSP). Rice

Table 1.2 Average rice production (in Millions of tons) of the main rice producers in the Mediterranean region in 2017/2018 [150]

Country	Average production (millions of tons)
Egypt	4.9
Italy	1.5
Turkey	0.94
Spain	0.81

research activities in the Mediterranean region are mainly focused on rice diseases [100] and secondly on grain quality [101].

Rice Breeding for grain quality improvement is fundamental because unlike wheat, mainly processed into food, rice is primarily eaten as whole cooked grain [102]. However, determining the genes associated with rice grain quality is complicated because each grain quality trait is explained by numerous quantitative traits loci (QTLs) with small phenotypic effects and is affected by environmental conditions. Additionally, there is still a lack of effective methodologies for measuring the grain quality traits [103, 104]. Consequently, progress in breeding of high-quality rice remains limited and improving rice quality is far behind improving rice yield. In the Mediterranean region, only Italy is engaged in research and studies on exploring the genetic basis of rice cooking and eating quality and the results are promising [97, 105]. Today, Italy is striving to retain competitiveness in the global rice market and this can be achieved by developing high-yielding varieties with a good quality that meets the standards of milling industry [105]. The Mediterranean basin lacks an important rice genetic resources in progenitor lines which limits the work of breeders in introducing new favorable traits. The existence of foreign germplasm limits the identification of agronomical or quality traits that adapt in a specific region [106]. Only Italy possess a genetically rich and diverse rice germaplasm which can play a key role in plant breeding for improved varieties however, the germplasm was not sufficiently explored and characterized on a genetic and molecular levels for quality traits [97]. The Italian rice germplasm was collected through a long history of recurrent introduction of foreign rice cultivars (mainly from North China, Japan, India and the US) to Italy to safeguard the rice production affected several times by blast disease and thus offering new sources of allelic variability. Local Italian varieties were exported over the years to many countries of North Europe and the Mediterranean region e.g. Greece and Turkey.

The rice grain quality parameters have been set by the European Union *"Regulation on the common organisation of the market in rice" (Council regulation No. 1785/2003)"* and include grain appearance (shape, color, integrity and fissuring) and cooking (gelatinization and pasting properties) and sensory properties (springiness, stickiness and chewiness). Milling quality and nutritional value are another important rice quality parameter to consider [107]. All the rice grain quality parameters (milling, appearance and cooking and sensory) are genetically associated with each other [108].

Based on the Length over Width ratio (L/W), four types of European rice are identified; Long A with L/W of 2–3, Long B with a L/W more than 3, Medium with a L/W below 3 and Round with a L/W below 2. In Italy, Round and Medium rice type were mainly cultivated before the introduction of the Long A rice varieties (developed using a US accession variety) the preferred ingredient to make the traditional well- known 'risotto'. However, Long B rice varieties were bred in the 1980s with the aim to be exported to North European countries. Modern varieties were developed using the Japonica gene pool and the contributions of the Asian and US germplasm [109].

Rice grain starch content (90% of the grain) and composition govern the cooked rice quality [111–113]. Other grain quality traits are seed storage proteins, vitamin E compounds and grain shape [113]. Rice grain starch is classified into linear amylose (20–30%) and highly branched amylopectin (70–80%). The higher the grain amylose content is, the firmer and less sticky the cooked rice is [114, 115]. Apparent amylose content (AAC) (the amylose content in addition to a detected amount of amylopectin measured with iodine-based assay) is responsible of the grain texture quality. Rice varieties can be classified based on their AAC content into waxy AAC (1–2%), very low AAC (2–9%), low AAC (10–20%), intermediate AAC (20–25%), or high AAC (>25%) types [116]. Cooking and sensory properties are the primary aspects of rice quality. Rice having a low to medium amylose content, a low gelatinization temperature and a low fragrance is mainly preferred by consumers. The rice genes, GS3, Waxy, ALK and fgr, encoding for grain length, amylose content, gelatinization temperature and aroma, respectively, govern most of the genotypic differences and are consequently considered key target traits for rice quality amelioration [108]. Three alleles governing the rice waxy gene were identified, Wx, Wxa and Wxb and they are associated with pasting properties (analyzed by the Rapid Visco Analyzer (RVA)) [97, 116, 117]. Usually Japonica rice varieties have a low AAC, while Indica rice varieties have a heterogeneous grain AAC with the majority of the varieties having a high AAC. Consequently, Japonica rice varieties are generally stickier than Indica rice varieties. When amylose concentration in the grain is the same among varieties, the lower the ratio of amylose over amylopectin is, the stickier the cooked grain becomes [114]. Soluble starch synthase (SS) is responsible of amylopectin synthesis and was reported to play a key role in determining starch quality and the gelatinization temperature (GT), a trait associated with the rice cooking time and quality [119–121]. A functional SSIIa (type of SS) was shown to increase the GT and degree of retrogradation by lowering the proportion of amylopectin short chains [121]. Indica varieties have been shown to have a higher GT as a result of an active SSIIa while most Japonica have low or no SSIIA activity [122]. Gel consistency (GC) and starch paste viscosity parameters are another quality parameter used to define the texture quality of cooked grain especially among varieties with same AAC content. Thus, breeding programs are encouraged to combine AAC content with favorable pasting characteristics and GC for improving the rice cooking quality. Both pasting profile and AAC were shown to be affected by environment. GC is influenced firstly by the amylopectin, and then by grain protein content, fat oxidation and rice flour particle size. *Waxy* and *SSIIa* markers showed consistent major effects on starch quality traits across studies. Thus, these markers should have priority for utilization in marker-assisted breeding [32, 97].

As rice is grown in an irrigated agro-system in the Mediterranean basin, its quality and productivity is particularly susceptible to climate change [123]. Rice research institutes are encouraging moving the rice cultivation toward ecological intensification, while conserving grain yield and improving the genetic diversity of the grain quality [123, 124]. In the Mediterranean region, France has already adopted the organic farming strategy in the late 1980s. Organic rice farming (labeled, Agriculture

Biologique) represented 10% of total rice farming area in 2014 and 16% of rice produced in Camargue [125, 126]. Spain failed to adopt organic rice farming due to fungal disease damaging the crop [127].

4 Conclusion

In the Mediterranean region, great interest from researchers, breeders and industrials was attributed to cereals crops. Breeding evolution resulted in varieties with higher yield and higher adaptability, yet such crops negatively impacted the soil and the environment. Grown under the rain-fed conditions of the Mediterranean basin, cereal production is constricted by rainfall and biotic and abiotic stresses and climate change outcomes are expected to intensify and to adversely affect crops yield and quality in the future, posing a major risk on food security especially in developing countries. New cultivars must be constantly produced and huge efforts are being led by research and cereal breeding centers to develop more resilient, productive and nutritious genotypes with improved technological quality. These favorable traits: can be found in the germplasm of cereal crops. Furthermore, reutilizing ancient and old genotypes is of high importance as a valuable alternative to be used in organic farming system and to ensure the preservation of biodiversity.

References

1. Salamini F, Ozkan H, Brandolini A, Schafer-Pregl R. Genetics and geography of wild cereal domestication in the near East. Nat Rev Genet. 2002;3:429–41.
2. Zohary D, Hopf M. Domestication of plants in the old world: the origin and spread of cultivated plants in West Asia, Europe and the Nile Valley. Oxford: Clarendon Press; 2000. p. 239–40.
3. Kilian B, Martin W, Salamini F. Genetic diversity, evolution and domestication of wheat and barley in the Fertile Crescent. In: Glaubrecht M, editor. Evolution in action. Berlin: Springer; 2010. p. 137–66.
4. Haas M, Schreiber M, Mascher M. Domestication and crop evolution of wheat and barley: genes, genomics, and future directions. J Integr Plant Biol. 2019;61(3):204–25.
5. Mefleh M, Conte P, Fadda C, Giunta F, Piga A, Hassoun G, Motzo R. From ancient to old and modern durum wheat varieties: interaction among cultivar traits, management and technological quality. J Sci Food Agric. 2018;99:2059–67.
6. Royo C, Soriano JM, Alvaro F. Wheat: a crop in the bottom of the Mediterranean diet pyramid. In: Fuerest-Bjelis B, editor. Mediterranean identities - environment, society, culture. London: Intechopen; 2017. p. 381–99.
7. Arzani A. Emmer (Triticum turgidum spp. dicoccum) flour and breads. In: Preedy VR, Watson RR, Patel VB, editors. Flour and breads and their fortification in health and disease prevention. London: Academic Press; 2011. p. 69–78.
8. Cooper R. Re-discovering ancient wheat varieties as functional foods. eJTCM. 2015;5:138–43.
9. Aboubacar A, Yazici N, Hamaker BR. Extent of decortication and quality of flour, couscous and porridge made from different sorghum cultivars. Int J Food Sci Technol. 2006;41:698–703.

10. Abecassis J, Cuq B, Boggini G, Namoune H. Other traditional durum-derived products. In: Sissons M, editor. Durum wheat chemistry and technology. 2nd ed. AACC International Press: Elsevier; 2012. p. 177–99.
11. Arendt EK, Zannini E. Wheat and other triticum grains. In: Cereal for food and beverage industries. 1st ed. Cambridge: Woodhead Publishing Series in Food Science, Technology and Nutrition; 2013. p. 1–66.
12. Matsuoka Y. Evolution of polyploid triticum wheats under cultivation: the role of domestication, natural hybridization and allopolyploid speciation in their diversification. Plant Cell Physiol. 2001;52(5):750–64.
13. Ceccarelli S, Grando S, Capettini F, Baum M. Barley breeding for sustainable production. In: Kang M, Priyadarshan PM, editors. Breeding major food staples. Ames: Blackwell; 2008. p. 193–216.
14. Royo C. From domestication to modern agriculture: wheat as a case study. Old and New Worlds: The Global Challenges of Rural History. International Conference, Lisbon, ISCTE-IUL, January 2016.
15. Feldman M. Origin of cultivated wheat. In: Bonjean P, Angus WJ, editors. The world wheat book. a history of wheat breeding. Paris: Tec & Doc/Intercept Ltd; 2001. p. 3–56.
16. Moragues M, Moralejo MA, Sorrells ME, Royo C. Dispersal of durum wheat landraces across the Mediterranean basin assessed by AFLPs and microsatellites. Genet Resour Crop Evol. 2007;54:1133–44.
17. Buerli M. Farro in Italy: a desk-study by Markus Buerli. Rome: Global Facilitation Unit for Underutilized Species (GFU); 2006. 20p.
18. Linares OF. African rice (Oryza glaberrima): history and future potential. PNAS. 2002;99(25):16360–5.
19. Foo KY, Hameed BH. Utilization of rice husk ash as novel adsorbent: a judicious recycling of the colloidal agricultural waste. Adv Colloid Interf Sci. 2009;152(1–2):39–47.
20. Abaye S, Kasai M, Ohishi K, Hatae K. Textural properties and structures of starches from indica and japonica rice with similar amylose content. Food Sci Technol Res. 2009;15(3):299–306.
21. Cai X, Fan J, Jiang Z, Basso B, Sala F, Spada A, et al. The puzzle of Italian rice origin and evolution: determining genetic divergence and affinity of rice germplasm from Italy and Asia. PLoS One. 2013;8(11):e80351.
22. Bodie AR , Micciche AC, Atungulu GG, Rothrock M, Ricke S. Current trends of rice milling byproducts for agricultural applications and alternative food production systems. Front Sustain Food Syst. 2019;13(47).
23. Maclean J, Hardy B, Hettel G. Rice almanac. In: IRRI, editor. Source book for one of the most important economic activities on earth. 4th ed; 2013. p. 283.
24. Muthukumaran S. Between archaeology and text: the origins of rice consumption and cultivation in the Middle East and the Mediterranean. Pap Inst Archaeol. 2014;24(1):14.
25. Dubcovsky J, Dvorak J. Genome plasticity a key factor in the success of polyploid wheat under domestication. Science. 2007;316:1862–6.
26. Sweeney M, McCough S. The complex history of the domestication of rice. Ann Bot. 2007;100(5):951–7.
27. Shewry PR. Wheat. J Exp Bot. 2009;60(6):1537–53.
28. Chen M, Zhao Z, Chen L, et al. Genetic analysis and fine mapping of a semi-dwarf gene in a centromeric region in rice (Oryza sativa L.). Breed Sci. 2013;63(2):164–8.
29. Xu Y, Jia Q, Zhou G, Zhang X, Angessa T et al. Characterization of the sdw1 semi-dwarf gene in barley. BMC Plant Biol. 2017;17(11).
30. Lantican MA, Dubin HJ, Morris ML. Impacts of international wheat breeding research in the developing world. CIMMYT, Mexico, DF. 2008:1988–2002.
31. Thomas SG. Novel Rht-1 dwarfing genes: tools for wheat breeding and dissecting the function of DELLA proteins. J Exp Bot. 2017;68(3):354–8.
32. Li H, Chen G, Yan W. Molecular characterization of barley 3H semi-dwarf genes. PLoS One. 2015;10(3):e0120558.

33. Venske E, Dos Santos RS, Busanello C, Gustafson C, Costa De Oliviera A. Bread wheat: a role model for plant domestication and breeding. Hereditas. 2019;156(16).
34. Dockter C, Hansson M. Improving barley culm robustness for secured crop yield in a changing climate. J Exp Bot. 2015;66(12):3499–509.
35. Wang SC, Wong DB, Forrest K, Allen A, Chao S, Huang B, et al. Characterization of polyploid wheat genomic diversity using a high-density 90 000 single nucleotide polymorphism array. Plant Biotechnol J. 2014;12:787–96.
36. Nadolska-Orczyk A, Rajchel IK, Orczyk W, Gasparis S. Major genes determining yield-related traits in wheat and barley. Theor Appl Genet. 2017;130(6):1091–8.
37. Scarascia Mugnozza TG. The contribution of Italian wheat geneticists: from Nazareno Strampelli to Francesco D'Amato. In: Tuberosa R, Phillips RL, Gale M, editors. Proceedings of the international congress "in the wake of the double helix: from the green revolution to the gene revolution". Bologna: Avenue Media; 2005. p. 53–75.
38. Zampieri M, Ceglar A, Manfron G, Toreti A, Duveiller G, Romani M, et al. Adaptation and sustainability of water management for rice agriculture in temperate regions: the Italian case study. Land Degrad Dev. 2019;30:2033–47.
39. Lakkakula P, Dixon BL, Thomsen MR, Wailes EJ, Danforth DM. Global rice trade competitiveness: a shift-share analysis. Agric Econ. 2015;46:667–76.
40. Zhou MX. Barley production and consumption. In: Zhang G, Li C, editors. Genetics and improvement of barley malt quality. Advanced topics in science and technology in China. Berlin: Springer; 2009.
41. Medimagh S, ElFelah M, ElGhazza M. Barley breeding for quality improvement in Tunisia. Afr J Biotechnol. 2012;11(89).
42. Blondel J, Aronson J. Biology and wildlife of the Mediterranean region. (Oxford: Oxford University Press). J Nat Hist. 1999;38(13):1723–4.
43. Pswarayi A, Van Eeuwijk FA, Ceccarelli S, Grando S, Comadran J, Russel JR, et al. Changes in allele frequencies in landraces, old and modern barley cultivars of marker loci close to QTL for grain yield under high and low input conditions. Euphytica. 2008;163:435–47.
44. Mefleh M, Conte P, Fadda C, Giunta F, Motzo R. From seed to bread: variation in quality in a set of old durum wheat cultivars. J Sci Food Agric. 2019;99:2059–67.
45. Carvalho F. Agriculture, pesticides, food security and food safety. Environ Sci Pol. 2006;9:685–92.
46. FAO. The Future of Food and Agriculture – Trends and Challenges. Rome; 2017. ISBN 978-92-5-109551-5.
47. Guarda G, Padovan S, Delogu G. Grain yield, nitrogen-use efficiency and baking quality of old and modern Italian bread-wheat cultivars grown at different nitrogen levels. Eur J Agron. 2004;21:181–92.
48. Newton AC, Akar T, Baresel JP, Bebeli PJ, Bettencourt E, Bladenopoulos KV, et al. Cereal landraces for sustainable agriculture: a review. Agron Sustain Dev. 2010;30:237–69.
49. Nazco R, Pena RJ, Ammar K, Villegas D, Crossa J, Moragues M. Durum wheat (Triticum durum Desf.) Mediterranean landraces as sources of variability for allelic combinations at Glu-1/Glu-3 loci affecting gluten strength and pasta cooking quality. Genet Resour Crop Evol. 2014;61:1219–36.
50. Eliazer Nelson AR, Ravichandran K and Antony U. The impact of the green revolution on indigenous crops of India. J Ethn Foods. 2019;6(8).
51. Godfray HC, Beddington JR, Crute IR, et al. Food security: the challenge of feeding 9 billion people. Science. 2010;327(5967):812–8.
52. Bajželj B, Richards K, Allwood J, Smith P, Dennis J, Curmi E, et al. Importance of food-demand management for climate mitigation. Nat Clim Chang. 2014;4:924–9.
53. Cheng SH, Min SK. Rice varieties in China: current status and prospect. Rice China. 2000;1:13–6.
54. Ben Hassen M, Cao TV, Bartholomé J, Orasen G, Colombi C, Rakotomalala J, et al. Rice diversity panel provides accurate genomic predictions for complex traits in the progenies of biparental crosses involving members of the panel. Theor Appl Genet. 2018;131:417–35.

55. Asseng S, Martre P, Maiorano A, Rotter GJ, O'leary GJ, Fitzgerald C, et al. Climate change impact and adaptation for wheat protein. Glob Chang Biol. 2019;25(1):155–73.
56. Giunta F, Bassu S, Mefleh M, Motzo R. Is the technological quality of old durum wheat cultivars superior to that of modern ones when exposed to moderately high temperatures during grain filling? Foods. 2020;9(778):1–15.
57. De Vita P, Li Destri Nicosia O, Nigro F, Platani C. Breeding progress in morpho-physiological, agronomical and qualitative traits of durum wheat cultivars released in Italy during the 20th century. Eur J Agron. 2007;26:39–53.
58. Kabbaj H, Sall AT, Al-Abdallat A, Geleta M, Amri A, Filali-Maltouf A, Belkadi B, Ortiz R, Bassi FM. Genetic diversity within a global panel of durum wheat (Triticum durum) landraces and modern germplasm reveals the history of alleles exchange. Front Plant Sci. 2007;18(8):1277.
59. Xynias IN, Mylonas I, Korpetis EG, Ninou E, Tsaballa A, Avdikos ID et al. Durum wheat breeding in the Mediterranean region: current status and future prospects. Agronomy. 2020;10(432).
60. Food and Agriculture Organization of the United Nations. FAOSTAT 2020. Production quantities of Wheat by country. http://www.fao.org/faostat/en/#data/QC/visualize. Last Update June 15, 2020. Accessed 4 August 2020.
61. Genesys database. https://www.Genesis-pgr.org/c/wheat. Accessed 15 Aug 2020.
62. Sissons M. Pasta. In: Wrigley C, Corke H, Seetharaman K, Faubion J, editors. Encyclopedia of food grains. 2nd ed. Oxford: Academic; 2016. p. 79–89.
63. Xue C, Matros A, Mock HP, Mühling KH. Protein composition and baking quality of wheat flour as affected by split nitrogen application. Front Plant Sci. 2019;10(642).
64. Furtado A, Bundock PC, Banks PM, Fox G, Yin X, Henry RJ. A novel highly differentially expressed gene in wheat endosperm associated with bread quality. Sci Rep. 2015;5(10446).
65. MacRitchie F. Physicochemical properties of wheat proteins in relation to functionality. Adv Food Nutr Res. 1992;36:1–87.
66. Wieser H, Gurser R, Von Tucher S. Influence of sulphur fertilization on quantities and proportions of gluten protein types in wheat flour. J Cereal Sci. 2004;40:239–44.
67. Giunta F, Motzo R, Pruneddu. Trends since 1900 in the yield potential of Italian-bred durum wheat cultivars. Eur J Agron. 2007;27:12–24.
68. Edwards NM, Gianibellib MC, McCaig TN, Clarke JM, Ames NP. Relationships between dough strength, polymeric protein quantity and composition for diverse durum wheat genotypes. J Cereal Sci. 2007;45:140–9.
69. Mariani BM, D'Egidio MG, Novaro P. Durum wheat quality evaluation: influence of genotype and environment. Cereal Chem. 1995;72(2):194–7.
70. Subira J, Pena RJ, Alvaro F, Ammar K, Ramdani A, Royo C. Breeding progress in the pasta-making quality of durum wheat cultivars released in Italy and Spain during the 20th Century. Crop Pasture Sci. 2014;65:16–26.
71. De Santis MA, Giuliania MM, Giuzio L, De Vita P, Lovegrove A, Shewry PR, et al. Differences in gluten protein composition between old and modern durum wheat genotypes in relation to 20th century breeding in Italy. Eur J Agron. 2017;87:19–29.
72. Horvat D, Drezner G, Sudar R, Magdić D, Španić V. Baking quality parameters of wheat in relation to endosperm storage proteins. CJSAU. 2012;4(1):19–25.
73. Ormoli L, Costa C, Negri S, Perenzin M, Vaccino P. Diversity trends in bread wheat in Italy during the 20th century assessed by traditional and multivariate approaches. Sci Rep. 2015;5(8574).
74. Fois N, Schlichting L, Marchylo B, Dexter J, Motzo R, Giunta F. Environmental conditions affect semolina quality in durum wheat (Triticum turgidum ssp. durum L.) cultivars with different gluten strength and gluten protein composition. J Sci Food Agric. 2011;91:2664–73.
75. Padalino L, Mastromatteo M, Lecce L, Spinelli S, Conte A, Del Nobile MA. Effect of raw material on cooking quality and nutritional composition of durum wheat spaghetti. Int J Food Sci Nutr. 2015;66(3):266–74.

76. De Vita P, Riefolo C, Codianni P, Cattivelli L, Fares C. Agronomic and qualitative traits of T turgidum ssp dicoccum genotypes cultivated in Italy. Euphytica. 2006;150:195–205.
77. D'Egidio MG, Mariani BM, Nardi S, Novaro P, Cubadda R. Chemical and technological variables and their relationships: a predictive equation for pasta cooking quality. Cereal Chem. 1990;67:275–81.
78. Dinelli G, Marotti I, Di Silvestro R, Bosi S, Bregola V, Accorsi M, et al. Agronomic, nutritional and nutraceutical aspects of durum wheat (Triticum durum Desf.) cultivars under low input agricultural management. IJA. 2013;8:85–93.
79. Pogna NE, Mazza M, Redaelli R, Ng PKW. Gluten quality and storage protein composition of durum wheat lines containing the Gli-D1/Glu-D3 loci. In: Wrigley CW, editor. Gluten 96. Sydney: Cereal Chem Div Royal Austr Chem Inst; 1996. p. 18–22.
80. Fadda C, Sanguinetti AM, Del Caro A, Collar C, Piga A. Bread staling: updating the view. Compr Rev Food Sci Food Saf. 2014;13:473–92.
81. Giunta F, Pruneddu G, Zuddas M, Motzo R. Bread and durum wheat: Intra- and inter-specific variation in grain yield and protein concentration of modern Italian cultivars. Eur J Agron. 2019;105:119–28.
82. Sissons M, Pleming D, Margiotta B, D'Egidio MG, Lafiandra D. Effect of the introduction of D-genome related gluten proteins on durum wheat pasta and bread making quality. Crop Pasture Sci. 2014;65:27–37.
83. Triboi E, Martre P, Girousse C, Ravel C, Triboi-Blondel A. Unravelling environmental and genetic relationships between grain yield and nitrogen concentration for wheat. Eur J Agron. 2006;25:108–18.
84. DuPont FM, Hurkman WJ, Vensel WH, Tanaka C, Kothari OK, Chung SB. Altenbach Protein accumulation and composition in wheat grains: effects of mineral nutrients and high temperature. Eur J Agron. 2006;25:96–107.
85. Carmona S, Alvarez JB, Caballero L. Genetic diversity for morphological traits and seed storage proteins in Spanish rivet wheat. Biol Plant. 2010;54:69–75.
86. Branlard G, Dardevet M, Amiour N, Igrejas G. Allelic diversity of HMW and LMW glutenin subunits and omega-gliadins in French bread wheat (Triticum aestivum L.). Genet Resour Crop Evol. 2003;50:669–79.
87. Turchetta T, Ciaffi M, Porceddu E, Lafiandra D. Relationship between electrophoretic pattern of storage proteins and gluten strength in durum wheat landraces from Turkey. Plant Breed. 1995;114:406–41.
88. Alvarez J, Caballero L, Ureña P, Vacas M, Martin LM. Characterisation and variation of morphological traits and storage proteins in Spanish Emmer Wheat germplasm (Triticum Dicoccon). Genet Resour Crop Evol. 2007;54:241–8.
89. Hawkesford MJ. Reducing the reliance on nitrogen fertilizer for wheat production. J Cereal Sci. 2014;59(3):276–83.
90. Martine-Moreno F, Solis I, Noguero D, Blanco A, Ozberk I, Nsarellah N, et al. Durum wheat in the Mediterranean rim: historical evolution and genetic resources. Genet Resour Crop Evol. 2020;67:1415–36.
91. MacKey J. Wheat, its concept, evolution and taxonomy. In: Royo C, Nachit M, Di Fonzo N, Araus JL, Pfeiffer WH, Slafer GA, editors. Durum Wheat breeding: current approaches and future strategies. New York: Haworth Press; 2005. p. 3–62.
92. Soriano JM, Villegas D, Aranzana MJ, García del Moral LF, Royo C. Genetic structure of modern durum wheat cultivars and Mediterranean landraces matches with their agronomic performance. PLoS One. 2016;11:e0160983.
93. Balfourier F, Roussel V, Strelchenko P, Exbrayat-Vinson F, Sourdille P, Boutet G, Koenig J, Ravel C, Mitrofanova O, Beckert M, Charmet G. A worldwide bread wheat core collection arrayed in a 384-well plate. Theor Appl Genet. 2007;114:1265–75.
94. Migliorini P, Spagnolo S, Torri L, Arnoulet M, Lazzerini G, Ceccarelli S. Agronomic and quality characteristics of old, modern and mixture wheat varieties and landraces for organic bread chain in diverse environments of northern Italy. Eur J Agro. 2016;79:131–41.

95. Dinu M, Whittaker A, Pagliaia G, Benedettellic S, Sofia F. Ancient wheat species and human health: biochemical and clinical implications. J Nutr Biochem. 2018;52:1–9.
96. Shewry P. Do ancient types of wheat have health benefits compared with modern bread wheat? J Cereal Sci. 2018;79:469–76.
97. Caffagni A, Albertazzi G, Gavina G, Ravaglia S, Gianinetti A, Pecchioni N, et al. Characterization of an Italian rice germplasm collection with genetic markers useful for breeding to improve eating and cooking quality. Euphytica. 2013;194:383–99.
98. Ferrero A. Challenges and opportunities for a sustainable rice production in Europe and Mediterranean area. Paddy Water Environ. 2006;4:1–12.
99. Guerrero M. No rain in Spain falling on the plain. Report by USDA GAIN: Oilseeds, Cotton, Sugar, Grain and Feed, Spain Rice Market in the Iberian Peninsula, Spanish Rice Producers to switch to Japonica Varieties. Madrid. Date: 2 February 2020.
100. Food and Agriculture Organization of the United Nations. FAOSTAT 2020. Production quantities of Rice, paddy by country. http://www.fao.org/faostat/en/#data/QC/visualize. Last Update June 15, 2020. Accessed 4 August 2020.
101. Titone P, Mongiano G, Tamborini L. Resistance to neck blast caused by Pyricularia oryzae in Italian rice cultivars. Eur J Plant Pathol. 2015;142:49–59.
102. Biselli C, Bagnaresi P, Cavalluzzo D, et al. Deep sequencing transcriptional fingerprinting of rice kernels for dissecting grain quality traits. BMC Genomics. 2015;16:1091.
103. Bergman CJ, Bhattacharyya KR, Ohtsubo K. Rice end-use quality analysis. In: Champagne E editor. Chemistry and technology. Minnesota: AACC; 2004. p. 415–72.
104. Anacleto R, Cuevas RP, Jimenez R, Llorente C, Nissila E, Henry R, Sreenivasulu N. Prospects of breeding high-quality rice using post-genomic tools. Theor Appl Genet. 2015;128:1449–66.
105. Custodio MC, Cuevas RP, Ynion J, Laborte AG, Velasco ML, Demont M. Rice quality: how is it defined by consumers, industry, food scientists, and geneticists? Trends Food Sci Technol. 2019;92:122–37.
106. Tesio F, Tabacchi M, Cerioli S, Follis F. Sustainable hybrid rice cultivation in Italy. A review. Agron Sustain Dev. 2014;34:93–102.
107. Domingo C, Andres F, Talon M. Rice cv. Bahia mutagenized population: a new resource for rice breeding in the Mediterranean basin. Span J Agric Res. 2007;5(3):341–7.
108. Chen LG, Song Y, Li SJ, Zhang LP, Zou CS, Yu DQ. The role of WRKY transcription factors in plant abiotic stresses. Biochim Biophys Acta. 2012;1819(2):120–8.
109. Zhou H, Xia D, He Y. Rice grain quality—traditional traits for high quality rice and health-plus substances. Mol Breed. 2020;40:1.
110. Faivre-Rampant O, Bruschi G, Abbruscato P, Cavigilio S, Picco AM, Borgo L, et al. Assessment of genetic diversity in Italian rice germplasm related to agronomic traits and blast resistance (Magnaporthe oryzae). Mol Breed. 2011;27:233–46.
111. Juliano BO. Criteria and tests for rice grain qualities. In: Juliano BO, editor. Rice chemistry and technology. Minessota: AACC; 1985. p. 443–524.
112. Bason ML, Blakeney AB. Grain and grain products. In: Crosbie GB, Ross AS, editors. The RVA handbook. Minessota: AACC; 2007. p. 31–47.
113. Zhang C, Zhu J, Chen S, Fan X, Li Q, Lu Y, et al. Wx lv, the ancestral allele of rice Waxy gene. Mol Plant. 2019;12(8):1157–66.
114. Juliano BO. A simplified assay for milled-rice amylose. Cereal Sci Today. 16:34.
115. Ayabe S, Kasai M, Ohishi K and Hatae K. Textural properties and structures of starches from indica and japonica rice with similar amylose content. Food Sci Technol Re. 2009;15:299–306.
116. Mutters RG, Thompson JF. Rice quality handbook. California: University of California, Agriculture and Natural Resources.S. Dept. of Agriculture, Federal Grain Inspection Service; 2009. p. 12–3.
117. Biselli C, Cavalluzzo D, Perrini R, Gianinetti A, Bagnaresi P, et al. Improvement of marker-based predictability of Apparent Amylose Content in japonica rice through GBSSI allele mining. Rice. 2014;7(1).1.

118. Xu F, Sun C, Huang Y, Chen Y, Tong C, Bao J. QTL mapping for rice grain quality: a strategy to detect more QTLs within sub-populations. Mol Breed. 2015;35(105).
119. Preiss J, Sivak M. Starch synthesis in sinks and sources. In: Zamski E, Schaffer AA, editors. Photoassimilate distribution in plants and crops: source-sink relationships. New York: Marcel Dekker; 1996. p. 139–68.
120. Umemoto T, Terashima K. Activity of granule-bound starch synthase is an important determinant of amylose content in rice endosperm. Funct Plant Biol. 2002;29:1121–4.
121. Umemoto T, Aoki N, Lin HX, Nakamura Y, Inouchi N, Sato Y, Yano M, Hirabayashi H, Maruyama S. Natural variation in rice starch synthase IIa affects enzyme and starch properties. Funct Plant Biol. 2004;31:671–84.
122. Umemoto T, Horibata T, Aoki N, Hiratsuka M, Yano M, Inouchi N. Effects of variations in starch synthase on starch properties and eating quality of rice. Plant Prod Sci. 2008;11:472–80.
123. Nakamura Y, Francisco PB Jr, Hosaka Y, Satoh A, Sawada T, Kubo A, Fujita N. Essential amino acids of starch synthase IIa differentiate amylopectin structure and starch quality between japonica and indica rice varieties. Plant Mol Biol. 2005;58:213–27.
124. ElBasiouny H, Elbehiry F. Potential soil carbon and nitrogen sequestration in future land use under stress of climate change and water deficiency in northern Nile Delta. Egypt Agric. 2019;3(4):111–2.
125. Sharma N, Khanna R. Rice grain quality: current developments and future prospects. In: Shah F, Khan Z, Iqbal A, Turan M, Olgum M, editors. Recent advances in grain crop research. London: Intechopen; 2019.
126. Comoretto L, Arfib B, Talva R, Chauvelon P, Pichaud M, Chiron S, et al. Runoff of pesticides from rice fields in the Ile de Camargue (Rhône river delta, France): Field study and modeling. Environ Pollut. 2008;151(3):486–93.
127. Quiedeville S, Barjolle D, Mouret JC, Stolze M. Ex-post evaluation of the impacts of the science-based research and innovation program: a new method applied in the case of farmers' transition to organic production in the camargue. Econ Innov New Technol. 2017;22(1):145.
128. ImproRICE: Improvement of Spanish organic rice production facing climate change effects. Programme: Sustainable Field Crops. starting date: January 2019, end date: December.

Chapter 2
The Evolution of Milling Process

Aleksandar Fišteš

Abstract Emergence of cultivated cereals and consequently milling of cereals is essentially connected with the beginning and development of human civilization. It is almost impossible to imagine a human nutrition without cereals. Flour milling is considered by many to be the man oldest continuously practiced industry. The evolution of milling process was slow and gradual, from early beginnings at the dawn of civilization up to today's fully automated and computerized process. Flour milling was and still is deeply connected with many spheres of human life both socio-political and technical-technological. The following chapter aims to provide a relatively brief overview of the evolution of milling process by pointing out the most significant steps which led to the modern approach in flour milling technology.

Keywords Cereals · Traditional milling · Roller mills · Modern machinery

1 Cereals and Civilization

Without exaggeration, it can be said that cereals are one of the most important resources in the history of the human race and that they are crutial in the foundation of human civilization.

According to the Bible, man lived in paradise in the abundance of the fruits of the earth, which he used for food [1]. The reason why man decided to give cereals an advantage over other plant species lies in the fact that he recognized its nutritional benefits and saw the way to provide himself with a food source in the long run. By learning to select seeds of superior plants people started to modify natural vegetation, thereby improving the production and wheat, barley, lentils and many other crops of edible seeds were planted.

Wild cereal grasses were the early food grains [2, 3]. They were domesticated by early primitive humans of the ancient farming communities in the Fertile Crescent region in the Middle East, which covers the area of modern day Iran, Iraq, Syria,

A. Fišteš (✉)
Faculty of Technology, University of Novi Sad, Novi Sad, Serbia
e-mail: fistes@uns.ac.rs

F. Boukid (ed.), *Cereal-Based Foodstuffs: The Backbone of Mediterranean Cuisine*, https://doi.org/10.1007/978-3-030-69228-5_2

Lebanon, Israel, Palestine, Jordan and Egypt. It has been rightly called the cradle of civilization. In this region, approximately 10,000 BC people shifted from hunter-gatherer societies and first began to settle in permanent communities. The geography and climate of the region were favorable for agriculture and people started to cultivate wild grasses such as barley, emmer and einkorn wheat (three of the so-called Neolithic founder crops). In this way, they were able to support themselves from the land. The origins of rice and millet farming date to the same Neolithic period in China. Sorghum and millets were also being domesticated in sub-Saharan West Africa. Maize-like plants appear to have been cultivated at least 9000 years ago, while the first directly dated corn cob dates to around 5500 years ago [4].

In fact, from the moment when people were able to satisfy their energy needs through plant-based food that they planted and harvested (together with animal domestication), instead of hunting and foraging, civilization begins. Agriculture allowed enough food for increased population, which led to larger societies and the development of first cities. Also, agriculture led to the accumulation of material goods, division of labor, social stratification and finally need for organization of political power [5]. The very word "cereal" originates from the name of the Roman goddess Ceres. Her particular responsibility was the food-giving plants, and for that reason the food grains came to carry her name [6]. Ceres was credited with giving the gift of agriculture to humankind. She is also called the law-giver because before, man had subsisted on acorns, and wandered in woods without settlement or laws. After grains were found out they divided and tilled the land and this was the origin of government and laws [7] proving the inseparable link between civilization and cereals.

2 Historical Overview of Flour Milling Process

Flour milling is man oldest continuously practiced industry. No other activity can be explored as far in its history of development as milling [8]. It began with the rise of civilization and followed its development both in the sphere of socio-political relations as well as technical-technological development. Early Neolithic villages show evidence of the development of processing grain so the beginning of milling should be sought in prehistoric times, gradually evolving throughout slave system in ancient times, followed by middle ages and feudalism, than industrial revolution and capitalism and all the way to modern times with completely automated and computerized production system.

The Greek philosopher Poseidonius said that the first man, when he used his teeth to crush food, was in fact the first miller. It is certain that the grains for food were not used as a whole, but were previously crushed. Soon man found the first primitive mechanical means to crush the grains of various fruits, instead of teeth, with his hands with the help of stones [8, 9]. The evidence of flour milling exists in the Book of Genesis when Patriarch Abraham said to Sarah "Quick! Knead three measures of fine flour and make bread."

Regardless of the stage of development, whether as a primitive or developed technological process, flour milling always consists of two basic operations—grinding and sifting. According to the method of grinding, i.e. the type of mill, the history of milling can be divided into several periods [7]:

- the period of the hand stone,
- the period of the slave and cattle-driven mill,
- the period of the single water-wheel (Greek),
- the period of the water-wheel geared to several stones (Romans),
- the period of windmills,
- the period of steam-driven mills,
- the period of cleaning the wheat and dressing the product through silk,
- the period of the perfecting of the millstone dress and extension of the dressing process and
- the period of the introduction of rolls and plansifters and a multitude of intermediate processes before the flour was finished.

2.1 Milling in Ancient Times

According to ancient inscriptions and excavations, the process of flour milling can be dated back to ancient civilizations of Babylon, Assyria and Egypt and it was done between two stones. Millstones dominated the process and were used until the nineteenth century, only over time its shape and drive changed, according to the development of technology [10, 11].

The mortar and pestle (Fig. 2.1) probably were the first grinding instruments to be used. Most probably it comes from the idea of pounding and rubbing the grains between two stones. During the process, people came to know that hard seeds are crushed easier in deeper and wider containers so the bottom stone was hollowed out to form a mortar while the upper stone was made to form a sort of pestle [7, 10, 12].

Hieroglyphics of the ancient Egyptians indicate that this milling technique was implemented by using first the wooden, and later on stone mortars in which the grain was crushed by the strike from pestles. Kozmin [1] gave a relatively detailed description of this process based on pictures from walls in the Egyptian town of Thebes: the loading was done by one man who poured the grain out of the vessel into the mortar; two man have done the grinding by the blows from pestles; one is empting the crushed grain out of the mortar into a sieve; finally the last man is sifting. Also, it is presumed that ancient Egyptians heated the grain before grinding because dried grains were broken more easily by the blows from the pestle.

This description suggests some very important facts. Even in this primitive stage the flour milling was a process that consisted of several operations (conditioning, loading, milling, and sifting), performed by several people, and partly reminiscent of operations that still exist today in modern milling. At this early stage sifting was also a part of the process, and even some kind of preparation of grain for grinding was done.

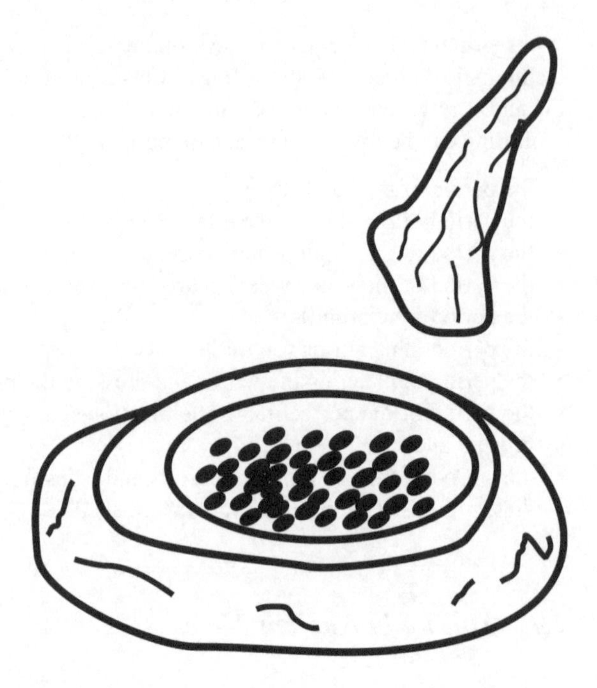

Fig. 2.1 Mortar and pestle

Similar type of this primitive mill with mortar and pestle was also found on the drawing on the excavated vases from ancient Greece. This type of grinding machine was in use in ancient China where mortar and pestle were used for polishing the rice, i.e. for removing the coating from the inner part of the rice kernel. Here, a pestle was attached to one end of a lever and mill was supplied with a foot drive on the other end. This type of mill is reported to be used by the Native Americans for corn grinding. Here, a lever, with pestle was attached to one end, and a box to the other, is placed on the fork. When the box is filled with water it outweighs the pestle and moves it to the higher position. When water runs out of the box the pestle falls into the mortar crushing the grain [1].

For thousands of years, stone mortar and pestle were the primary grinding instrument. The invention of saddlestone was the next and very important step in the development of flour milling because it involved thinking about the efficiency of the process itself. It was noticed by the people that the strikes of the pestle wear out the mortar and the pestle itself.

The saddlestone (Fig. 2.2) is consisted of lower larger stone where grains were crushed by the small stone [1]. These stones were in the shape of a trough adjusted to each other so that the lower one is firmly planted stone block hollowed out from above, and the second one is cylindrical and adjusted to the first. The above stone was shaped to adapt to that recess, so that it can be moved back and forth. Between the stones, seeds were placed and the upper stone was moved back and forth by the hands. In this way, the seeds were crushed and ground [9, 12]. Wooden handles were later inserted into this stone to facilitate the process [13].

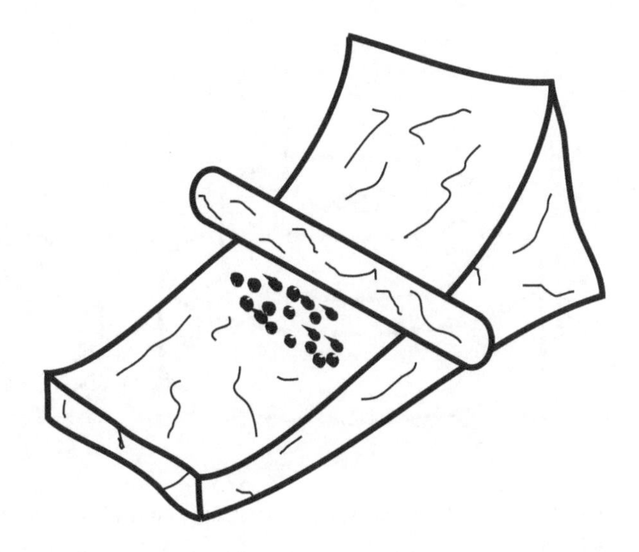

Fig. 2.2 Saddlestone

It is supposed that the mills consisting of two grindstones appeared in ancient Egypt but the archaeological finds from 1000 BC at the Boden Lake in modern Switzerland and around Dresden and Leipzig in modern Germany also reported similar grinding stones [7, 12].

Very important feature of this grinding instrument is a shift from impact (mortar and pestle) to frictional force as a source of stress that cause comminution. Even today the milling equipment is still classified according to working principle and the source of predominant stress that cause size reduction [14]. Although still primitive in its nature this grinding instrument where grains were crushed between two stones paved the way for further progress [2] and could be referred to as a transitional type from mortar and pestle to two-stones rotary mills [1].

The mills consisted of grindstones probably can be attributed to ancient Egypt. During exodus from Egypt around 1300 BC, the Hebrews carried portable mills with them as evidenced in the fifth book of Moses where it was stated that "no man shall take the nether or the upper millstone to pledge". Also, the fourth book of Moses mentions the heavenly manna which had to be ground: "people went about and gathered it in mills, or beat it in mortar" [1, 7, 8]. It is obvious that these grinding instruments were considered as an integral part of the household. It is likely that each household owned its own mortar and pestle or millstone to prepare daily supplies of grinded grain [2, 10]. On the other hand, it shows that mortars and pestles were still used and not fully replaced by millstones. Grinding mills are also mentioned in Homer's Odyssey (around 1100 BC) so the probable route of the double stone mill was from the Egypt through Greece to Rome [1, 12].

The basic principle of hand driven double stone mills (Fig. 2.3a) is similar wherever they were constructed or used. The grain was fed through a hole in the center of upper stone which was rotated with a stick fixed into its edge. By the action of the grinding surface of the stones assisted by the centrifugal force, the grain is milled and meal moves towards the outlet of the lower stationary stone [7, 10]. Similar type

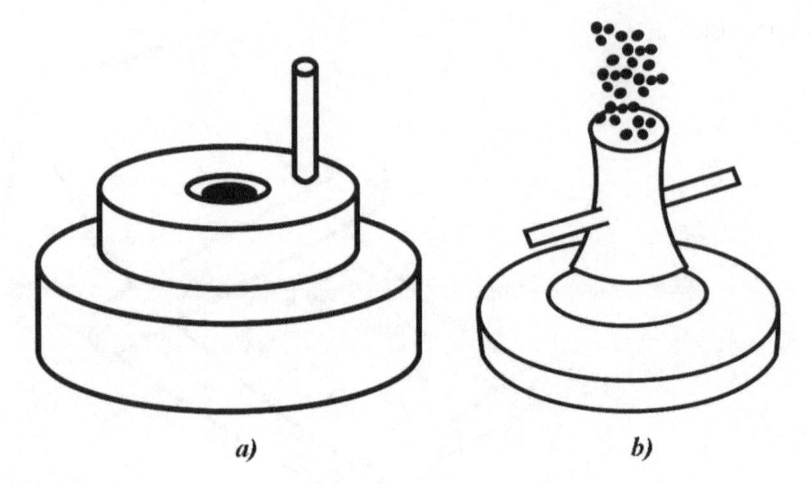

Fig. 2.3 Hand driven millstones: (**a**) stone with a stick fixed into edge and (**b**) stone with the handles

of mill associated with Roman period (Fig. 2.3b) consisted of the vase-shaped upper stone rotating on a lower conical fixed stone. The grain was poured from above and the slaves turned the top stone with the handles [8]. The Chinese rice mills were of similar construction but they were intended to remove the cover of the kernel and not for grinding [1]. Hand-powered rotary mills superseded saddlestones and they were used for centuries to come [2].

These type of hand mills are associated with ancient times. They were used as family mills and as a part of household while usually milling was performed by women. Also, they are associated with slave owning societies and milling was performed by slaves or convicts as a part of their punishment [1].

2.2 Animal Driven Mills

Very important step in the evolution of the flour milling process was a shift from milling to satisfy personal-family needs to milling to satisfy the market needs. Moreover, the increase in population and the greater need for flour forced man to replace the former exclusively human drive with a partial or complete animal power [9].

Significant progress in the development of flour milling was made during the time of the Romans. Actually, the word mill itself is of Latin origin—*molinum*. Most of the European languages for milling use the derivate of this Latin word (German—*mühle*; French—*moulin*; Hungarin—*malom*; Polish—*mlyn*; Dutch—*molen*; Spanish—*molino*, Russian—*мельница*, Italian—*mulino*; etc.). The animal-driven rotary mill appeared around the 200 years BC in Italy. These Roman mills (Fig. 2.4) with their shape, especially the narrow structure, resembled hourglass.

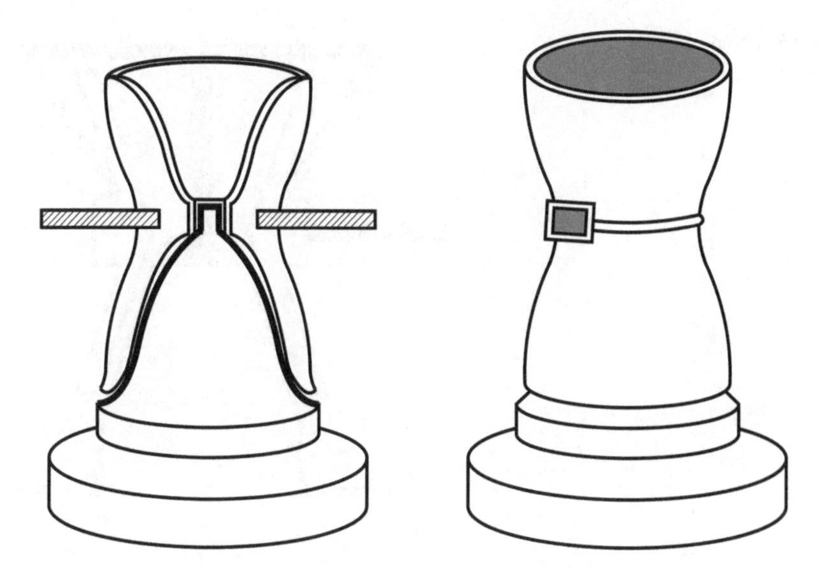

Fig. 2.4 Roman hourglass mill

The mill consisted of two parts: the lower (nether) fixed stone—*meta* and the revolving upper stone—*catillus*. The foundation of this mill consists of cylindrically built wall. To this foundation is fixed conically shaped stone with a twisted lead groove placed on the top. The upper stone was in the shape of a double hollow cone, it has two "bell" shaped hollows much resembling an hourglass. In the place where the tops of the bells are joined an iron cross beam is fixed. In the middle of this beam, the journal (sliding surface between a rotating shaft and a hole) is inserted. It provides just enough space between the lower bell and the cone for grain to be grinded. The grain is poured into a narrow opening in the middle of the upper bell of the *catillus* and it falls into the space between the grinding surfaces of the stone. The rotation took place with the help of levers (wooden poles) inserted in the narrowest place of the *catilus* in two or four quadrangular (or rectangular) openings. The obtained flour was discharged through ring-shaped lead groove of the foundation [1, 9].

This type of hourglass mills (Fig. 2.5) were found well preserved, along with rotary-hand millstones, during excavations in the ancient city of Pompeii which have been buried by the Somma-Vesuvius volcanic products during the 79 AD eruption. The grindstones of the Pompeian mills were made of volcanic rocks [15, 16].

In the beginning, crushed grains were used as a whole (something like whole grain flour). By sifting the crushed material the flour was separated from larger products. These coarse products are again ground on a stone and sifted so this can be considered as the beginning of a gradual reduction process (basis of the modern flour milling process) with various types of flour produced. The use of animal driven mills facilitated the grinding and the advances in screening techniques allowed to differentiate the quality of flour [17]. According to Roman author Pliny, at that time

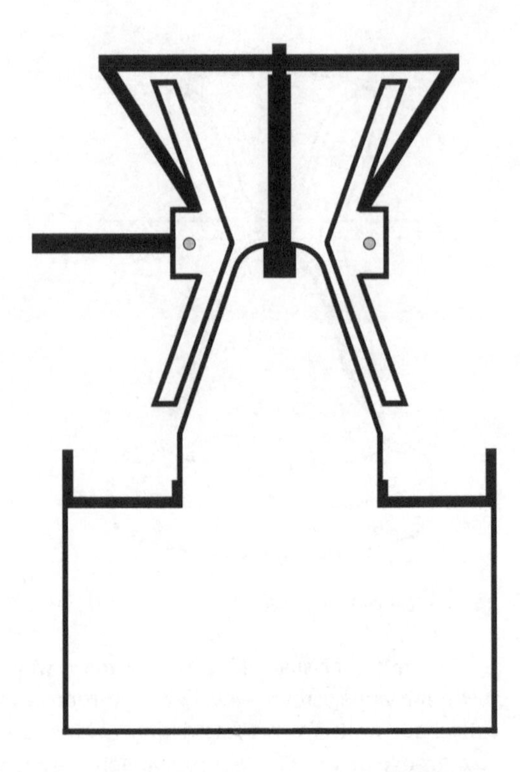

Fig. 2.5 Pompeian hourglass mill

horsehair and linen cloth sieves were used to separate the bran from the flour, and even flour of different quality were found in the stores [1, 8].

It is known that the ancient Romans ground several types of flour: fine flour (*siligo*), common fine flour (*farina*) and black flour (*farina secundaria*) [12]. Kozmin [1] states that Pliny mentioned number of different flour grades: flour of finest quality (*pollen*), flour of medium quality (*similago*), flour of semolina first quality (*farini tritici*), flour of semolina second quality (*secondarii panis*), flour of semolina third quality (*cibari panis*) and bran (*furfur*). Accordingly there were numerous types and of bread. With flour of superior quality (*siliga*) was produced the *panis siligineus.* Starting from the way in which flour was sifted there were the *panis cibarius, secundarius, plebeius, rusticus, militaris castrensis* (for soldiers), *nauticus* (for sailors) [17]. According to these facts, it is assumed that among the ancient Romans, during the glory and power of the Roman Empire, milling was at a high level, because they separated the bran from the flour, and classified the flour into several types.

The animal-driven rotary mill saved humans from hard work and significantly increased the flour output since heavier stones worked with greater efficiency while donkeys, or horses could drive the mill for hours. They originated in Rome and very quickly spread throughout the Roman Empire, in Africa, Asia and especially in Arabia where they were driven by camels [12]. In fact, this type of mills could be seen as hand mills with a more sophisticated drive, which is especially reflected in the higher number of revolutions of the stone, which is achieved through gear wheels. The smaller gear is mounted on the shaft of the stone, and the large one is driven by animal power.

2.3 Water Driven Mills

Probably, in ancient times, people already came to idea that the manual power to move the millstones could be replaced by water power. Looking at the banks of the river, the flow of water, they noticed that wood and planks float on the water, in some places faster, and in others slower. They assumed that there is a certain force in the water which moves lighter objects. Based on their observations, the millers transferred the mills to their households near rivers and streams. In the direction of the water flow, they placed a water wheel and as the water flowed or fell from a height, it turned the wheel. Therefore, instead of manual power, the millstones were driven by water power.

The stones were placed by the water or on the water itself under the roof of the object. Usually in the place where the water flow was the strongest, they set up a water wheel, the axis of which was extended to the axis of the stones (Fig. 2.6).

At that point, two axles are connected by a pair of conical gears, which transmit rotation at an angle of 90°. The big water wheel actually gave the whole drive. The mill wheel was larger, usually 2 m in diameter, and paddles of appropriate shape and

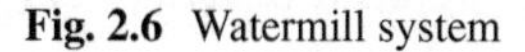
Fig. 2.6 Watermill system

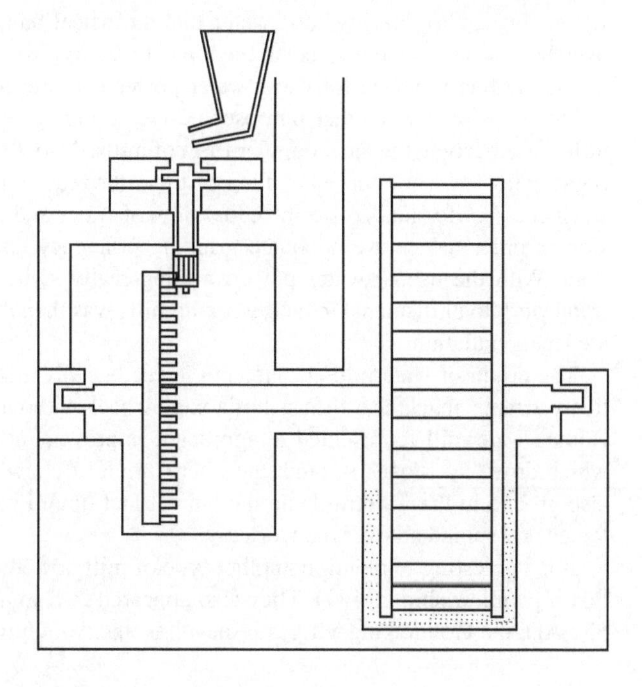

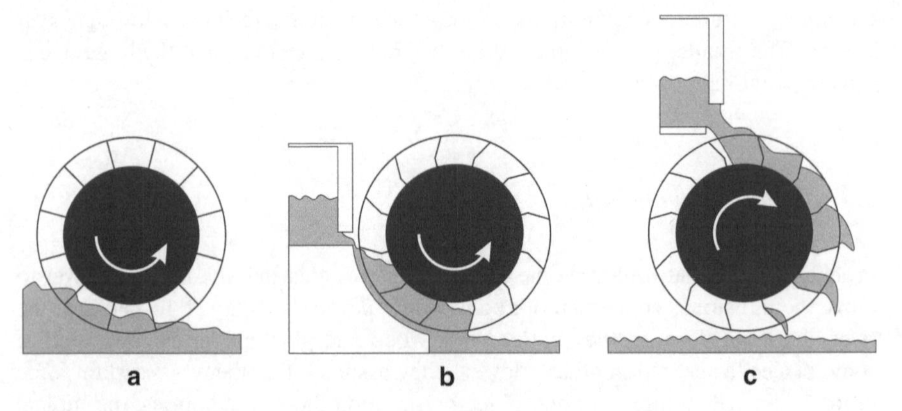

Fig. 2.7 Waterwheel design: (**a**) bottom, (**b**) breast-shot and (**c**) overshot

size were fixed at small intervals on the wheel, so as the water fell on the blades it moved the mill wheel, which moved over its axis by gears [12].

Among the Romans, the mill wheel took on a different shape. Its shaft is placed vertically, a mill wheel is placed on its lower part, and the upper end is directly connected to the stone. The mill wheel consists of radially placed blades, which are specially curved to better absorb the kinetic energy of water. Such a wheel could move two stones and in that case a suitable gear was placed on the upper end of the shaft. There are various designs of waterwheel, depending on the water supply available, including undershot—water hits the wheel paddles at the bottom of the wheel (Fig. 2.7a), breast-shot—water hits the wheel half way up (Fig. 2.7b) and overshot—water hits the wheel at the top (Fig. 2.7c) [18].

This ordinary device for using water power is a predecessor of the later water turbine used for many other purposes besides grinding of grains. Soon the watermills have become the most popular form of mills. Also, these mills required proper organization of work while handling of the mills required special skills because it was necessary to standardize the equal flow of water and use its power for propulsion in order to achieve adequate grinding efficiency and appropriate quality of flour. With the use of water power, milling really switched from the household framework to craftsmanship, and work in mills was then standardized by appropriate legal regulations [12].

The origin of watermills is ancient Greece but this first watermills were a relatively simple machines with a horizontal wheel without a gearing system. The Roman watermill represented a significant improvement having a vertical wheel which drove the stones through the intervention of cogwheels [7]. The Egyptians also ground in the watermills on the Nile at that time. Later, these mills were perfected and spread all over the world.

It is interesting to mention another type of mill powered by water, the so-called floating mill or ship mill [7]. They also appeared in Roman times. In the winter of 537 AD, the Gothic king Vitiges besieged Rome, which was captured and held by the Byzantine (East Roman) military leader Belisarius. The Goths closed all roads to Rome to force Belisarius to surrender. However, Belisarius came up with the idea of transferring the mills to the Tiber and placing them on ships which were anchored. Belisarius succeeded and on the Tiber he ground flour to the besieged people and saved them from starvation. The Goths could not capture Rome, so they left it after a year of siege. Ship-mills are actually watermills, because they are driven by water power by means of a water-wheel. Similar type of mill appeared in France, the mill was placed on boats and a shaft was pulled on both sides of the mill with water wheel attached to it. In Germany, the mills were set up on the river bank while the water wheel was put at the end of the very long shaft which lay in the water and powered the whole plant. This type of mill domesticated on the rivers of Danube, Tisza, Rhine, Marne, Audrey and Volga [12].

2.4 *Wind Driven Mills*

As it was mentioned in previous section numerous water mills were built throughout the world. However, where power of the water was not available, wind was used [2]. A windmill converts the power of the wind into energy through the rotation of a wheel made up of adjustable sails (blades). This movement is transmitted through a system of gears and shafts to the millstone to grind grain into flour.

Windmills were used throughout the high medieval and early modern periods The horizontal windmills first appeared in Iran during the ninth century (originally used for pumping water then adapted for grinding grain), while the vertical windmills in northwestern Europe in the twelfth century [19] and remained until nineteenth century [2] when they were replaced by steam power as a source of motion. Before the development of steam power, the windmills were the primary source of energy for many productions such as processing of spices, oil seeds, wool, dyes, saw mills, etc.

There are two basic designs of windmills depending on the position of the shaft. It is reasonable to assume that man has used wind to power machines for centuries. Although there are some examples of using wind driven wheel to power the machines in Roman Egypt and China [20, 21], the first practical windmills were designed in Persia (Iran) in the ninth century as being operated in Khorasan (Eastern Iran and Western Afghanistan) [20]. These so called horizontal windmills (Fig. 2.8) had vertical sails made of lightweight wood or some cloth material that rotated in a horizontal plane, around a vertical shaft. These windmills were used to grind grain or draw up water [22].

The vertical-axle windmill had reached parts of Southern Europe at eleventh century. However it is debatable whether or not these Middle Eastern horizontal windmills influenced the creation of European vertical windmill [20]. As it was mentioned earlier in the text, the horizontal-axis (instead of vertical) or vertical windmill is believed to originate from northwestern Europe: France, England and Flanders, from the end of the twelfth century [23]. The reason for this is that

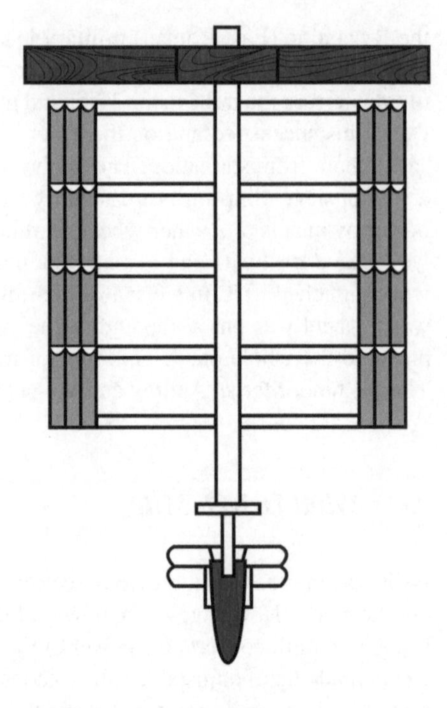

Fig. 2.8 Horizontal windmill

Fig. 2.9 Vertical windmill

Europeans for centuries were acquainted to watermills which are based on the horizontal axis design.

In general, these mills had four blades mounted on a central post and these sails are carried on the horizontal shaft [12]. The gear system inside a windmill (Fig. 2.9) transferred the power from the motion (rotary) of the sails to millstones (or any other machinery). They had a cog on horizontal shaft which translated the horizontal motion of the central shaft, by means of horizontal gearwheel on the top end of the vertical upright shaft, into vertical motion for the millstone (directly) or wheel which would then be used for grinding grain.

Through centuries, the European engineers, especially from the Netherlands, England, Germany, Denmark, were able to significantly improve the windmill technology. These improvements made it possible for the sails to be adjusted to collect the most wind no matter what direction it is blowing. Also the main body structure of the windmill became more permanent, larger and taller and finally ended with multi-story towers, known as a tower mill so more equipment could be housed in it, it became more resistant to weather changes, the sacks of grain could be stored at the foot of the building and even provided the millers a place to live.

All previously described, types of mills (hand mill, animal driven mill, watermill, and windmill) are primarily associated with millstone as a device for flour milling. The invention of steam power triggered the industrial revolution which had its consequences on flour milling as well by enabling the easier transition from millstone to roller mills as a primary milling device. In addition it triggered the invention and use of other equipment for wheat cleaning and condition, sifting, conveying, auxiliary equipment, etc. Also, it enabled complete mechanization and later on complete automation of the process which led far higher processing capacities and establishment of flow sheet of modern flour milling process.

2.5 Steam Engine in Flour Milling

Significant progress in flour milling was made with the invention of the steam engine because it enabled further increase and improvement of the mill plant and the transition of milling from crafts to industry, as it was with all other branches of industry [10]. This enabled the reduction of production costs, mechanization of work, increase of production capacity, improvement of product quality, all of which resulted in better economic effects [12].

With the invention of the steam engine and its application in flour milling, the energy of water and wind gave way to steam power and subsequently to electricity as the mill's energy source [2]. Although flour mills were among the first to employ steam, the steam engine of the first steam mill did not directly move the mill, but served to pump the water to a high reservoir (mill-dam), from which the water flowed to the water-wheel which powered the main shaft of the mill. In year 1785, the first directly driven steam mill was built in London and it became operational in 1786 having 10 millstones for grinding while cylindrical sieves were used for sieving. Actually, the milling was still performed by the millstones which dominated the process until the invention of the roller mills [1, 7, 8].

For centuries and almost until the beginning of the nineteenth century the grinding procedure which was mostly practiced involved scarce cleaning, the grain is washed or moistened and it was ground between two stones and the product is sifted through bags [8]. This basic drawback of this one-pass system is that the outer layers of the grain kernel are ground as finely as the endosperm so the separation of flour from bran particles is difficult [24]. Even when people started to mill several times on consecutive millstones, the stones was always set low and goal was to get as much flour as possible on the first pass through the stone. Certainly, the millstones have improved over time, as well as the types of stone used. Sandstone, basalt, granite, were replaced by high-quality freshwater quartz, known as Champagne or French stone [8]. The surface of the stone was also improved over time, with upper and lower stones having identical cuts [2] which required occasional sharpening and roughing because of wear of the surface (Fig. 2.10).

One of the fundamental change in flour milling technology was developed by French and Hungarian millers in the seventeenth and eighteenth centuries. This new milling system linked together several sets of stone mills, the gap between each pair of stones being set slightly closer than the one before [24]. This milling system introduced intermediate sieving which meant that after each grinding pass the obtained flour was sieved and some bran separated, before the remaining material was sent to the next set of stones. Also, in contrary to previous practice where stones were set "low", in this new process the millstones were set "higher" so they can perform a gentle grinding action resulting in more efficient dissociation between flour and bran [11, 25]. This was the actual beginning of the gradual reduction flour milling process which still exists in the modern flour mills. However, this milling system was perfected after the introduction of roller mills.

The invention of the steam engine triggered the industrial revolution but this period (the end of the eighteen century) is also characterized by significant changes in the political, social and economic spheres primarily manifested in French Revolution and America war for independence. The economic structure of feudal Europe was ended and industry turned to the principles of organized production in order to make a profit. As Kozmin [1] stated America was the first country to adopt the principles of capitalism in flour milling which led to the next significant

Fig. 2.10 Millstone cut

progress in the development of milling process—automated mill. Here, mechanical sifters in the shape of cylindrical and polygonal sieves are included in the process, instead of the sifting bags used in Europe. In addition, the flaxen tissue in sifting bags was replaced by wool, then by wire, and finally by silk. At the beginning of the nineteenth century, the bucket elevators and screw conveyors were included into US mills so the flour milling process became mechanized and continuous process [2]. These improvements have had a significant influence on milling in European countries and in general.

2.6 The Emergence of New Machines in Flour Milling: Roller Mills, Plansifters and Purifiers

The first mention of a kind of roller mill was given at the end of the sixteenth century by Rameli, military engineer of the Italian Renaissance. The first roller consisted of a grooved (sharpened) log and a grooved steel coating, which could be tightened against the log if necessary, which caused stronger or weaker grinding of grain [10, 12].

The invention of the roller-mill dates to period around year 1821–1822 and can be attributed to work of Helfenburger (Rohrschach), Ballinger (Vienna) and Kollio (Paris). In 1823, Müller (Lucerne) began building (Warsaw, Trieste, Frauenfeld) steam driven mills operated by iron rolls instead of millstones. However, the shortcomings of these first roller mills were eliminated around 1833–1834 by the Swiss engineer Sulzberger (Zurich) who invented the first successful system of grinding by rolls. Soon, in 1837 Sulzberger type of mills were constructed by in Mainz, Milan, Munich, Leipzig, and Stettin, and in 1839 Budapest [1, 8, 10, 26]. The flour obtained from these mills gained popularity because of appearance and high quality. Despite the fact that the millers achieved good results, rollers did not become domesticated in flour milling until the seventies of the nineteenth century [12]. In fact, an era of struggle between the rollers and millstone began. During forties and up to seventies of nineteenth century roller mills struggled against millstones. Of course there was a lot of conservatism, keeping the old, but the fact is that millers were acquainted to well establish so called French process in automated stonemills. In addition, one of the main argument against rollers was the expense of the roller mills themselves, the first rollers consumed a lot of energy, broke down, required repairs and caused frequent interruptions in the operation of the mill [1].

The breakthrough considering the construction of roller mills happened in the seventies of the nineteenth century. Swiss engineer Wegmann constructed roller mill made of porcelain whose patent was purchased by the Ganz company in Budapest. Mechwart, mechanical engineer and chief executive of the Ganz Works, significantly improved the Wegmann roller mill by changing the porcelain rolls to wear-resistant, hard cast, diagonally grooved grinding rolls and patented it [8, 27]. These perfected roller mills became world famous and this can be marked as the beginning of the period when roller mills began to replace millstones on massive scale. The roller mills increased the capacities of flour mills, as well as amount, diversity and quality of flour and speed up the milling process compared to millstone ground operation. As it was mentioned earlier in the text the gradual reduction milling system was widely practiced in France and Hungary. During the seventies of the nineteenth century English engineer Simon (Manchester) perfected the gradual reduction system whose main features were the use a large number of process stages and exclusive use of roller mills for grinding. The principle of gentle grinding and intermediate sifting between two grinding operation develop to break, purification and reduction system [11] which still are, after almost 150 years, the backbone of the modern flour milling process.

The success of the roller mills can be attributed to the fact that they provide a controlled and selective grinding action, i.e. the best possible dissociation between the inner part of the grain kernel—endosperm (flour) and outer layers (bran) and economy of operation [28]. The grinding of the grain should be as selective as possible, i.e. the grinding of the endosperm should not be accompanied by the same degree of size reduction of outer layers (pericarp, aleurone layer and germ), which should be separated as completely as possible in the form of bran. The control of the grinding action is due to the fact that roller mills are once-through type of mills where retention time of the particles in the mill is short, measured in [ms] [29, 30]. Also, roller mills offer the possibility of controlling the grinding action by controlling the type and the intensity of stress imposed on particles in the grinding zone. This is achieved by the adjustment of a set of roll parameters.

In roller milling, the feed material is passed between two counter-rotating rolls with either a corrugated or a smooth finish [28]. Very important feature of the roller mills is that there is a difference in speed (or rpm—revolution per minute) between two rolls, fast and slow rolls, and this difference is defined by differential. If two rolls were to rotate at different speeds, shear stress would be induced in the particles [29]. Figure 2.11 shows the particle being drawn into the grinding zone of the rolls which starts from the point where a particle is initially engaged between two rolls to the point at which the gap is at minimum.

Particles passing through the grinding zone of the roller mill are subjected to shear forces from contact between points on the particles and the roll surfaces, and compressive forces on the particles as a whole. Shear force from the side of the fast roll (F_F) tends to draw particle into the roll nip while shear force from the side of the slow roll (F_S) tends to eject it because of the different speeds of the rolls. Particle is also subjected to the compressive forces (F_{F1} and F_{S1}) reaching the maximum at the line connecting the centers of the cross section of the rolls [31, 32]. Once the particle is drawn into the roll nip, the strain increases as particle goes toward the roll nip, and the particle is crushed. The particles are held on the slow roll [33] while the rotation of the fast roll causes both compressive and shear deformation. The set of roll parameters: the roll gap, the feed rate of stocks to rolls, the roll speeds, the roll differential (the ratio of speeds of the fast and slow rolls), and the type and condition of roll surface (corrugated or smooth), influence the magnitude of the stress and the relative contributions of compressive and shearing forces which determine the

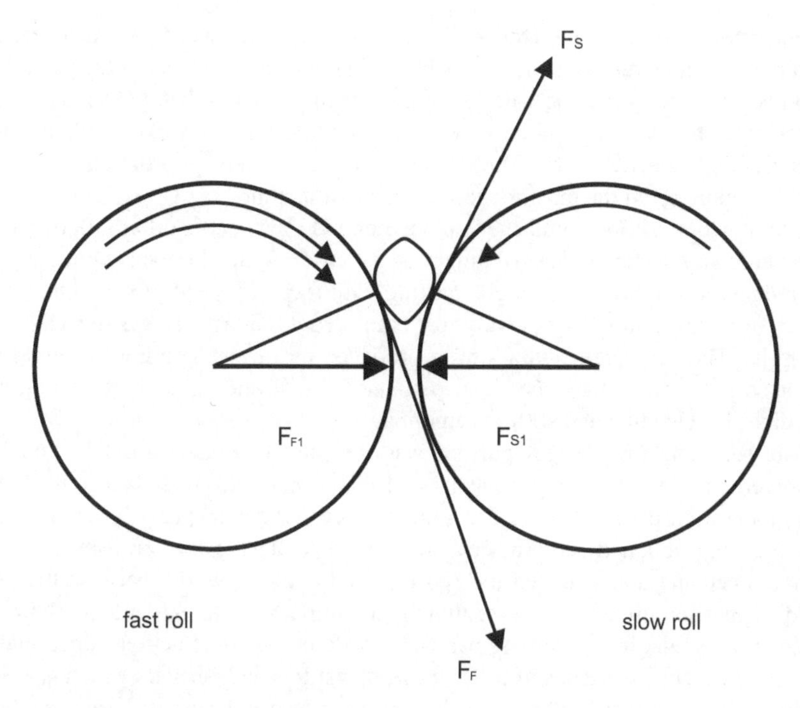

Fig. 2.11 Grinding zome of the roller mill

degree of particle size reduction, energy required for grinding and bran contamination of the flour [34].

The gradual reduction system which was adopted by most millers required higher capacities and greater area of sifting surface. The horizontal separating reel and centrifugal sifting machines could not meet these requirements (only 1/3 of the sieving surface is actually usable) and soon they were supplemented by the plansifters [2]. Invention of the plansifter, by Hungarian engineer Haggemacher in 1887, represented an enormous improvement in sieving technic [35]. This sieve consisted of several frames—flat sieves, one above the other, with bolting clothes of various size openings which enabled easier separation of the crushed material into products of various particle sizes and qualities [10]. It revolutionized bolting methods in flour mills because it took up less space and required less power than machines previously used for same functions.

Haggemacher got an idea by watching how housewives sifted flour on a hand sieve where the motion of the sieve is gyratory in horizontal plane. He noticed that heavier and smaller products fall through the sieve, and lighter and larger ones remain on the surface of the sieve. He made the first plansifter with 10 frames placed one above the other while hanging on iron rods. The material is introduced at one end and pushed to the other end by means of laterally placed blades (lamellas). Besides numerous advantages, the first Hagenmacher's plansifter had some shortcomings, mainly considering the balance, and attracted the attention of many milling engineers. Around 1890–1892, Luther and later Konegen made certain repairs in the construction considering balancing, and created a perfected plansifter. Important feature of the construction of the hanging sifter is that it always remains in balance. This system required very little driving power and worked with no interference. The capacity of the sifter is relatively large, takes up little space, and is therefore considered the most modern creation of its time [12].

Combination of Haggenmacher's plansifter and Mechwart's roller mill triggered a revolutionary change in the grinding process. By applying these machines, a gradual reduction process, i.e. a "high" milling with focus on yield of semolina (large pieces of endosperm with relatively few bran particles attached) was introduced to the mills. The best quality flours were produced by further grinding of semolina. Advances in sifting have been accompanied by inventing and perfecting the machines for cleaning and sorting semolina.

The semolina (middlings) purifier was invented in France around 1860 [2]. However, only in 1896 Haggenmacher and Wall, millers from Budapest, made the first modern sense semolina purifier with two sieves and a fan [12]. These machines classified the obtained semolina by size and separate the fine bran by air flow. Purifiers become domesticated in flour mills all over the world, because the high yield of pure semolina was an essential precondition for obtaining quality flour.

Over the years, huge progress has been made in the construction, functionality and technical characteristics of roller mills, plansifters and purifiers which has contributed into a significant increase in their performance and capacity. However, their basic working principles remained the same and they still are the principal machines in the modern flour milling process. Increased output of these machines required equally capable conveying system. Increased need for streamlined, cleaner and more efficient handling system has been met with pneumatic conveying (transport of powdered or granular material through pipes by air). It represents the most significant development in this area whose origins date back to the mid-twentieth century [2]. Over the years, bucket elevators have been gradually replaced with the pneumatic conveying system that are used exclusively today.

In the second half of the last century, progress in flour milling technology included further improvement of existing machinery, changes in flow sheet to accommodate auxiliary machines, on-line quality control, automatization, remote control, computerization and adoption of information-based technologies [2, 11, 28].

3 Modern Flour Milling Process

This section aims to provide a brief overview of the process applied in today's flour mills represented on the simplified block diagram (Fig. 2.12).

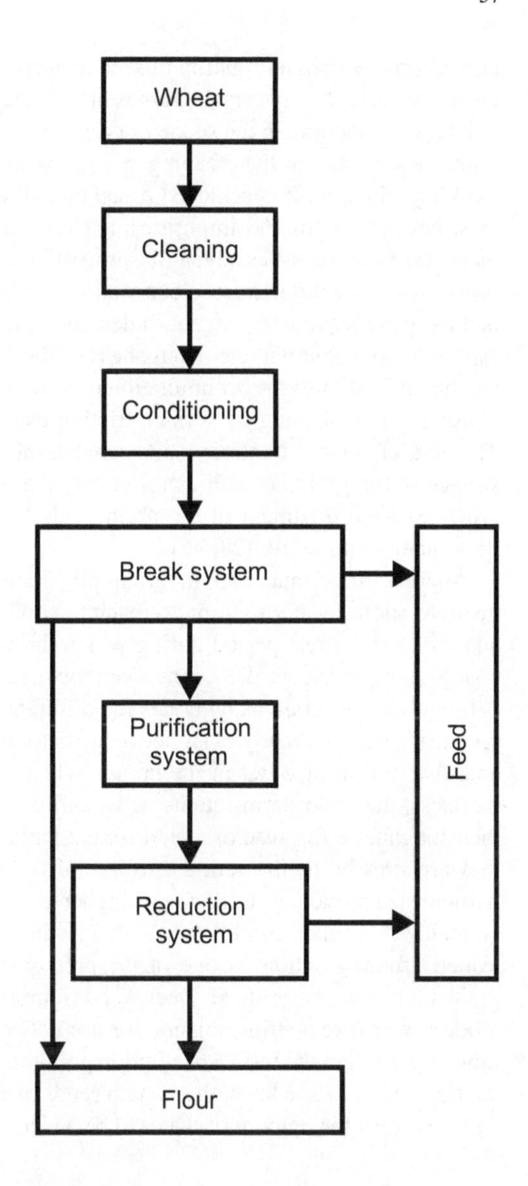

Fig. 2.12 Simplified block diagram of flour milling process

3.1 Wheat Cleaning and Conditioning

Wheat grain sent to mill for processing always contains, to a greater or lesser extent, foreign material such as straw, wood, stones, metal, weed seeds, etc. They reach the grain in different ways, mostly in the field and their amount depends on farming and environmental conditions, such as: adequate crop rotation, level of applied agricultural techniques, quality of seed material, the adjustment of machines during the harvest, etc. Upon arrival, grain is pre-cleaned before storage and later subjected to intensive cleaning in the scope of preparation for milling. The separation of these impurities is done in the cleaning system by multiple machines based on different working principle. It is achieved based on the differences in certain properties that exist between grain and impurities, such as: size (width, length, volume), shape, magnetic properties, aerodynamic properties (air resistance), elasticity, frictional properties, color differences, electrostatic properties, etc. The usually used machines include sieve separator, magnet, indented cylindrical separator—trieur, dry stoner, aspirator, optical sorter, etc. Roughness of the grain surface, the presence of crease and brush facilitates the accumulation of impurities. A very important aspect of this phase is grain cleaning by surface friction usually achieved by horizontal scourer. The goal of surface treatment is to remove the impurities and microflora from the surface of the grain, as well as to remove the peripheral layer of the husk and the brush. Surface treatment of the grain, and cleaning in general), greatly improves flour quality and safety [28, 36].

Another important step in grain preparation for milling is conditioning. It involves adding water to grain to reach the optimum milling moisture (between 14 and 17%) and a rest period during which the water penetrates into kernel. Proper conditioning weakens the connection between the anatomical parts of the grain. Adding water to grain exaggerates the differences in the structural and mechanical properties which already exist between anatomical parts. With the optimal content and distribution of water in the kernel, which should be in accordance with grain hardness, the endosperm retains its friability, while germ and bran layers increase their toughness. Because of that, different parts of the grain reacts differently when they are stressed by the action of roller mills, i.e. there is a difference in degree of particle size reduction. In this way, higher yield of flour with a defined ash content is achieved. Proper conditioning, along with gradual reduction system and controlled grinding action, is one of the prerequisites for selective milling. Optimal grain moisture also reduces energy consumption during milling [28, 36].Today, modern wheat tempering mixers are used. The system continuously measure the flow of grain and the moisture of grain going in our out of the tempering. Based on the targeted moisture level, the system sends back a signal to water dosage regulator which control the amount of water to be added.

3.2 Grinding Process

As it was mentioned in the text, flour milling is a gradual reduction process consisting of sequential and consecutive size reduction and separation. The grinding passage is a combination of at least one pair of rollers and one section of a plansifter which process the stock following the rollers. More pairs of rollers and several sections of plansifter can be engaged on the same grinding passage, provided that they process the same material. After every grinding step the ground material is sieved and the undersize material removed before regrinding. This means that the grinding passages throughout the flour milling process differ in characteristics of the input material to the rolls, as well as in the set of roll and sifting parameters which are adjusted according to the characteristics of the input material [37].

The break system is the most complex and most important phase in flour milling process. The success of the entire process of flour production depends on the management of this phase. The objective of the break system is to open the wheat kernel and remove the endosperm and germ from the bran coat with the least amount of bran contamination, and at the same time, obtain a granulation distribution of maximum large middlings with a minimum of flour and small middlings. However, some shattering of the bran occurs and results in a mixture of endosperm and bran in the released middlings. This is achieved by multi-stage milling with up to four or five grinding passages. The first three break passages, called the primary or head break system, release relatively pure endosperm particles. Practically they determine the yield and quality of flour that will be obtained by grinding pure middlings (semolina) in the reduction system. Secondary or tail break system (4th and 5th Break) cleans up the bran and releases smaller pieces of endosperm together with fine pieces of bran and germ [28]. After every roller milling operation, the stock is sifted and classified according to size into: overtails—sent to next break, sizings, coarse and fine middlings (particles of endosperm that yet to be ground to flour)—sent to sizing or purification system, and flour. Remaining material from the last break passage is discharged as a by-product to feed [11]. Break passages employ corrugated rolls with differential of 2.5. Going from 1st to 5th break, the number of corrugations increases, while the roll gap decreases.

As it was mentioned, middlings released from the break system represent mixture of endosperm and bran particles. Actually, middlings obtained by sorting on plansifter, in addition to the particles of pure endosperm, also contains particles of endosperm with adherent bran particles and pure bran particles. Separation of these particles, which cannot be achieved by sieving on plansfiter (separation is based only on difference in size of the particles), is performed on purifiers where sieves together with air-flow separate particles of different size, specific gravity and shape [38, 39]. Also, it grades the pure endosperm into particle size ranges. The clean endosperm from the purifier is sent to the head end of the reduction system while the branny materials to the tail break system [11, 28].

The main goal of sizing system is to reduce the size of coarse middlings or sizings and produce clean midllings, while minimizing flour production and shattering of bran. Reduction system represents the final stage of wheat flour milling process. The objectives are to maximize the yield of quality flour, control starch damage and minimize the amount of fine bran particles in flour. The best results in this phase are achieved by using smooth rollers together with small differential of 1.25 while the roll gap is gradually reduced from the head to tail-end reduction passages (around eight grinding passages). The grinding zone is dominated by compressive forces which are effective in fracturing endosperm particles. This creates more flour while flattening the tougher branny particles, which would not pass into the flour. Bran, being tough and fibrous, is more prone to fracture imparted by shear forces [34, 40,

41]. After grinding on smooth rollers, the stock is sifted on plansifters, whereby 2 or even 3 flours of different granulation are obtained. The overtails depending on composition, predominantly endosperm or bran particles, are directed to the following passages in the reduction system or to the appropriate passage of the break system, respectively. Remaining material from the last reduction passage is discharged as a by-product to feed.

3.3 *Auxiliary Machines in the Process*

It is fair to say that the milling industry is very conservative. After more than 100 years, roller mills and plansifters still remain the primary machines used in the process. On the other hand, millers always sought out for possibilities to simplify the process and make it more efficient in terms of reducing the investment, operating and maintaining costs as long as the quantity and quality of the finished products are not affected [42].

Over the decades, roller mills and plansifters have been redesigned to such an extent that it has been possible to multiply the throughputs of these machines but flour milling technology has not changed fundamentally [43]. The emphasis in this area has been on improving the effectiveness of these machines so the short surface mill is now the norm [11]. The term roll surface is expressed as the unit of roll length per day capacity of the mill (24 h). By improving the roller mill construction and design it was possible to increase roll velocities which made possible to increase loads to the roller mill and reduction of roll surface [28]. The efficiency of plansifters has been increased by making them larger, with up to 30 stacked sieves, more space efficient and by increasing sieving rates [11]. Over the years, rationalization of the process has been achieved by increasing the efficiency of the roller mills and plansifters as well as by supporting and supplementing the rolls with more auxiliary machines. By introducing bran finishers, drum detachers and impact (pin) mills it was possible to reduce the number of rolls and therefore the number of process steps making the shorter roll surface a reality [28, 44].

Bran finishers are used for additional scraping of floury endosperm remained on bran particles. Usually they are included on the third break passage (and other following break passages) for the treatment of overtails which are directed from the corresponding section of the plansifter to the rolls of next break passage. Some materials from tail-end reduction passages also can be treated by bran finishers. The machine consists of steel bars attached on the rotating shaft enclosed in a perforated metal cylinder. Endosperm particles are released by friction (should prevail) and impact and they are forced through the screen by the action of the beaters [28].

Drum detachers are designed for the gentle detaching of the stock flakes following the grinding on reduction passages. At tight roll gaps, together with smooth surface of the rolls and small differential of 1.25, the compressive forces could be so intense that fractured endosperm particles create flakes. Flakes would reduce the flour yield and commercially flake disrupters are commonly used before sieving to overcome this. They are positioned immediately after the roller mills or just before plan sifter inlet (after the air locks of the pneumatic conveying system) [45]. The material to be detached is fed through the inlet directly to the rotor and is caught by the inner surface of the jacket, which is installed with number of impact bars. The pitch of beaters directs the material from the feeding point to the discharge point of the machine. The use of these machines should increase flour yields without disintegrating any bran and germ particles. In order to do that the rotary speed is limited to a maximum 14–16 [m/s] [46]. They allow the sifter passage to perform more effectively and prevent losing good stock to by-products.

In impact mills the particles are struck by a high–velocity impeller or accelerated and thrown against a wall and crushed by the force of the impact [47]. The internal parts can be rotating blades or pins. However, they are not selective in grading and they are used to produce fine bran powder when used in the break system or for grinding of clean middlings at the head of the reduction system [28]. In this way they are reducing the number of roller mills in the process [11].

3.4 New Machines in the Process

As it was mentioned in previous section the emphasis considering the developments of flour milling process has been on improving the effectiveness of these machines rather than on new types of machine [11]. However, introduction of some new machines in the flow sheet such as debranning machine, eight-roller mill (also known as two high or double high roller mill), and near infrared reflectance (NIR) instruments, can be marked as the most notable ones.

The use of debranning machines to remove the bran ahead of the first break can simplify break and reduction steps of conventional milling [48, 49]. It is an operation in which the bran layers are removed by applying friction and abrasion. Over the years, a number of debranning machines have been developed whose work principle is based on successive exposure of grain to friction and abrasion or a system where the mentioned phenomena (friction and abrasion) take place simultaneously. Friction and abrasion can occur in separate devices, different parts of the same device, or they can occur simultaneously in the same device [50–52]. Abrasion occurs as a consequence of grain contact with abrasive parts of the device (abrasive discs or stones) and slotted screens. Friction occurs in friction chambers, while grain moves with the help of paddles, as a result of kernel-to-kernel friction and friction between the kernels and the special screen [28]. By including debranning machines in flour milling process, the capacity of the mill is increased by the fact that the bran is removed from the surface of the kernel and the endosperm is exposed. Because of that the number of grinding steps (grinding passages) for the same processing capacity is reduced. It results in lower capital investments because the mill flow sheet is simplified, less equipment is required in the process and greater compactness of the plant is achieved [48]. Also, debranning reduces the number of molds and bacteria (located in the outer pericarp) in flour [51, 53]. On the other hand, by debranning the kernel physical properties are significantly changed so the mill flow for grinding debranned wheat differs from conventional one [28, 52]. In this way a relatively large amount of pure middlings (semolina) are obtained relatively quickly while the first and second break passages function more like sizing stages than like conventional brakes [28]. Once purified, the semolina can be immediately sent to the reduction system to be grind the appropriate particle size [54].

The traditional wisdom in flour milling is that after every grinding step the ground material should be sieved and the undersize material removed before regrinding. Double grinding of intermediate streams, without intermediate sifting after every roller milling operation, has been one of the most notable process developments in flour milling [11]. The eight-roller mill (8 rolls in one housing), also known as two high or double high roller mill, provides two grinding passages without any intermediate sifting. Many flour mills introduced eight-roller mill in order to reduce capital cost. Compared to a conventional flour milling process, the introduction of the eight-roller mill into the milling flowsheet offers numerous advantages such as: smaller number of roll stands, reduction of sifter surface, fewer pneumatic suction lifts (conveying the stock from the roll to the sifter) resulting in lower material and installation costs, lower pneumatic system air requirements resulting in lower energy costs and lower filter surface requirements for cleaning the pneumatic conveying air, less spouting and auxiliary components, lower space requirements for equipment installation, less cleaning and maintenance [28, 42–44, 55–57]. On the other hand, with twin stage grinding, coarse material is not separated from the fines and therefore the conditions for controlled milling are less favorable. The decision as to how many double grinding passages can be applied in the flowsheet depends directly on the finished products to be made. It is not always possible to equip mill exclusively with eight-roller mills [43, 57]. Disadvantages of the eight-roller mill concept are mainly coming from the fact that the undersize material produced on the upper pair of rolls is not removed before regrinding on the lower pair of rolls which are not design to mill them [58]. With twin passage, the particle size distribution of the stock will not be the same as with conventional system [42, 57, 59]. Appropriate adjustment of the roll and/or sifting parameters is needed regardless of whether the eight-roller mill is used on breakage or reduction passages [41, 60–63].

The main advantage of NIR technology is that provides rapid measurements and reliable results. NIR instruments are used for monitoring the characteristics of end-products such as moisture, ash and protein content. Also, NIR instruments are used for on-line control enabling millers to observe differences in intermediate stock quality. Any deviation from targeted range electronically activates machine adjustment. In this way, it is used as a part of decision making system in the mill in cases of changes that can influence milling performance [28].

References

1. Kozmin PA. Flour milling (translated by Falkner M and Fjelstrup T). New York: D. Van Nostrand Company; 1917.
2. Bass EJ. Wheat flour milling. In: Pomeranz Y, editor. Wheat: chemistry and technology, vol. 2. 3rd ed. American Association of Cereal Chemists: St.Paul, Minnesota; 1988. p. 1–68.
3. Serna-Saldivar SO. Cereal grains: properties, processing, and nutritional attributes. Boca Raton: CRC Press, Taylor & Francis Group; 2010.
4. National geographic society: the development of agriculture. 2016. https://web.archive.org/web/20160414142437/https://genographic.nationalgeographic.com/development-of-agriculture/. Accessed 25 Jul 2020.
5. Mazoyer M, Roudart L. A history of world agriculture: from the neolithic age to current crisis. New York: Monthly Review Press; 2006.
6. Merriam-Webster: Cereal. In: Merriam-Webster.com dictionary. (n.d.). https://www.merriam-webster.com/dictionary/cereal. Accessed 25 Jul 2020.
7. Halliwell W. The technics of flour milling. London: Straker Brothers Limited, The Bishopsgate Press; 1904.
8. Stürzinger E, Künzlo A. Vermahlungs-und Betriebskunde. Interkantolale Berufsbildungskommisision für müller: Uzwil; 1971.
9. Krička T, Kiš D, Matin A, et al. Tehnologija mlinarstva (technology of milling). Poljoprivredni fakultet u Osijeku, Agronomski fakultet u Zagrebu: Osijek; 2012.
10. Šenborn A. Tehnologija brašna - Mlinarstvo (Technology of flour – Milling). Novi Sad: Tehnološki fakultet (Faculty of Technology); 1973.
11. Owens WG. Wheat, corn and coarse grains milling. In: Owens WG, editor. Cereals processing technology. Cambridge: Woodhead Publishing Ltd., 2001. p. 27–52.
12. Kristoforović I. Mlinarstvo (Milling). Trgovinska knjiga: Beograd; 1951.
13. Jacob HE. Sechstausend Jahre Brot. Rowohlt Verlag: Hamburg; 1954.
14. Haque E. Application of size reduction theory to roller mill design and operation. Cereal Food World. 1991;36:368–74.
15. Peacock DPS. The mills of Pompei. Antiquity. 1989;63:205–14.
16. Buffone L, Lorenzoni S, Pallara M, et al. The millstones of ancient Pompei: a petro-archaeometric study. Eur J Mineral. 2003;15:207–15.
17. Staccioli, L.: Bread and bakers in ancient Rome. (n.d.). http://www.cerealialudi.org/en/alimentazione/pane-e-panettieri-nellantica-roma/. Accessed 28 Jul 2020.
18. Local heritage initiative: history and technology of watermills. (n.d.). http://www.jesmond-deneoldmill.org.uk/mill/technology.html. Accessed 28 Jul 2020.
19. Glick TF, Livesey S, Wallis F. Medieval science, technology, and medicine: an encyclopedia. New York: Routledge, Taylor & Francis Group; 2006.
20. Lucas A. Wind, water, work: ancient and medieval milling technology. London, Boston: Brill Publishers; 2006.
21. Drachmann AG. Hero's windmill. Centaurus. 1961;7:145–51.
22. Wailes R. Horizontal Windmills. Transactions of the Newcomen Society. 1967–68;XL:125–45.
23. Braudel F. Civilization and capitalism, 15th–18th century, vol. I. California: University of California Press; 1992.
24. Catterall P. Flour milling. In: Catterall P, Young LS, editors. Technology of breadmaking. Dordrecht: Springer Science+ Business Media; 1999. p. 296–329.
25. Storck J, Teague WD. Flour for man's bread, a history of milling. Minneapolis: University of Minnesota Press; 1952.
26. Edgar WC. The story of a grain of wheat, vol. 4. New York: D. Appleton and company; 1903. p. 200.
27. Farkas J. Malomipari gépek és karbantartásuk. Mezőgazdasági Kiadó: Budapest; 1978.
28. Posner ES, Hibbs AN. Wheat flour milling. American Association of Cereal Chemists: Minnesota; 2005.

29. Fang C, Campbell GM. Stress-strain analysis and visual observation of wheat kernel breakage during roller milling using fluted rolls. Cereal Chem. 2002;79:511–7.
30. Fang C, Campbell GM. Effect of roll fluting disposition and roll gap on breakage of wheat kernels during first-break roller milling. Cereal Chem. 2002;79:518–22.
31. Fistes A, Tanovic G. Effect of smooth roll grinding conditions on reduction of sizings in the wheat flour milling process. In: Pletney NV, editor. Focus on food engineering research and developments. New York: Nova Science Publishers; 2007. p. 453–66.
32. Мерко ИТ. Совершенствование технологических процессоб сортовога помола пшеницы. Колос: Москва; 1979.
33. Scott JH. Flour milling and processes. 2nd ed. London: Chapman and Hall; 1951.
34. Scanlon MG, Dexter JE. Effect of smooth roll grinding conditions on reduction of hard red spring wheat Farina. Cereal Chem. 1986;63:431–5.
35. Baumgartner F. Handbook of mill construction and milling. D. & S. Loewenthal: Berlin; 1900.
36. Fišteš A, Tanović G. Praktikum iz Tehnologije mlinarstva (Handbook on milling technology). Tehnološki fakultet: Novi Sad; 2014.
37. Fistes A, Rakic D, Takaci A, et al. Solution of the breakage matrix reverse problem. Powder Technol. 2014;268:412–9.
38. Posner ES. Wheat. In: Kulp K, Ponte JG, editors. Handbook of cereal science and technology. New York: CRC Press, Taylor and Francis Group; 2000. p. 1–30.
39. Flamer R. Designing a purifierless millflow. Milling J. 2008;First Quarter:20–2.
40. Scanlon MG, Dexter JE, Biliaderis CG. Particle-size related physical properties of flour produced by smooth roll reduction of hard red spring wheat Farina. Cereal Chem. 1988;65:486–92.
41. Fistes A, Tanović G, Mastilović J. Using the eight-roller mill on the front passages of the reduction system. J Food Eng. 2008;85:296–302.
42. Baltensperger W. New development in the mill flow charts grinding process using eight-roller mills. Assoc Operative Millers Bull. 1993;December:6327–32.
43. Baltensperger W. State-of-the-art grain milling technology. Assoc Operative Millers Bull. 2001;January:7583–92.
44. Wanzenried H. Benefits and results with 8-roller mill, model MDDL. Assoc. Operative Millers Bull. 1991;December:5977–81.
45. Fišteš AZ, Tanović GM. The effect of using the drum detachers in the industrial wheat flour mills. Acta Periodica Technologica. 2013;44:49–56.
46. Cinquetti M. Molini da grano, vol. 2. Chiriotti Editori: Pinerolo; 2000.
47. Hibbs A, Pence R, Shellenberger J. Impact milling. Northwest Miller. 1947;232(10):2a.
48. Dexter JE, Wood PJ. Recent applications of debraning of wheat before milling. Trends Food Sci Tech. 1996;7:35–41.
49. Lin Q, Liu L, Bi Y, Li Z. Effects of different debranning degrees on the qualities of wheat flour and chinese steamed bread. Food Bioprocess Tech. 2012;5:648–56.
50. Tkac JJ. Process for removing bran layers from wheat kernels. US Patent 5,082,680; 1992.
51. Hemery Y, Rouau X, Lullien-Pellerin V, et al. Dry processes to develop wheat fractions and products with enhanced nutritional quality. J Cereal Sci. 2007;46:327–47.
52. Ranieri R. Wheat debranning: industrial applications on durum wheat. In: 6th annual IAOM Eurasia District, Florence, Italy. 2011. http://www.openfields.it/en/iaom-eurasia-report/. Accessed 21 Sept 2020.
53. McGee BC. A new rollermill and debranner for use in a compact mill. Assoc. Operative Millers Bull. 1996;January:6674–5.
54. Willis M, Giles J. The application of a debraning process to durum wheat milling. In: Kill RC, Turnbull K, editors. Pasta and semolina technology. Oxford: Blackwell Science; 2001. p. 64–85.
55. Baltensperger W. Particle size of semolinas for pasta production. Assoc Operative Millers Bull. 1997;March:6879–84.
56. Eugster W. Advances in process technology. Assoc. Operative Millers Bull. 2001;October:7706–07.

57. Tegeler VC. Eight-high roller mills vs. four-high roller mills the pros and cons. Assoc. Operative Millers Bull. 1999;February:7229–30.
58. Flamer R. Developments in flour milling. Milling J. 2008;Fourth Quater:14–6.
59. Handreck BL, Pötschke L, Senge C. Intensives Aufschroten von Weizen im Achtwalzenstuhl. Die Mühle + Mischfuttertechnik. 1999;26:818–26.
60. Zwingelberg H. Verschiedene Walzenstuhlbeschüttungen und deren Auswirkungen auf Produktanfall und Mineralstoffgehalt Teil 2: Versuche an nicht geputztem Grieß als Einzel - und Doppelvermahlung. Die Mühle + Mischfuttertechnik. 1998;20:649–54.
61. Zwingelberg H, Arzt B. Verschiedene Walzenstuhlbeschüttungen und deren Auswirkungen auf Produktanfall und Mineralstoffgehalt Teil 1: Versuche beim I. Schrot und als Doppelvermahlung beim I. Und II. Schrot. Die Mühle + Mischfuttertechnik. 1998;18:593–5.
62. Fistes A, Rakic D. Using the eight-roller mill in the purifier-less mill flow. J Food Sci Tech Mysore. 2015;52:4661–8.
63. Fišteš A. Comparative analysis of milling results on the tail-end reduction passages of the wheat flour milling process: conventional vs. eight-roller milling system. Hem Ind. 2015;69:395–403.

Chapter 3
Wheat Bread in the Mediterranean Area: From Past to the Future

M. Paciulli, P. Littardi, M. Rinaldi, and E. Chiavaro

Abstract Bread origins are closely linked with those of the first civilizations of the Mediterranean area. Today bread is spread all over the world and available in many types. The relationship between ingredients, processes and quality of the final products is very close. The use of additional ingredients beside those basic (i.e. wheat flour, water and yeast), and the advent of new technologies have allowed a wide diversification of bread products over the centuries. Despite the technological evolution, bread is still a very traditional product, rooted in the territories of origin and in the habits of some places. In this chapter, an overview on ingredients, technologies, and new trends in breadmaking are described, highlighting the strong link between tradition and innovation existing for this kind of products.

Keywords Leavened bread · Flat bread · Fortification · Ancient grains · Ingredients

1 Overview on Bread in the Mediterranean Area

Bread's birth is indissolubly tied to the history of cereals that seems to plunge one's roots between Palaeolithic and Neolithic eras, when some population discovered wild wheat and barley plants near rivers and started fed with sprouted, fermented, roasted and boiled grains. Subsequently, but long before the invention of agriculture, they began to grind/mill, knead with water and bake this raw material to produce the bread forefather. The plantation of cereals (e.g. *Hordeum vulgare* L., *Triticum monococcum* L. and *Triticum dicoccoides*) and the production of flour for breadmaking took hold about 4000 years Before Christ during the Uruk civilization in the Fertile Crescent (between Tigris and Euphrates rivers) and gradually expanded in all the Mediterranean Basin [1].

M. Paciulli (✉) · P. Littardi · M. Rinaldi · E. Chiavaro
Department of Food and Drug, University of Parma, Parma, Italy
e-mail: maria.paciulli@unipr.it

F. Boukid (ed.), *Cereal-Based Foodstuffs: The Backbone of Mediterranean Cuisine*, https://doi.org/10.1007/978-3-030-69228-5_3

One of the first civilization producing bread was the Egyptian: initially they produced it with only wheat flour and water in the form of unleavened bread. Also, when Israeli clans began the exodus, leaved home with unleavened bread, that is easy to preserve from deterioration. This kind of bread has a flat shape which enable a rapid baking and easy preservation from spoilage and extended shelf life and consequently it facilitates the nomadic life in the desert. As converse the leavened bread, as we usually know it with an aerated crumb, is typical of settled peoples, as the ancient Greeks and Romans; the latter even establishing the use of public ovens.

Starting from the ancient Mediterranean population and over the centuries, bread became the symbol of the survival, due to its simple and cheap ingredients (only wheat flour and water). It assumed different characteristics considering the different geographic areas, where it was produced. Nowadays, bread is considered one of the first food produced after planning and designing by mankind, not simply harvested, or hunted. A large array of bread types is available including knead or not, with or without fat, salted/unsalted or sweet, flat, or leavened, big, or small, thick or thin, starting from a dough or a batter. In addition, different proofing method (brewer's yeast, sourdough, baking powder, ecc) can be used for bread making. Obviously, all this kind of bread present different structural, sensorial, and nutritional characteristics based on recipes and processing. To fill in the blanks, in the recent years, new trends in the baking sector emerged considering the use of innovative ingredients such as pseudocereals, pulses and also in the optic to produce enriched bread with functional ingredients from different origin and giving particular nutritional characteristics to the final product.

2 Main Ingredients in Breadmaking

Flour and water are the most significant ingredients in a bread recipe and most of the bread is traditionally produced from wheat flour. Besides, more ingredients can be added playing different roles as summarized in Table 3.1.

2.1 Wheat Flours

Durum (*Triticum turgidum* L.) and soft/bread (*Triticum aestivum* L.) wheat are two cereal species well adapted to Mediterranean environments, with durum wheat mainly used to produce semolina for the pasta industry, whereas bread wheat flour is mainly used for bread production. Durum wheat is traditionally assigned to lower-yielding, more stressful environmental conditions compared with bread wheat. Bread and durum wheat also differ in terms of grain protein percentage—usually higher in durum than in bread wheat [19]. The use of durum wheat in breadmaking has been restricted because durum wheat gluten proteins generally lack the elastic strength of some of the strongest bread wheats. New cultivars of durum wheat with

Table 3.1 Main ingredients for breadmaking

Ingredient	Type	Role	References
Flour	Wheat flour Ancient grains Other cereals Pseudocereals	Provides structure	[2] [3] [4] [5]
Water		Involved in the formation of gluten and in modification of the dough rheology	[6]
Yeast	Sourdough Compressed yeast	Produces carbon dioxide which causes dough rise	[7] [4]
Salt		Enhances flavours, strengthens the gluten and act on yeast for controlled expansion of the dough	[8]
Sugar		Provides sweetness and helps yeast to begin producing gas	[9]
Fats	Flour lipids Shortening surfactants	Increase dough machinability, increase loaf volume, tenderise crumb structure, extends bread shelf life	[10]
Improvers	Oxidising and Reducing agents, Enzymes, Hydrocolloids	Compensate flour quality deficiencies	[4, 11]
Antistaling agents	Enzymes, Hydrocolloids, Emulsifiers	Prevent starch retrogradaton during storage	[12–16]
Preservatives	Propionic acid, Ethanol, Antioxidants, Biopreservatives	Inhibit growth of moulds and thermophilic bacteria	[4, 17, 18]

stronger gluten properties were introduced in the United States, Canada, and Italy around 1970s. Results of this agronomic research questioned the belief that durum wheat was suitable for high quality pasta production but not for breadmaking. In the Mediterranean area, and particularly in the southern of Italy, durum wheat has been, and continues to be, used in the formulation of several types of bread [20]. In the Middle East and North Africa, local breadmaking accounts for about half of the durum wheat consumption [21].

Wheat flour consists mainly of starch (ca. 70–75%), water (ca. 14%) and proteins (ca. 10–12%). In addition, non-starch polysaccharides (ca. 2–3%), in particular arabinoxylans (AX), and lipids (ca. 2%) are important minor flour constituents relevant for bread production and quality [2], as illustrated in Table 3.2. Starch is present in the flour in the native state where it appears as distinct semi-crystalline granules. The major components of starch are the glucose polymers amylose and amylopectin. Amylose is an essentially linear molecule, consisting of α-(1,4)-linked D-glucopyranosyl units with a degree of polymerization (DP) in the range of 500–6000 glucose residues. In contrast, amylopectin is a very large, highly branched polysaccharide with a DP ranging from 3×10^5 to 3×10^6 glucose units. The amylose/amylopectin ratio differs between starches, but typical levels of amylose and amylopectin are 25–28% and 72–75%, respectively. At room temperature and in

Table 3.2 Wheat flour components and functionality for bread making [2]

Components	Properties	Functionality
Starch	The glucose polymers amylose and amylopectin are the major components In flour, it is in the form of semi-crystalline granules	– Absorb water and in presence of heat during baking gelatinize and swell. Upon cooling it forms a continuous network that act as structural element
Proteins	Between 80 and 85% are represented by the gluten proteins: monomeric gliadins and polymeric glutenins Albumins is a minor class	– During dough mixing, gluten proteins are transformed into a viscoelastic network. It is involved in retaining the carbon dioxide produced during fermentation, influencing loaf volume and crumb structure of the resulting bread
Lipids	Consist predominantly of triglycerides and other non-polar lipids	– Contribute into dough handling properties – Increase the stability and retention of gases in the dough, it affects loaf volume
Fiber	Arabinoxylan	– Increase dough consistency and decrease mixing time – Reduce the diffusion rate of carbon dioxide out of the dough

sufficient water, starch granules absorb up to 50% of their dry weight of water. When the starch suspension is heated above a characteristic temperature (the gelatinization temperature), it undergoes a series of changes which eventually lead to the irreversible destruction of the molecular order of the starch granule. This process is termed gelatinization. During dough preparation, starch absorbs up to about 46% water. Due to the combination of heat, moisture and time during baking, the starch granules gelatinize and swell. Upon cooling, the solubilized amylose forms a continuous network, in which swollen and deformed starch granules are embedded and interlinked. Because of its rapid retrogradation, amylose is an essential structural element of bread and is a determining factor for initial loaf firmness [2].

The proteins present in wheat comprise albumins, globulins, glutelins (glutenins) and the prolamines (gliadins). Between 80 and 85% of total wheat protein are represented by the gluten proteins. They are the major storage proteins of wheat. Two functionally distinct groups of gluten proteins can be distinguished: monomeric gliadins and polymeric glutenins. The properties of gluten proteins allow wheat flour to be transformed into a dough with suitable properties for breadmaking. These properties cannot be found in other cereals, even closely related to wheat, such as barley and rye. During dough mixing, wheat flour is hydrated and because of the mechanical energy input discrete masses of gluten protein are disrupted. The gluten proteins are transformed into a continuous cohesive viscoelastic network. During mixing, the resistance of dough mixing first increases, then reaches an optimum and finally decreases during what is called 'over-mixing'. The gluten protein network in the fermenting dough plays a major role in retaining the carbon dioxide produced during fermentation and during the initial stages of baking. Gas retention properties in turn determine loaf volume and crumb structure of the resulting bread. The gliadin/glutenin ratio of the gluten proteins influences the quality of the final products. The continuous network formed by glutenin polymers provides strength and elasticity to the dough. On the other hand, gliadins provide plasticity/viscosity to wheat flour doughs. Up to a certain limit, higher dough strength increases loaf volume. However, dough rise is hindered with flours that are too strong. An appropriate balance between the two class of molecules is thus required. During baking, dramatic changes occur in the gluten proteins that are probably a combination of changes in protein surface hydrophobicity and formation of new disulphide cross-links. As a result of these heat-induced changes as well as those of the starch, the typical foam structure of baked bread is formed [2].

Among the non-starch polysaccharides, arabinoxylan, especially the water extractable fraction, play a relevant functional role in breadmaking. These molecules increase dough consistency and decrease mixing time. Water extractable—arabinoxylans seems to act somewhat as gluten during fermentation as they slow down the diffusion rate of carbon dioxide out of the dough. However, they lack elastic properties. This class of polysaccharides is thus reported to improve bread characteristics such as crumb firmness, structure and texture, loaf volume. Water Unextractable-Arabinoxylans, on the other hand, enhance gas cell coalescence and decrease gas retention, resulting in poorer bread quality [2].

Lipids in wheat flour originate from membranes and organelles and comprise different chemical structures. They consist predominantly of triglycerides, as well as of other non-polar lipids. The fatty acid pattern of the flour lipids is dominated by linoleic acid (C18:2) with lower amounts of palmitic (C16:0) and oleic acids (C18:1). It is well known that flour lipids, significantly affect the breadmaking quality and in particular the loaf volume. Presumably, they form lipid monolayers at the gas/liquid interphase of the gas cells, increasing their stability and retention in the dough. Furthermore, polar flour lipids positively contribute into dough handling properties as well. The majority of flour lipids are related with starches. As a result, amylose–lipid complexes have been found after gelatinisation of cereal starches and in native starches as well. The oxidation of polyunsaturated fatty acids, by wheat lipoxygenase, can activate chain reactions leading to the oxidation of other constituents, such as proteins and carotenoids, thus affecting dough rheological properties and crumb colour [2].

2.1.1 Ancient Grains

Since the beginning of twentieth century, bread and durum wheats were the object of intensive breeding programs aimed to the development of modern cultivars. The main results of this revolution were the development of new cultivars characterized by higher yield, reduced susceptibility to diseases and an increased tolerance to environmental stresses. Simultaneously, a substantial improvement in technological quality has been achieved. After the World War II the replacement of autochthonous wheat landraces and old varieties with the modern cultivars was completed [22, 23]. Despite the fast decline of landraces and old varieties, some of them survived until today on small acreages located in marginal areas. In recent years, the revival of the cultivation and use of ancient grains contributed into the preservation of biodiversity. Moreover, it permitted the development of a local micro-economy, which is continuously growing, and which allows local producers differentiating their products and increasing their remuneration. In addition, old grains are good sources of proteins, lipids, minerals, and polyphenols [3, 24]. Therefore, foodstuffs based on old cultivars grew in response to consumers request for healthy and sustainable foods. Nevertheless, their breadmaking aptitude was expected to be lower than those produced using new cultivars (since 1970s), where breeders have focused on the optimization of the production of bread in terms of flour quality and production process. Current industrial processing and machinery are indeed more adaptable for handling new cultivars and not older grains due to a less standardization of the matrix in the latter. Subsequently, bakers are currently facing a challenging situation to achieve the perfect balance between producing old cultivars-based bread with high technological quality and desirable sensory attributes. It was previously reviewed the breadmaking performance of einkorn, Khorasan wheat (Oriental wheat) spelt and emmer as ancient wheat species [3]. Also, old *Triticum aestivum* L. varieties are sometimes reported as ancient grains [25]. The presence of gluten in these old wheat species make them possibly exploitable for breadmaking. In general, in comparison to common wheat, the flours of ancient wheat species yield softer doughs with low elasticity and high extensibility because of the poor gluten quality [26].

Einkorn flour is generally described as not suitable for breadmaking because of its sticky dough and poor rheological properties, however some authors reported high breadmaking quality [27, 28]. Brandolini et al. [29] conducted a survey of 65 einkorn samples by studying their pasting properties and concluded that einkorn had higher final viscosity than modern wheats. The differences are probably related to the smaller size of einkorn starch granules as well as to the lower amylose percentage of einkorn flour [28]. Regarding bread colour, einkorn has a lighter colour than common wheat and durum wheat suggesting that einkorn undergoes lower heat damage than modern wheat during baking because low α- and β- amylases limit the degradation of starch in reducing sugars, leading to lower Maillard reactions [28]. Low lipoxygenase activities in einkorn dough also limits the degradation of carotenoids [30].

Of particular interest, among Khorasan wheat is the QK-77 variety, registered by USDA in 1990 and known by the commercial name of Kamut [31]. Pasqualone et al., [32], studying Kamut bread, showed good sensory properties and loaf volumes, highly resembling to those of bread obtained from modern wheat. Moreover, Balestra et al. [33] found that kamut was more suitable than durum wheat for the fermentation processes at acidic conditions as an increase in the bread volume and the metabolic heat production by yeast were observed.

Korczyk-Szabò and Lacko-Bartosovà [34], making bread with different spelt varieties, found acceptable sensory scores with significant differences among the varieties. They concluded that spelt might be suitable for breadmaking, but it remains closely related to the choice of the variety. Another study compared common and spelt bread; the latter had high crumb elasticity, but low crumb cell homogeneity, probably due to its special dough rheological attributes [35]. Nutritionally, these breads had less total starch, more resistant starch, and less rapidly digested proteins in comparison to bread made with modern wheat flours [36]. Coda et al. [37] reported that spelt and emmer sourdoughs had slightly higher pH values than wheat ones but titratable acidity, concentration of free amino acids, and phytase activity higher than in common wheat sourdough. Specific volume and crumb of spelt breads showed higher resemblance to those of wheat breads than emmer. Sensory analysis also revealed that spelt and emmer can be used for producing acceptable bread products. Geisslitz et al. [26], compared a set of eight cultivars each of common wheat, spelt, durum wheat, emmer and einkorn in terms of protein content and composition, gluten aggregation, dough and bread properties. They found that, despite the good gluten content of the ancient grains, their baking performances were generally lower than that of the modern cultivars, because of the gluten composition. Spelt showed the best performances in comparison to the other ancient grains, being however ekinorm the more different from controls. Emmer showed intermediate behaviors between the other two. Based on a robust dataset, the authors concluded that at least one spelt, emmer and einkorn cultivar each with a favorable gluten protein composition for good baking performance was identified. More recently, Boukid et al. [38] studied flour properties and breadmaking quality of seven cultivars of "old" Italian wheat (*Triticum aestivum* L. ssp. *aestivum*) in comparison to a new one. Interestingly, some old cultivars presented superior texure, volume, colour, and crumb grain properties than those of new cultivar. Physicochemical and sensory evaluations resulted aligned, enabling the identification of some old cultivars with interesting breadmaking quality that can be potentially used and further improved with an optimized breadmaking process for advanced breeding programs.

2.2 Other Flours

In the past, rye, corn, barley, oats, sorghum, millet, and rice were more often used as ingredients for bread products. More recently, in countries where these other cereals are commonly grown, they have been utilized in breadmaking to reduce the proportion of wheat imports or the need to use indigenous materials.

Of all the non-wheat breads, those based on rye are the most common. Rye has similar protein contents to wheat but a distinctly limited ability to form gluten, mostly because of the high levels of pentosans present in rye flours. Moreover, the overall enzymic activity in rye flours is high compared with that of wheat flours. In order to restrict amylolytic activity and breakdown of starch during baking, acidification of rye bread doughs has become common. Acidification of rye doughs improves their physical properties by making them more elastic and extensible and confers the acid flavour notes so characteristic of rye breads [4].

Triticale has some potential in breadmaking. Early strains of triticale gave flours with poor baking performance, in part because of higher α–amylase activity and lower paste viscosity than wheat flours and in part because of weaker protein qualities. Later developments of triticale varieties and the use of bread improvers have brought the baking performance closer to that of wheat flours [4].

The interest of using oats for breadmaking is mainly due to the health benefits associated with this cereal. In particular, it contains high levels of β-glucans that are associated with the reduction of blood cholesterol levels [39, 40]. Oat flours are not capable of forming a strong gluten network able to retain large quantities of gas and thus products are usually based on a mixture with wheat flour. The final loaf has generally a bland taste and firm texture. Oats also have a high lipid content, reported to be around 5.5 and 9.7% for hulled and naked oats, respectively [39]. For this reason, oats are prone to rancidity and long-term storage requires inactivation of enzymic activity. For these reasons oats are used in bread recipes only when their nutritional contribution can override their adverse effects on bread quality or where their unique flavour can provide positive benefits [4].

The use of barley for breadmaking is attracting interest and as with oats as a nutritional supplement to wheat flour. Barley, indeed, contains higher levels of β-glucans than wheat, even though significantly lower than those in oats. The use of malted barley grains is founding also some interest in the baking industry, because of the distinctive flavours of these ingredients that are carried through to the final product. After the various stages of the malting process, a milled barley malt will have a high diastatic (enzymic) activity. The enzymic activity is usually dominated by α-amylase and proteolytic enzymes. The former can be beneficial in improving the gas retention abilities of flours, although such enzyme-active malt flours should be used with caution in breads which are later to be sliced, because of the high levels of dextrins which will be produced during dough processing and baking. The proteolytic activity can also have both beneficial and adverse effects since they cause softening of the dough. Some softening can improve dough machinability, but too much may cause dough to stick to equipment surfaces during moulding and intermediate proof [4].

Other cereals, which are not normally considered for the production of bread, because they lack proteins capable of forming gluten, are even used in 'composite' flours, blended with wheat. The use of rice, corn, sorghum, millet and teff, or pseudocereals such as buckwheat (*Fagopyrum esculentum* Mönch), amaranth (*Amaranthus caudatus* L.) and quinoa (*Chenopodium quinoa* Willd), and possibly even starchy roots such as cassava (*Manihot esculenta* Crantz) flours are used in countries wishing to produce bread with local flours or to reduce the importation of wheat. The absence of gluten makes these flours also interesting for people suffering from celiac disease. The application of these alternative cereal flours in breadmaking will be discussed in other sections of the chapter.

2.3 Water

Without water, the most changes in bakery products would not occur. In breadmaking, the added water level does not usually exceed the flour weight. According to 100% flour, the use of approximately 50% water results in a finely textured, light bread, even though most artisan bread formulas contain anywhere from 60 to 75% water. Firstly, the addition of water in the 'right' quantity essential for the formation of gluten and for modifying the rheology of the dough [6]. The optimum water level is affected by the properties of the flour and some other ingredients that may be present in the formulation. When an excess of water encounters the flour particles there is a gradual uptake of water. Water is taken up by the wheat gluten proteins, which at about 16% moisture content, pass from a hard glassy material to a soft rubbery one. Starch granules, especially the ones damaged during milling, are also able to absorb water. The whole starch granules present in wheat flour are largely insoluble in cold water, but when heated in an aqueous medium they begin to absorb water and swell (gelatinization). The water soluble pentosans have a significant water-absorbing capacity, estimated to be around seven times their own weight, and can form viscous solutions. The fibrous components of the wheat grain, which derive from the bran skins, also absorb water during mixing but more slowly than other flour components. The optimum water level is usually set according to the required viscosity and other dough rheological characters that will result in the 'right' product qualities. If the dough has too little water, it will have a high viscosity, and it will be difficult to change the shape of the dough during dividing, handling, and moulding. In contrast, if the added water level is too high in the dough it will have a low viscosity, and while it will be easier to change its shape during moulding, it may not retain its shape and is likely to flow during proof [6].

The optimum water level for a given dough varies according to the bread variety required, the breadmaking processes employed, and the methods by which it will be handled and processed, especially moulded. Mathematical models have been developed to predict water absorption from flour properties. More recently, dough mixing tests are used to make a direct measurement of the optimum water level, based on assessment of dough rheology, either during mixing or afterwards. The most common of these methods is based on the use of Brabender Farinograph; 500 BU (Brabender Units) is typically used to determine optimal dough development [41]. As well as playing a key role in dough formation, water also plays a significant role in the control of dough temperatures and in turn in the activity of the yeast present in the formulation. The level of moisture remaining in the baked product is also a major contributor into the characters of breads. In general, the crust of bread has a lower moisture content than the baked crumb and is therefore expected to have a hard, crisp eating character compared with the softer, chewier character of the crumb. Within certain limits, the higher the crumb moisture content, the fresher the bread will be perceived by the consumer. Gradually, depending on the storage conditions, the moisture held within the crumb migrates to the areas of lower moisture content closer to and at the crust surface. This phenomenon causes the crumb to become firmer and drier, and the crust to lose its crisp eating character and become soft. Changes in texture and eating quality are a result of loss of moisture from the crumb and the intrinsic firming of the structure associated with staling The variations in the crumb moisture content, the ratio of water to starch and the ratio of water to gluten in bread will clearly have an effect on the physicochemical changes that occur during storage. We commonly refer to this phenomenon as staling, that leads to loss of product freshness [6].

Water, above certain levels, also supports the mould growth in bread: the major microbial spoilage agent for many this kind of products. Controlling the free water levels or using anti-mould agents are essential to ensure a mould-free shelf- life of the products [6].

2.4 Salt

The general term salt in bread formulas refers to sodium chloride. The flavour-enhancing function of salt is well known. Omitting salt from the formula results in baked products that are quite tasteless. Another important function of salt in bread-making is its stabilizing effect on fermentation [8]. In dough made without salt, the yeast ferments excessively resulting in gassy, sour dough and baked products with open grain and poor texture. Salt inhibits or "controls" fermentation rate by decreasing the rate of gas production, which results in longer proof times. This appears to be the result of increased osmotic pressure and the action of the sodium and chloride ions on the membrane of the yeast cells [8].

It is well known that salt lengthens the mixing time of dough and decreases water absorption. It has a strengthening effect on dough. These effects are explained as follows. In a flour-water system at a normal pH (~6.0), the gluten protein has a net positive charge. These positive charges repulse each other. This allows the gluten hydrating faster (shorter mixing time) and keeping the protein chains from interacting with each other, resulting in a weaker dough. Low levels of salt shield the charges allowing the protein chains to approach each other. This causes the flour to hydrate more slowly (longer mixing time) and allows the protein chains reacting more tenaciously to form a stronger dough. Although salt increases dough strength, levels of salt above the optimum level of 1.5–2% for bread do not necessarily improve loaf volume. A substantial decrease in volume of loaves at elevated salt levels has been reported. Also salts other than sodium chloride had a deleterious effect on loaf volume [8].

Salt also affects the water phase of the dough system. Increased salt concentration generally increases the ordering of water structure. The increase in ordered water structure allows proteins to interact with each other through hydrophobic interactions. Moreover, different salts have widely different effects on dough rheology. However, the effect of various ions on yeast activity and on human health as well as the effect of the various salts on the taste of the final product, should be considered. Salt contains 39% sodium, and this has spurred interest in reducing the sodium chloride level or completely or partially replacing sodium chloride in baked product formulations with alternative salts. Small incremental decreases in sodium have shown to be effective, as consumers may not be able to detect gradual reductions in sodium up to a one quarter reduction [42]. Sodium chloride replacers such as potassium chloride, calcium chloride and magnesium sulfate have been used to replace or enhance salt taste in several foods. While these compounds contribute with a certain salty taste quality, they may also provide undesirable after tastes such as bitter, metallic and astringent tastes, which has limited their current use in food manufacturing [43]. Sodium chloride also limits the growth of yeast and enables the gluten structure in bread to develop. The reduction of sodium chloride in bread may therefore result in an increase in the growth of yeast and an undeveloped gluten structure, which has a negative effect on the texture and volume of bread [43].

2.5 *Yeast*

Leavened bread is obtained from fermentation of wheat flour sugars liberated from starch by the action of natural flour enzymes. Fermentation is caused by yeasts. Due to fermentation, sugar is converted to moisture and CO_2. As water vapor and CO_2 expand due to high temperature, yeasts act as an insulating agent preventing high rate of temperature rise of bread crumb and the possibility of excessive moisture evaporation. Sugar is often added for initiation of fermentation. Leavening is produced only if the gas trapped in a system that will hold on it and expand along with it. Therefore, much of baking technology is the engineering of food structures through formation of the correct dough to trap leavening gases and the fixing of these structures by the application of heat [9].

The oldest-known bread-leavening agent is the spontaneous sourdough. It is prepared with a mixture of flour and water that is then fermented by spontaneous lactic acid bacteria (LAB) (several species of *Lactobacillus* sp., *Leuconostoc* sp., and *Weissella* sp.) and yeasts (mainly *Saccharomyces cerevisiae* and *Candida* sp.). Fermentation by endogenous microorganisms takes place in the dough, producing metabolites that affect the characteristics of the dough. The repeated addition of new flour and water to the dough, known as *backslopping*, allows a composite ecosystem of yeasts and LAB to develop inside the dough, giving it its typical sour taste. The application of this technology involves the continuous transfer of microbial populations, under specific environmental conditions, resulting in their long-term genetic adaptation and gene differentiation. The yeasts are mainly responsible for the production of carbon dioxide (CO_2), whereas LAB are mainly responsible for the production of lactic acid, acetic acid, or both; both produce the aromatic precursor compounds of bread. Furthermore, the technological performance of the dough and the nutritional properties, aroma profile, shelf life, and overall quality of the bread are greatly affected by the metabolic activity of the sourdough microorganisms [7]. The quality of some sourdough breads is indeed not always consistent. This is mainly due to the complexity of microbiota, the influence of flour, and the environment of propagation. In the population dynamics leading from flour to mature sourdough, lactic acid bacteria and yeasts outcompete other microbial groups contaminating flour, and interact with each other at different levels. Several factors may interfere with the persistence of species and strains associations that are typical of a given sourdough: metabolic adaptability to the stressing conditions of sourdough, nutritional and antagonistic interactions among microorganisms, intrinsic robustness of microorganisms, and existence of a stable house microbiota [44].

It is reported that the production of sourdough bread gradually spread from ancient Egypt throughout Greece and the Roman Empire and has continued ever since. During the Barbarian migration period in Europe (AD 300–700), industrial bread manufacturing disappeared. The technology of sourdough bread survived in the monasteries until the twelfth century. After the Middle Ages, breadmaking technology saw the advent of new progress, especially in northern Europe, where baking went together with brewing. A substitute for sourdough is indeed the beer yeast *Saccharomyces cerevisiae* commonly known as baker's yeast or compressed yeast. Although evidence indicates the use of beer yeast in breadmaking in ancient Egypt and other communities familiar with the brewing process, it was only after the Middle Ages that it became a widely used agent for the leavening process. Nevertheless, the culturing of baker's yeast for breadmaking purposes on the industrial scale only occurred in the nineteenth century. During the twentieth century, it began to replace the use of sourdough in bread production almost completely, particularly in industrial bakeries. For many years, sourdough was employed exclusively by artisanal bakeries and home bakers. The rapid expansion of baker's yeast was due to its greater capacity to meet the requirements of innovative bakeries, whose bread production became mechanically assisted. Thus, the rapid and simple leavening process associated with baker's yeast replaced the sourdough-baking process, which was much more time consuming due to its long fermentation time and laborious management.

In the last 20 years, the popularity of sourdough bread has grown worldwide for several reasons, but mostly due to the renewed interest of both consumers and bakers in this product. Consumers are attracted to its pronounced flavour, high nutritional value and healthy properties, prolonged shelf life, use of fewer additives, and, finally, its traditional aspects.

The use of selected microbial strains (commercial starters) together with constant technological parameters in the usual environment, can today help to attenuate the variations in the sourdough performances [7, 44].

3 Additional Ingredients

3.1 Fats

Lipids exhibit important functional properties in breadmaking, although they are present in lower levels (<5%) than starch or proteins. The three main sources of lipids in a typical bread formula are wheat flour lipids, shortening and surfactants.

The role of wheat flour lipids in breadmaking has already been discussed in Sect. 1; in this paragraph the role of added lipids such as shortening and surfactants will be discussed. Bread formulations generally contain low levels of added oil or fat.

The term "shortening" is used to define a group of solid lipids formulated especially for baking applications. Shortenings are made from vegetable and/or animal crystalline lipids and oils and consist of nearly 100% lipid. The typical level of shortening added in a bread formula is about 2–5% on flour basis. The main roles of shortening in breadmaking are to plasticise and lubricate dough, as well as to increase dough rise, oven spring and loaf volume. Shortening also affects crumb structure by tenderising the baked bread and extending the shelf life or keeping quality. In particular, during dough mixing shortening plasticises dough. Higher levels of shortening in dough require lower levels of water to reach an equal dough consistency [10]. Chin et al. [45] noted that the crystalline lipid (partially) coats the gluten proteins and starch granules, thereby reducing water absorption during dough making. A further function of shortening in bread dough is lubrication. Added solid lipids, uniformly distributed within the gluten network between the starch granules, lead to formation of thin and easily expandable gluten films with less tendency to contract after expansion than when no lipid or liquid oil is added. Shortening also has a major impact on air incorporation during mixing, which in turn affects loaf volume. At the end of mixing, the interface surrounding each crystal coalesces with the gas-liquid interface of the bubble, such that the TAG are directly exposed to the gas phase lowering the surface tension of the oil-water interface. Moreover, lipid crystals attach to proteins surrounding the gas cells and, hence, increase the stability of the protein-lipid film surrounding these cells in a way that liquid lipids cannot. During dough fermentation shortening does not affect carbon dioxide production. Addition of shortening containing sufficient solid lipid to dough recipes strengthens dough and, therefore, improves gas retention. By contrast, dough containing no added shortening suddenly exhibits loss of part of the gas already during earlier stages of the fermentation. During Baking in the absence of added shortening, neighboring bubbles coalesce, resulting in a coarser crumb structure. During baking, the lipid crystals of shortening melt and provide a source of extra (liquid) material at the gas cell surface. This allows the gas cells to expand without rupturing, which leads to a large number of relatively small gas cells and, hence, produces large volume bread with a fine crumb structure. Molten shortening can also coat gluten and contribute into the formation of a lipid-protein film. When shortening is added as a dough ingredient, the bread has a larger loaf volume than control bread [10]. It was already noted that freshly baked bread made with shortening has a softer and more uniform crumb. Also, the breads have a crispier crust as shortening reduces moisture migration from the centre of the loaf to the drier outer crust region which otherwise causes the latter to lose its crispness [45]. Longer shelf lives have also been observed for shortening containing breads. While Smith and Johansson [46] referred to amylo pectin complex formation with TAG as a reason for the lower firming rate of TAG supplemented breads, this seems highly unlikely for steric reasons. A more probable explanation may well be that the shortening also forms a barrier to moisture migration in the crumb [10].

Surfactants are surface-active agents: molecules with amphiphilic properties, i.e. they contain both hydrophilic and hydrophobic parts. When concentrated at the interface between two phases (liquid, solid or gas), they decrease the interfacial tension. In breadmaking, surfactants are generally divided into dough strengtheners that mainly interact with gluten, and crumb softeners or anti-firming agents that can complex gelatinised starch [10]. In breadmaking, surfactants are commonly used at a level of 0.3–1.0%. MAG and DAG, typical examples of crumb softeners, are the most often used surfactants in many processed foods. Other well-known surfactants are derived from acylglycerols such as diacetyl tartaric acid esters of mono- and diacylglycerols (DATEM) (i.e. a typical dough strengthener), and succinylated MAG. Polysorbate fatty acid esters or lactated esters of MAG, such as sodium and calcium stearoyl lactylate are also commonly used in breadmaking [10] as dough strengtheners. Galactolipids, lecithin and its different phospholipids are naturally occurring surfactants.

The impact of added surfactants during breadmaking is like that of shortenings. Indeed, it has been proposed that they can replace (part of the) shortening in some bread formulas [10]. This may be due to the fact that they increase dispersion of the added shortening lipids and thereby increase their action. The impact of added surfactants on dough and bread properties cannot be underestimated as, depending on their nature, they improve dough handling and increase gas retention, and consequently oven spring and loaf volume. They also improve crumb structure and slicing and extend the shelf-life of bread. During dough mixing the fine crumb grain in the final product obtained upon addition of surfactants to a breadmaking formula can be attributed to increased air incorporation during dough mixing, to smaller cells formed during mixing, or to a combination of both. The generation of smaller air cells is logically assumed to result from the beneficial effect of lowering surface tension. Therefore, it is currently assumed that surfactants stabilize the mixed protein-lipid interface lining the gas cells either directly by modulating surface tension, or indirectly by sequestering certain components (e.g. proteins or lipids), turning the mixed interface into a more homogenously stabilized one (i.e. stabilized by lipids/surfactants or proteins, respectively). The amphiphilic nature of surfactants contributes into the gluten network strength while the addition of lecithin lowers dough stability, in general, anionic surfactants such as SSL and CSL strengthen dough [47]. Surfactants also improve gas cell retention during dough fermentation. Surfactant use, in general, increases dough height during proofing. The presence of surfactants also increases the proof time needed to reach maximum dough height, probably by strengthening the gluten network. During baking surfactant increase in bread loaf volume. Observed that one of the major differences between MAG and shortening (mainly TAG) is that MAG can complex amylose, whereas shortening cannot. The two lipid classes therefore work through different mechanisms, with shortening suggested to interact with proteins to increase the loaf volume. The addition of 0.25% GMS to a bread formula improves the crumb structure which is much finer and more homogeneous. Surfactants that are classified as crumb softeners interact with starch. MAG are the best known crumb softeners. Surfactants that can complex amylose are more efficient in reducing bread firmness than those that do not form complexes.

3.2 Improvers

Flour quality is a key factor for achieving an acceptable product and is mainly related to gluten protein quality and quantity. Since deficiencies in flour quality must be compensated to obtain an acceptable product, the addition of flour improvers is a common practice in breadmaking. Moreover, the use of enzymes helps to improve dough handling properties, increase quality of fresh bread, and extend the shelf-life of stored bread. An increasing number of dough improver, as summarized in Table 3.3, is now available for both large- and small-scale bakeries.

Table 3.3 Improvers types and functionalities in breadmaking

Components	Type	Functionality	References
Oxidizing agents	Potassium bromate Potassium iodate, calcium iodate Azodicarbonamide Ascorbic acid	Form new cross-links between protein chains and improve the gas retention abilities of the dough	[4]
Reducing agents	L-cysteine	Soften the dough by breaking the cross-links between amino acids in the gluten network	[4]
Enzymes	Amylases	Increase fermentable sugars and gas production, leading to loaf volume increase and crumb hardness decrease	[48]
	Hemicellulases	Hydrolyse non-starch polysaccharides resulting in improved gluten matrix and breadmaking ability	
	Proteinases	Attack protein chains weaken the gluten structure. It leads to improved softness and dough-handling	
	Oxidases	Form of protein–protein bonds that strengthen the gluten network and stabilize the dough	
	Lipoxygenases	Have similar functions to oxidases. Bleaching effect on pigments, resulting in a whiter crumb	
	Lipases	Hydrolyse acylglycerols, contributing to form lipid films entrapping gases. Increase dough handling and stability, improve oven spring and crumb structure	
	Transglutaminases	Cross-linking effect on proteins that improves the resistance of the dough	
Hydrocolloids	Gums Modified cellulose Pectins Galactomannans Exopolysaccharides	Affect gluten hydration and interfere in starch gelatinization and retrogradation Improve bread volume, crumb porosity and texture. Retard staling	[49]

3.2.1 Oxidizing and Reducing Agents

The role of oxidants in breadmaking systems is mainly improving the gas retention abilities of the dough. With the addition of suitable oxidizing materials, it is possible to obtain a better gluten development and reduce the time for doughs formation. This function is possible thanks to the ability of these molecules to form new cross-links between protein chains. Oxidizing agents have been known for over 50 years and many different types are being used around the world. Slow acting ones such as potassium bromate have been used widely and are common throughout Europe. The faster acting ones such as potassium iodate, calcium iodate, and azodicarbonamide are more widely used in USA. However with a greater awareness of food additives by the general population, and a greater understanding of their function, a number of changes to legislation on Europe have been made with the result that ascorbic acid (or Vitamin C or E 300) is today the sole oxidizing agent allowed to use in the baking industry. Typically, it is used at levels between 100 and 300 ppm flour weight. However, high levels of ascorbic acid lead to the lack of doughs elasticity and loss of bread quality. These oxidizing agents are commonly added as part of the bread improver, but they can also be added to flours by the miller [4].

Reducing agents make dough more extensible. They are deliberately added to 'weaken' structure in specific products. The major material used in bread dough is L-cysteine. It can only be used at low levels in improvers but by reducing dough resistance to deformation it helps in moulding and shaping, such as rolls and baps, without structural damages. L-cysteine can also be used in pan breads at low levels where its ability to reduce resistance can help reduce streaking caused by moulding faults. Other ingredients such as deactivated yeast and proteases have a similar effect. Reducing agents soften dough by breaking the cross-links between amino acid chains in the gluten network rather than by breaking the chains themselves. This reaction is finite and so the process is inherently more controllable than that using enzymes. A wide range of recipes use this technology and reducing agents are sometimes used in conjunction with enzymes [4].

3.2.2 Enzymes

Among the enzymes used in food applications, those used in bakery industry constitute nearly one-third of the market. Baking enzymes are used as flour additives and they are used in dough conditioners to replace chemical ingredients and to perform other functions in a label-friendly way [48].

Amylases can act only on damaged or gelatinized starch since these are susceptible to enzymatic attack. The amount of damaged starch is dependent on wheat variety and especially on milling conditions. α-Amylase catalyzes the hydrolysis of α-1,4-glycosidic bonds of starch polymers, producing low molecular weight polysaccharides and dextrins. β-Amylase decreases the molecular size of these polysaccharides by cleaving the disaccharide maltose from the non-reducing end. Unlike higher glucose polysaccharides, maltose is fermentable by yeast. The resulting increase in fermentable sugars has a positive effect on yeast fermentative activity, which along with gas retention is a fundamental element of bread production. An increase in fermentative gas production, combined with the ability of the dough to retain that gas, leads to an increase in loaf volume and decrease the crumb hardness. However, extensive degradation of damaged starch due to too high levels of α-amylase leads to sticky dough [48].

The other widely used enzyme are hemicellulases, also known as pentosanase and xylanase. Hemicellulases include any enzyme that catalyzes the hydrolysis of non-starch polysaccharides present in the cells walls of bran and germ that are one of the reasons for the poor breadmaking quality of whole wheat flours. During dough mixing, arabinoxylans compete with gluten for water. Xylanases hydrolyze the backbone of arabinoxylans, reducing their molecular size and water-holding capacity. This allows a greater gluten hydration, which results in better gluten matrix development and breadmaking ability. Another effect ascribed to xylanases is to offset reduced gluten coagulation caused by pentosans by hydrolyzing them. These gluten-linked pentosans were considered to have a steric hindrance effect on gluten coagulation [48].

Proteinase is a group of enzymes that directly attacks protein chains, in doing so they increasingly and irreversibly weaken the gluten structure. Proteases have a long history in breadmaking and were traditionally used to treat doughs resulting from overly strong and too elastic flours. Originally, the aim of protease addition was to improve softness, dough-handling properties, and dough machinability. However, proteases have more functional effects. Functional effects of proteolytic enzymes are reduction of mixing time; improvement of dough machinability; improvement of gas retention due to better extensibility; improved pan flow in bun and roll production; improvement of grain and crumb texture; improved water absorption; improved colour; and improved flavour [48].

Oxidase are used in breadmaking with the aim to help rebuilding the gluten network, to improve texture, volume, freshness and also dough machinability and stability. Dough conditioners are specifically meant for gluten strengthening. Gluten strengthening results in improved rheological and handling properties of the dough. This effect is induced by the formation of protein–protein bonds that strengthen the protein network and thereby strengthen and stabilize the dough. For example, in presence of oxygen, glucose oxidase catalyses the oxidation of the beta glucose and in doing so produces hydrogen peroxide. The ability of hydrogen peroxide to oxidise thiol groups to form disulphide bonds often forms the basis of the claims for the improving effect of glucose oxidase in breadmaking [48]. Among oxidase, also lipoxygenase from soya bean flour has been used for decades in breadmaking not only for its bleaching effect, resulting in a whiter crumb, but also for its improving effect on dough rheology, loaf volume and gluten stability. It converts polyunsaturated fatty acids to fatty acid peroxyl radicals. The formed hydroperoxides react with the naturally occurring yellow carotenoid pigment in wheat flour, leading to a reduction of the yellow colour. Furthermore, it has been claimed that lipoxygenase has a direct oxidizing effect on gluten formation probably because of the oxidation of gluten proteins through the co-oxidation of accessible thiol groups of the gluten protein by the enzymatically oxidized lipids [48].

Lipases hydrolyse ester bonds of acylglycerols, yielding mono- and diacylglycerols, free fatty acids and, in some cases, also glycerol. Lipases usually function at lipid-air or lipid-water interfaces and their activity is sharply increased by the presence of organized lipid structures, which are normally found at such interfaces. There is evidence that a lipid film surrounding the gas cell contributes into the gas cell stability. Specific lipases are claimed to improve dough-handling properties, to increase dough strength and stability, to improve dough machinability and to increase oven spring. Besides this, such lipases also improve crumb structure and crumb whiteness [48].

Besides the discussed enzymes, several other classes of enzymes have been investigated for their effects in breadmaking and several other types mentioned above have been reported to have beneficial effects on one or more characteristics of dough or bread. Among them transglutaminase improves the resistance of dough [50], particularly yeast dough from weak wheat flour, in a manner comparable to potassium bromate. It is known that transglutaminase has a cross-linking effect on proteins independent of the redox system of the dough, not involving the thiol groups and disulphide bonds in the dough [48, 51].

3.2.3 Hydrocolloids

These compounds, commonly named gums, are water binding agents, indeed, their use in breadmaking requires a recipe with increased amount of water. In this class of compounds it's possible to find exudate gums (e.g. Arabic gum), gums from seaweeds (e.g. Sodium alginate, Agar agar, Carrageenans), modified celluloses (e.g. microcrystalline cellulose (MCC), hydroxypropylmethylcellulose (HPMC), Hydroxypropyl cellulose (HPC), carboxymethyl cellulose (CMC)), pectins, galactomannans from leguminous seeds (e.g. Guar gum, Locust bean gum) and exopolysaccharides from microbial fermentation (e.g. Xanthan gum, Dextran, Gellan gum) [49]. Hydrocolloids are able to modify gluten and starch properties, mainly affecting the gluten hydration and interfering in the starch gelatinization and retrogradation. Through specific interactions, particularly with gluten proteins, they can positively or negatively modify dough rheology, which depends on their structure, concentration, and the interactions with other components [52]. Hydrocolloids can improve bread volume, crumb porosity and texture, leading to products with enhanced technological quality. Several studies investigated the effect of hydrocolloids on breadmaking performances. Xanthan gum, alginate and locust bean gum have been reported to affect the moisture content, softening and retrogradation enthalpy of withe bread crumb [13]. CMC and HPMC have been found to improve the quality of white and whole wheat breads [12], and also the effect of sodium alginate, κ-carrageenan, xanthan gum and HPMC on bread specific volume, hardness and moisture content have been evaluated [53].

3.3 Anti-Staling

When stored at room temperature, most types of breads undergo a progressive and often rapid deterioration of quality, commonly known as 'staling'. The mechanism of bread staling involves the re-crystallization (retrogradation) of gelatinized starch, responsible for the texture changes that take place during bread storage. During baking, amylose and amylopectin tend to separate and to accumulate within the starch granules and in the intergranular space in the form of double helices. After baking, amylose retrogrades very quickly, stabilizing the initial structure and forming a more rigid, insoluble network. The subsequent increase in bread firmness is due to further physicochemical changes affecting the starch components, especially the amylopectin fraction. Moisture loss, starch-gluten interactions or development of macromolecular entanglements and water migration are also involved in bread staling [4, 15].

To prevent these phenomena, anti-staling agents such as enzymes, hydrocolloids and emulsifiers are used in breadmaking, as they interfere with the re-association of amylose, amylopectin, or both. In particular, the enzymes most frequently used are α-amylases. These enzymes, already described as improvers (Sect. 4.2) because they increase bread volume, improve crumb grain, crust and crumb colour, and contribute into flavour development, positively act on bread texture by producing low molecular-weight dextrins that, in turn, interfere with the amylopectin retrogradation and with the protein-starch interactions occurring during bread storage. Also lipases have been reported to positively influence bread staling. In fact, they act on the lipid fraction by increasing the number of molecules with emulsifying properties (monoacylglycerols and diacylglycerols) [54].

Emulsifiers are indeed used in bread as means of maintaining crumb softness for longer by retarding the process of staling. The anti-staling ability of emulsifiers is mainly due to their interaction with starch. Emulsifiers complex with linear amylose, and they may do some complexing with the outer linear branches of amylopectin. It appears that the formation of an emulsifier-amylose complex contributes into a decrease in the initial firmness of the crumb, while a complexing with amylopectin results in a distinct reduction in the rate of firming during storage. The emulsifiers that are widely used for their ability to reduce the staling of the bread crumb are distilled, saturated monoglycerides. Other types of emulsifiers that are effective as crumb softeners are sodium or calcium stearoyl-2-lactylate (SSL and CSL) and diacetyl tartaric acid esters of monoglycerides (DATA esters). Also lecithins, among the emulsifiers, have been reported to reduce staling in bread, acting as crumb softeners [55]. It is a group of complex phospholipids especially extracted from soybean. Soya lecithin hydrolysate complexed effectively with starch amylose and amylopectin retarding wheat starch crystallization.

Hydrocolloids, other than as improvers, have also been assayed as staling inhibitors [49]. One of the mechanisms involved in the anti-staling effect of gums is the increased moisture retention during storage, thus preserving crumb softness [14]. Moreover, inhibition of starch-gluten interactions or the development of macromolecular entanglements in the amorphous part of the crumb have been suggested among the anti-staling effect of hydrocolloids [16].

3.4 Preservatives

Preservatives are intended to inhibit the growth of moulds and thermophilic bacteria. The preservatives most commonly used in bread to prevent or minimize microbial growth are the propionates, i.e. propionic acid and its salts, acting by distorting the pH equilibrium of microorganisms [4]. The anti-microbial activity of propionates is mainly against moulds and the bacteria responsible for the development of rope in bread. Because their effectiveness on yeasts is minimal, propionates can be used in bread without disturbing the leavening activity in the dough [4]. The usual choice from this group is calcium propionate; its maximum levels today used in the EU is 3000 ppm by weight of flour. At this level the mould-free shelf life can be extended by 2–3 days. There is a very significant loss in propionic acid level during baking. Post-baking or post-cooling it is possible to use preservatives sprayed onto the surface of the bread. This avoids the adverse impact of the preservatives on yeast fermentation. New methods which have been suggested for mould control [17] include synthetic anti-oxidants (such as butylated hydroxytoluene), essential oils (such as clove, cinnamon, thyme) and biopreservatives (based on lactic acid bacteria). Another substance with effective preservative action in bread is ethyl alcohol. The addition of ethanol at levels between 0.5 and 3.5% of loaf weight leads to a substantial extension of the shelf-life of bread [18]. Increases in shelf life have been also obtained when ethanol was sprayed over all surfaces of the loaf prior to packaging and sealing as when it was added to the base of a bag of the same size before adding the product and sealing. This finding confirms that ethanol acts as a vapour pressure inhibitor. Despite the high cost of duty paid, ethanol has real potential as a bread preservative not only because of its anti-microbial activity but also its ability to delay staling. However, in some parts of the world the use of ethyl alcohol cannot be used for ethical or religious reasons. Alternatives to E-number listed preservatives are being offered. A common approach is to exploit the natural antimicrobial activity of fermented wheat flour, such as sourdough, attributed to the production of organic acids which operate synergically with other antimicrobial compounds. Such products may well contribute organic acids such as propionic acid at analytically detectable levels and may well involve questions regarding the lack of declaration of such compounds on the product label [56]. It is proposed as natural preservative instead of the chemical additives, such as sorbate, propionate, and benzoate, frequently mixed into the dough to control the growth of undesired microorganisms.

Irradiation can also be used to destroy mould spores which may be present on the surfaces of bread. The types of irradiation of primary interest in bread preservation are ultraviolet, microwave and infrared. Due to the poor penetrative capacities, UV light limits its use to surface applications. On the other hand, microwave and infrared radiations are more penetrating and can be used even on the packed products [4].

Modified atmosphere packaging became of interest in the late 1970s for bakery products, mainly in Europe because new labelling regulations demanded that the presence of preservatives should be declared. In contrast, carbon dioxide and nitrogen do not need to be declared. In addition, there was a need for greater increases in shelf-life than were possible using propionates. Packaging in carbon dioxide has the advantage that it can increase the shelf life of bread without affecting its flavour, aroma or appearance [4].

4 Types of Bread

Bread is a popular staple food throughout the world since thousands of years. Traditionally, breadmaking was a long process, but it has been drastically shortened with the mechanization of the bakery. In the European Union, by volume, 61% of the breads come from industrial production, 34% from artisan bakers, and 5% from in-store retailers and others. Industrial bread is sold mostly in Northern European countries, whereas artisanal breads prevail in the Mediterranean. Different baking technologies were developed to respond better to new market demands, as summarized in Table 3.4.

Table 3.4 Bread types and properties

Type	Ingredients	Processing	References
Leavened bread	– *Basic*: Flour, water, yeast, salt – *Optional*: Sugar, fats, improvers, Antistaling, preservatives	Mixing; fermentation, moulding, proofing, baking – *Straight dough*: all ingredients are mixed together – *Sponge and dough*: part of the flour is mixed with part of water and total amount of yeast. This preferment is called sponge or biga. The remaining ingredients are then added and mixed to form the dough	[4, 57, 58]
		– *Sourdough*: similar to the sponge and dough method, it is characterized by spontaneous fermentation with microbes originating from the flour. The repeated addition of new flour and water to the dough (*backslopping*) allows a composite ecosystem of yeasts and LAB, giving it its typical sour taste	[7]
Flat breads	– *Basic*: flour, water, salt – *Optional*: fats, with or without yeast	Mixing; leavening (which may be absent); shaping (by pouring the batter on a griddle or by sheeting and eventually punching the dough); baking	[59]
Enriched breads	Addition to the basic and optional ingredients of: – Pulses – Pseudocereals – Alternative flours – Fibers – Antioxidants – Food by-products – Novel ingredients	They can be prepared as leavened or flat breads In the formulation, the additional ingredients are added in percentages suitable for balancing the functional and the sensory aspects	[60–65] [5, 66–71] [4, 5, 39, 72] [73, 74] [75–80] [81–89] [90–101]

Leavened Bread

lmaking starts by adding water (and other ingredients) to flour and applying ic energy by mixing thereby forming extensible dough. Traditionally the dough formed by hands, requiring long mixing times (30–45 min); today mixing ıines are commonly used, which allow a mixing time of 10–20 min. Among the ng machines spiral or twin arms mixers are generally used. Twin arms mixers uce a better dough aeration, providing a higher volume increase and oxygen-ı of the dough, and consequently higher volume of the final product. Conversely, spiral mixer generates lower gas occlusion, and higher temperature increase ng the mixing phase (i.e. 9–10 °C vs 4–6 °C) [57]. After mixing, bread dough is nented, moulded and proofed.

3read is produced mainly according to two methods: the straight dough and the nge and dough method. In straight or direct dough baking all ingredients are ‹ed together at the same time; some bakers use to put all the ingredients together he kneader at the same time, while others follow a precise order: a gradual addi-n of the water amount or a gradual addition of the flour amount [58]. The gradual dition aimed to achieve a progressive flour hydration during the mixing step, a gher water absorption and a better dough development [57]. The bulk fermenta-ɔn lasts from 0.5 to 3 h, depending on the quality of the flour, yeast level, dough mperature and humidity, and variety of bread produced. The dough is usually mixed or punched once or several times during fermentation. After dividing and ıaping, dough is proofed once more for approximately half an hour and then baked. he resulting breads have fine and uniform crumb structures, but lower volume and arder crumb than breads produced by indirect methods [4] (Fig. 3.1).

In the indirect sponge and dough method, one third or a half of the total amount ɔf flour is mixed with part of water and total amount of yeast. It is allowed to fer-nent for 3–20 h at 21–27 °C. The preferment is called sponge or biga. After the first ermentation is ended, the remaining ingredients are added and mixed to form the dough. The dough is allowed to rest/proof for 15–30 min. After dividing and shap-ing, the dough is proofed for another hour. Baked bread has a very soft and porous crumb, with many bubbles of irregular size and distribution. Its flavour is distin-guished and rich [58]. This method differs from sourdough (Fig. 3.2) because this latter is characterized by spontaneous fermentation with microbes originating from the flour. Sourdough is fermented up to 24 or even 48 h at 22–30 °C until reaching pH $\leq$ 4, or until a total titratable acidity of 10–13 mL for 0.1 M NaOH is reached. Dough becomes acidic due to the production of organic acids by lactic acid bacteria. During fermentation, the yeast converts glucose into (mainly) carbon dioxide and ethanol [58]. At that point, an initial increase in dough volume can be observed. Then, after continued fermentation, the dough is moulded and panned. The proofing process is a final fermentation during which the dough volume increases in the pan. Gas cells are retained in dough during fermentation and baking. The more tradi-tional breadmaking practices requires a wood oven for baking. In this kind of oven, the temperature rises to around 300 °C. Other oven typologies used are heat-cycle

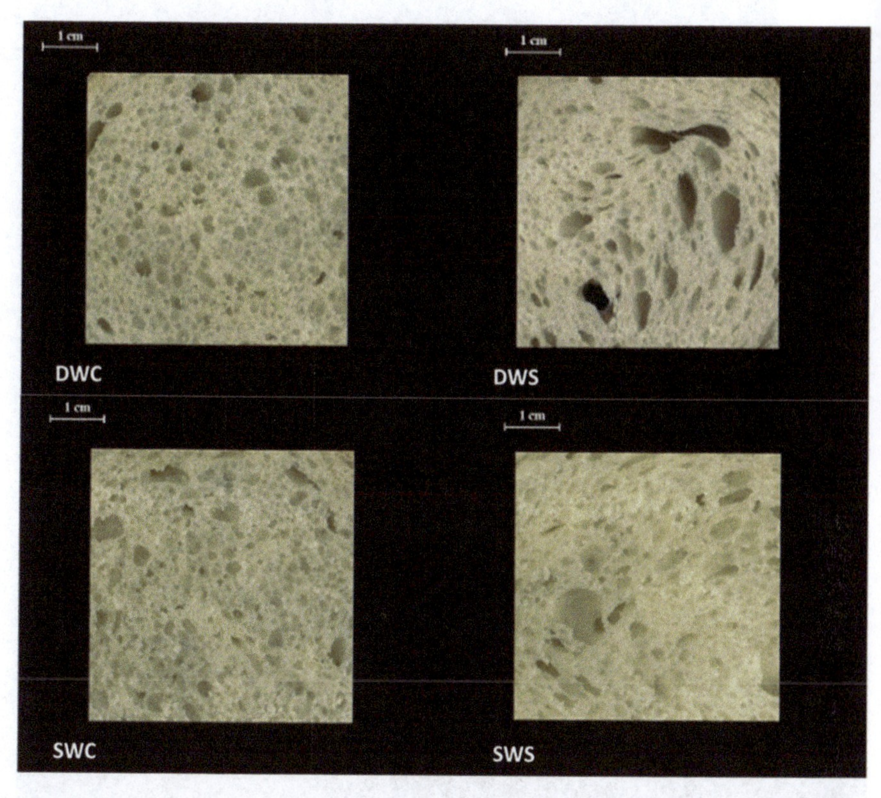

Fig. 3.1 Crumb structure of Durum Wheat Straight Dough bread (DWC); Durum Wheat Sourdough bread (DWS); Soft Wheat Straight Dough bread (SWC); Soft Wheat Sourdough bread (SWS). (Modified by [102] with the kind permission of Springer Nature)

oven, gas oven, electric oven [57]. Baking temperatures in these kinds of oven typically range from 220 to 250 °C, aiming to achieve a core temperature of approximately 92–96 °C by the end of the baking period [58]. The use of different baking temperatures affects the crust-crumb ratio [103]. High temperatures limits the extent of the oven-rise because the gases do not have enough time to complete their potential for expanding cells. On the other hand, long baking times tend to decrease crumb softness. The common use of higher temperatures at the beginning of the baking phase, followed by a temperature decrease, is probably linked to the necessity of obtaining a rapid crust development in order to create a thick crust and a soft crumb. Moreover, in this way the loaf leavened structure is locked, without allowing any further expansion of its inner gas [4]. The duration of the baking phase is reported to be 30 min for 350 g dough with oven temperatures of 200 °C or 235–275 °C; and 35 min for 600 g dough at 225 °C [57]. During baking, a further expansion of the dough is observed. The expansion in the oven, or so called oven spring, results from continued yeast action, with more carbon dioxide diffusing into the cells, expansion of the carbon dioxide by heating, and vaporisation of ethanol

. 3.2 Production stages of a traditional durum wheat sourdough bread (Altamura like bread). Wood oven burning leaves and small branches of the olive tree; (**b**) Sourdough *backslopping*; Dough fermentation; (**d**) Dough moulding; (**e**) Dough proofing; (**f**) Baking. Photos taken in a ery in Laterza (TA), Italy and water. As the bread bakes from the outside to the inside, the crumb is baked and the surface drying with consequent crust formation and browning. Some bakers use to inject steam in the oven; it delays the evaporation of water at the dough surfaces by condensation of water from the oven atmosphere onto the dough surface. Moreover, this practice produces a barrier on the dough surface against the release of carbon dioxide, resulting in a higher bread volume development and in a thinner, softer, and lighter crust [58]. On the other hand, when baked in a dry atmosphere, as in wood ovens, bakery products have a dull surface appearance and a rather harsh colour [58]. During baking the dough is transformed into the final product with a characteristic texture and desirable eating properties.

High-quality bread is characterized by a high volume, soft and elastic crumb with uniform appearance, and a relatively extended shelf life. Bread quality is reflected in its sensory attributes, physical properties, chemical composition, safety, and shelf life. It is the result of a good recipe, the ingredients quality and appropriate processing. The number of features in a type of bread is assessed during product development and quality control using sensory and instrumental methods. Bread volume is a result of crumb structure. It is brought about by correct dough mixing, fermentation, and moulding. Crumb porosity and bread volume strongly influence crumb texture. Fermentation is an essential step for bread's cellular structure, crust crispness, and aroma formation [104]. In general, the longer the fermentation time, the better the aroma and texture of bread. Pagani et al. [105] showed that after 24 h, the hardness of crumb produced by straight dough method is 50% higher than that of a sponge and dough loaf. Shittu et al. [106] showed that varying combination of baking temperature and time leads to significant differences in the quality of bread in terms of volume, crust colour, crumb moisture, density, porosity, and softness. In general, breads baked at higher heating rates have a lower moisture content compared with breads baked at lower heating rates [107]. The type of flour used, and the structure of the crumb influence the colour [108]. The texture of the crust and crumb is completely different. Crust should be of the proper thickness, colour, and glossiness typical for that bread type, without cracks and detachment from the crumb. Crust colour is influenced by the baking time and temperature, and the amount of sugar in dough [107]. The flavour of the bread is the most important. It is composed of approximately 200 compounds, among which are alcohols, aldehydes, ketones, esters, and pyrazines [109]. Bread volatiles are products of proteolysis, lipid oxidation, and Maillard reactions that occur during fermentation and baking. Quality, together with eating habits and price, influences consumers in purchase choices.

In recent years, different baking technologies, e.g., frozen dough baking or partially baked bread have been developed to better respond to market demands, and to drive the production of bread with an extended shelf life. Breadmaking procedure from frozen dough involves dough preparation, freezing of the prefermented dough, thawing and baking [110]. Partially baking involves initial baking encompassing approximately 75% of the total baking time, followed by freezing and storage. Final baking can be done in baking stations or households by nonskilled personnel, making freshly baked bread available at any time.

4.1.1 Sourdough Technology

The sourdough-baking process generally concerns artisanal bakeries, and it is relatively commonplace in the production of Protected Designation of Origin (PDO) and Protected Geographical Indication (PGI) breads that have been baked in a specific geographical area for at least 30 years and that are protected by European regulations, such as Council Regulation (EC) N. 510/2006 [111]. For example, Pane di Altamura, Pagnotta del Dittaino, and Pane Toscano are all PDO breads produced in Italy; the first two are produced using semolina from durum wheat harvested in a specific area, and the latter is produced using whole-grain flour from soft wheat grown in Tuscany and characterized by the absence of salt. Other sourdough breads produced in Italy are the PGI breads Pane di Matera, Pane di Genzano, and Coppia Ferrarese. The first is semolina based, and the others use bread flour; sourdough is the main leavening agent in all three. Four PGI sourdough breads are produced in Spain using bread flour: Pan de Cea, Pa de Pagès Català, Pan de Alfacar, and Pan de Cruz de Ciudad Real. The first bread only uses sourdough, the second two breads use both sourdough and the yeast *S. cerevisiae*, and for the last bread, the regulation reports the use of natural yeast as a leavening agent [7].

The differences between sourdough bread and baker's yeast–leavened bread have been documented by many authors. Sourdough fermentation affects the properties of both the dough and the resulting bread: it extends shelf life through the inhibition of spoilage fungi and bacteria [112], improves the bread's flavour [113], enhances nutritional properties [114], increases loaf volume, has a positive effect on texture, and delays staling [115]. Acidification is the most evident effect of LAB metabolism in the dough. The pH of a ripe sourdough ranges from 3.8 to 4.5, depending on endogenous factors (microbial composition and the nature of the flour) and exogenous factors (temperature and time of fermentation and dough yield). The acidification of bread dough affects the activity of microbial and cereal enzymes, the rheology of the dough, and the flavour of the bread. The acidity acts on the gluten network, improving the softness and extensibility of the dough and favouring the retention of CO_2 produced in the fermentation process [116]. Excessive acidity and hydrolysis of gluten proteins—for instance, as a consequence of a long fermentation time—results in a softer, less elastic dough, reduces the loaf volume, and increases staling and bread firmness [117]. Consumer appreciation of sourdough bread has increased in recent years, not only due to better knowledge of its nutritional properties, but mainly because consumers enjoy the stronger flavour of sourdough bread compared to yeasted bread [118].

4.2 Flat Breads

The "flat" breads include a multitude of bread types different from each other but are always relatively thin, ranging from a few millimeters to a few centimeters in thickness (Fig. 3.3). These breads, whose origin is very ancient, as demonstrated by

Fig. 3.3 Different types of flat bread. (**a**) Flat bread baking on a gas-fueled metal griddle. These metal griddles are named *saj* in Palestine, Jordan, Lebanon, Syria, and Iraq, *sadj* in Sudan and Ethiopia, and *saç* in Turkey. (**b**) Iranian*Barbari* bread. (**c**) Sardinian (Itaky) *Carasau* bread. (**d**) Greek *Pita* bread

the findings from Mesopotamia, ancient Egypt, and the Indus civilization, were probably the first processed foods. In the review of Pasqualone [59] the several pluses of flat breads over leavened breads are well described: (1) they can be obtained from cereals other than wheat, such as pseudocereals or legumes, allowing the use of sustainable local productions from marginal lands; (2) they do not necessarily require an oven to be baked; (3) they can serve as a dish and as a spoon/fork; (4) they can be dehydrated by a second baking process, preventing the growth of moulds and extending the shelf life; (5) they are transported with little encumbrance. Although flat breads originated in a rural society, their strong points make them still very popular, even beyond the areas of origin. Flatbreads are currently produced worldwide using a variety of primary materials and as both leavened and unleavened bread.

The production steps of flat breads are not different from those of leavened breads: kneading of ingredients (flour, water, salt, sometimes little amounts of fatty ingredients, with or without yeast according to the specific bread type); leavening (which may be absent); shaping (by pouring the batter on a griddle or by sheeting and eventually punching the dough); and baking. Depending on the gluten content of the starting flour, flat breads can be obtained from either compact and elastic dough, which requires strong and extensible gluten, or from a semi-fluid batter, which, on the contrary, can be obtained from an array of gluten-free flours.

Flatbread is quite popular in the Mediterranean area, in southern Italy (Sardinia), North Africa, and Spain, and bread wheat (*Triticum aestivum* L.) or durum wheat (*Triticum turgidum* subsp. *durum* L.) flour is used. According to the traditional habits, the production process of flat bread can include a leavening phase or not. Turkish (yufka), Lebanese (marquq), Italian, (piadina) Algerian (kesra) and Tunisian (malsouka) unleavened flatbread are well known [59]. When leavened, flatbreads are traditionally made using sourdough fermentation. The Moroccan Khobz El-daar is a durum-wheat single-layered flatbread, sourdough fermented. Baladi is double-layered flatbread produced in Egypt with bread flour and sourdough fermentation. In southern Italy, more precisely in Sardinia, an island in the middle of the Mediterranean Sea, different types of flat breads are produced. They are all leavened and double layered, and their production traditionally involves the use of semolina from durum wheat and sourdough fermentation. The most representative are the double-layered Spianata and Carasau breads. The former is a soft flatbread; the latter is a crisp flatbread obtained by baking the bread twice; in fact, after the first round of baking, the two layers are separated and then toasted, to obtain crispy and very thin (about 1–2 mm) sheets of bread. Both can be made in either a circular or rectangular shape [7]. Given the reduced thickness and mass, all flat breads are cooked very quickly (a few minutes) and remain in a rather pale colour, with almost no crust. The simplest baking methods of flat breads is under a layer of sand, embers, and ash. Most flat breads are produced in the same way as they were prepared thousands of years ago, in domes or vertical ovens. In domed ovens bread is put on a horizontal floor, while vertical ovens are cylindrical clay structures where bread is baked stuck onto the vertical inner walls [59]. Fully automated industrial lines are today also today available [119]. It will help to increase the appreciation and spread this kind of breads far beyond the area of origin.

5 New Trends: Enriched Breads

Composite breads formulated with mixtures of wheat flour and flours from other sources, such as pulses or pseudocereals; or breads containing whole grain or other functional ingredients are becoming increasingly available in the market. These products, other than being characterized by unique sensory features, are also appreciated for their healthiness. Bread, indeed, although providing carbohydrates and energy, lacks some essential amino acids and bioactive components, which are instead provided by these additional ingredients.

5.1 Composite Breads with Pulses, Pseudo-Cereals and Other Flours

Pulses, declared in 2016 by FAO (Food and Agricultural Organization of the United Nations) the theme of the year, are an affordable source of carbohydrates, dietary fiber, vitamins, minerals, phytochemicals, and particularly proteins. The most common pulses are dried peas, beans, lentils and chickpeas. Despite their dense-nutrient composition, they contain antinutritional compounds, which can negatively affect bioavailability of nutrients and digestibility [120], as well as impart bitter or unacceptable taste. These antinutritional molecules can be either thermolabile (e.g. protease inhibitors and lectins), or thermostable (e.g. phytic acid, raffinose, tannins and saponins), thus processes, such as dehulling, soaking, cooking and extruding, are needed before use [60]. Single or blended pulses' flours can be mixed with wheat flour for making bread. Regardless of pulses type, the resulting blends-based breads have increased the protein, fat, dietary fiber, and mineral contents compared to 100% wheat flour based-bread [61]. Moreover, also the physical and functional properties of the final products resulted affected by the presence of legume flours [61]. In this regard, many authors reported appreciated technological and sensorial traits up to 10–15% pulses' flour supplementation. At higher levels (> 20%), the aspect (volume and colour) and the texture (crumb elasticity and firmness) of breads were negatively impacted [62]. Colour of crust and crumb became darker as the level of pulse supplementation increased, due to Maillard reaction [61, 62]. The higher water holding capacity of pulses flour, associated with increased total protein and pentosan content, or the dilution of wheat gluten content or/and possible interactions among fiber, water and gluten, may be responsible for dough rheology, volume and texture changes [61, 62]. Addition of vital gluten or structuring agents, which can mimic gluten properties such as emulsifiers, hydrocolloids and enzymes has been proposed as a good alternative to overcome these technological flaws [63–65]. Moreover, pulses flour pre-treatment (e.g. germination and fermentation) may provide additional benefits such as increasing nutrients availability and decreasing antinutrients [60].

A pseudocereal is a plant grown to produce starchy grain suitable for human food. The term "pseudocereal" combines "cereal", referring to grains of grass species, with the prefix "pseudo" meaning "false". Within nutrition and food processing they are used like cereals. At present the most known representatives are amaranth (amaranthus spp.), quinoa (chenopodium quinoa) and buckwheat (fagopyrum esculentum). Pseudo-cereal flours can be excellent sources of proteins, vitamins, minerals, fibre, and other important nutrients [66], and show antioxidant, antinflammatory, and anticarcinogenic activities [67]. Baking characteristics of amaranth, quinoa and buckwheat have been assessed in gluten-free matrices [68], achieving breads with superior nutritional features and acceptable sensory scores. In wheat flour matrices, some studies demonstrated the feasibility of partial/low replacement of wheat flour with pseudocereals for processing baked goods [5, 69–71]. Iglesias-Puig et al. [69] demonstrated that quinoa could partially replace wheat flour in bread, increasing its nutritional value in terms of dietary fibre, minerals, proteins of high biological value and healthy fats. At 25% of flour substitution a small depreciation in bread quality in terms of reduction of loaf specific volume, crumb hardening and decrease of total acceptability was however observed. When buckwheat flour was used to substitute 15% wheat flour for producing bread, crumb hardening, and increased gummy and chewy textures were observed. Texture became even more affected when husked buckwheat flour was added [70]. On the other hand, the 15% substitution didn't interfere with bread specific volume. Buckwheat bread showed less lightness and higher redness and yellowness, despite this difference was not recognised in sensory evaluation. Buckwheat breads resulted better in flavour and mouth feel sensory attributes, other than with enhanced antioxidant activities [70]. Whole amaranth flour has been used as a partial replacement for wheat flour in bread formulations [71]. It increased the product's nutritional value and raised dietary fibre, mineral and protein levels. When amaranth flour was used in proportions between 10 and 20%, an increase in crumb hardness and elasticity were observed; moreover, the tristimulus colour values were significantly affected when the amaranth concentration was raised [71].

A unique flour, used to enrich both wheat and gluten free breads, is chestnut flour [121]. The European chestnut (Castanea sativa Mill.), has been a staple food in southern Europe and Turkey for millennia, largely replacing cereals where these would not grow well, if at all, in mountainous Mediterranean areas. That is why chestnut tree is also called "fruit of the bread tree". Chestnut flour has a promising potential for the production of diverse high-quality food products [122, 123]. Chestnut flour consists of high quality proteins with essential amino acids (4–7%), relatively high amount of sugar (20–32%), starch (50–60%), dietary fiber (4–10%), and low amount of fat (2–4%), also some important vitamins such as vitamins E, C, B group as well as minerals. Besides, it has a low fat content, but rich in essential fatty acids and also it has a low amount protein having high-quality essential amino acids. The enrichment of breads with chestnut flour is reported to be acceptable up about 20%–30% substitution. The addition of chestnut flour creates a darker and often harder product with lower volumes [72, 102], as visible from Fig. 3.4 where 0, 20 and 50% chestnut flour substitution were used in a soft wheat bread recipe [72]. However, chestnut flour has been reported to improve both the dough machinability, consistency, colour, and aroma, with even better performance during the shelf life, especially for gluten-free breads [124]. Besides, antioxidant capacity of bread, mainly attributed to the Maillard reaction compounds, increase with the supplementation of chestnut flour [72] and this is probably related to the presence of phenolics, tocopherols, tocotrienols and vitamin largely present in chestnut fruit and integument. Dall'Asta et al. [72] also reported a richer volatile profile for breads supplemented with chestnut flour, probably due to the marked increase of furans, with their toasty and nutty notes, and of phenolic compounds, with their woody and smoky notes.

5.2 *Wholegrain Breads*

Wholegrain flours from wheat and other cereals have gained considerable attention as breadmaking ingredients, due to their nutritional and health benefits [125]. Whole grain intake has been linked to health benefits such as decreased risk of chronic diseases including cardiovascular disease, diabetes, cancer, and obesity [126]. These beneficial effects are due to the presence in whole flours of higher levels of

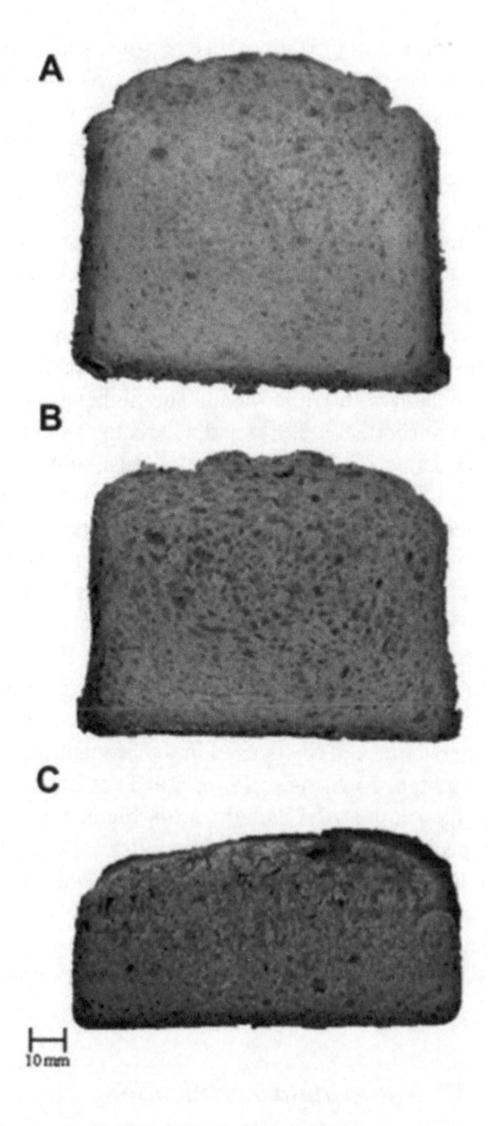

Fig. 3.4 Crumb slice images for 100% SW (**a**), 80% SW/20% CH (**b**) and 50%SW/50%CH (**c**). Abbreviations: *SW* soft whest, *CH* chestnut. (Modified by [72] with thew kind permission of Elsevier)

vitamins, minerals, fibers, antioxidants, and other phytochemicals such as carotenoids, flavonoids, and phenolic acids, in comparison to the corresponding refined flours [127]. Barriers to increasing whole grain consumption are often texture and sensory related, but also include higher cost of whole grain products and lack of knowledge regarding the health benefits of whole grain consumption [128]. Effects associated with whole grains bread production include low loaf volume, increased crumb hardness, coarse texture, darker colour, and distinctive flavour and aroma [73]. Reasons suggested for these effects are fiber-gluten interactions; dilution of gluten protein by the bran and protein; competition for water by the fiber constituents leading to insufficient hydration of gluten proteins and starch; physical effects of bran particles, fiber, and arabinoxylans on the gluten network; and higher levels of ferulic acid [73]. For these reasons, some formula and process modifications are necessary for whole bread production as compared to white bread. Water absorption must be increased. Vital wheat gluten, dough conditioners such as oxidizing agents, emulsifiers, and enzymes, as well as shortening and mould inhibitors are often added or their concentration is increased compared to white bread formulations [11]. Whole wheat dough is more susceptible to overmixing due to the physical action of the bran on the gluten. To reduce this effect, process adjustments, such as longer mixing times at lower speed, shorter total mixing time, or lower dough temperature are used. To avoid over fermentation, common for whole wheat doughs, lower temperatures and decreased fermentation times help to minimize this problem. Longer baking times and lower baking temperatures are also often needed compared with white bread. The higher water activity of whole wheat breads can lead to shorter shelf life and necessitates the addition of mould inhibitors [11].

Despite its ability to supply considerable amounts of DF, the market share of bread made with wholemeal flour is smaller than that of bread baked with white flour. Consequently, there is much interest in the development of white bread with elevated DF content. Incorporation of ground dietary fibers in a white bread recipe is an alternative to the use of wholegrain flours. Fiber with the most appropriate physicochemical properties should be chosen to prevent permanent protein matrix disruption, thus, to avoid excessive weakening, particularly in highly substituted flour systems. In addition to wheat which contains a mixture of soluble and insoluble pentosans in endosperm and the insoluble hemicelluloses available in bran, other sources of fibers used in the production of bread included rice bran, soy fiber, corn bran, powdered cellulose, and oat fiber. The criteria used in selecting the type of fiber to be used include color, dietary fiber content, cost, and water absorption capacity, along with water retention capacity following baking [129]. In general, reducing the particle size of the fiber powder allows for an increase of water-holding capacity. In the bread production process, this feature is extremely important, because water is involved in the processes of dough formation, starch gelatinization, protein denaturation, flavor and colour development, and staling kinetics [74].

5.3 Antioxidant Fortification

In recent years, there has been a global trend towards the use of the natural substances present in the food as a source of antioxidant and functional ingredients. Wheat bread is a good source of energy and nutrients for the human body; however, it is a food with a low antioxidant capacity. Due to the bread widespread consumption, it can be considered the best vehicle for functional supplements [75]. Among the regular breads, the ones made by durum wheat have been found to have higher antioxidant levels than breads made with bread wheat [130]. Ancient wheats have also been reported to be a good source of antioxidant compounds for bread fortification, as well as barley, oats, rye and pseudocereals flours [75]. Breads made by whole grain flours, or directly enriched with bran, are also rich sources of fiber and bioactive compounds [131], however, this bread type is not acceptable everywhere, and besides this the bran can include many impurities, such as heavy metals, pesticides and microorganism [132]. Supplementation of wheat flour with other sources of antioxidants such as spices [76], herbs and parts of green plants [77], seeds [78], fruit or vegetable products [79], and waste products from the food industry [80], is a new approach towards the enrichment of white bread. These raw materials are often cheap and a very good source of antioxidants, especially phenolic acids. Phenolic acids retain their antioxidant activity after the baking process, which has potential health benefits for consumers. From the literature data, a compromise between nutritional value and sensorial quality is generally achieved via the enrichment of wheat flour by 10–30% with other cereals and/or pseudo cereals and up to 5% of functional components (e.g. onion skin, green coffee, dietary fiber) [75]. From the consumer point of view, one of the important factors is the sensory value of bread, because taste, smell and flavour significantly influence consumer preferences towards cereal products. Another very important factor, often overlooked in studies of bread enrichment, is the bioaccessibility and bioavailability of functional compounds conditioning their activity in living organisms.

5.4 Food Industry by-Products for Bread Fortification

Functional ingredients obtained from food industrial by-products (BP) are a promising vehicle for the nutritional improvement of bread. Food BP are rich sources of functional ingredients, such as fibre, minerals, and phytochemicals, among others. Finding alternative uses for those BP is of major interest for both environmental and economic issues. Food industrial BP differ on their origin and include a wide range of components, such as: peel, stem, leaf, seed, shell, bran, kernel, pomace, oil cake, etc. The incorporation of functional ingredients obtained from different BP has impact on technological, nutritional, and health promoting properties of final products.

Food BP have been often used to increase the bread fiber content. In particular, fruit and vegetables BP may be an important source of pectin (in the case of fruits), cellulose (from peels), hemicellulose (from pomace) and lignin (from peels and seed coats). Bhol et al. [81] and Soares Júnior et al. [82] added respectively 15% pomegranate bagasse and 30% rice bran to bread formulations. They observed an increase of the dietary fiber content in the final products, making them eligible for the nutritional claim "high in fibre" (≥6 g fibre/100 g food product). On the other hand, bread formulations with 20% of brewer's spent grain, resulted in a final dietary fibers content higher than 3 g fibre/100 g, getting the nutritional claim "source of fibre" [83]. Salgado et al. [84], adding 3 and 6% cupuassu peel to bread recipes, reached the claim "source of fibre", while breads with 9% cupuassu peel resulted as "high in fibre".

The presence of bioactive compounds in food BP has also been exploited for bread fortification. Grape pomace, rich in antioxidant polyphenols compounds [85] or tomato skins and seeds, rich in lycopene and other bioactive compounds [86], are only two examples of by-products that have been added in bread formulations, to increase proportionally the anti-radical activity or the concentration of specific bioactives. Despite the fact that thermal treatments during baking may alter phenolics, different studies have indicated that a significant amount of polyphenols still remain in bread after processing. Moreover, some studies reported also a nutritional improvement (*in vivo*) after the consumption of breads containing such molecules. As an example, the use of dried powdered grape skins resulted in lowering total cholesterol, low-density lipoprotein cholesterol and also lipid peroxidation of breads [87]. In another study, Sudha et al. [88] observed improved free radical scavenging as well as cyto/DNA protective properties for buns enriched with apple pomace. Borczak et al. [89] demonstrated that the addition of certain freeze-dried fruits with an elevated content of polyphenols, such as elderberry, can reduce the glycemic response, potentially making it a useful tool to prevent hyperglycemia-related diseases.

The utilization of food BP, in most cases, results in a loss in food acceptability, attributed mainly to changes in flavour and appearance. Thus, food manufacturers should select the best formulation and process conditions to mitigate the negative effects, maximizing the nutritional benefits, of the incorporation of food BP in bread recipes.

5.5 Novel Ingredients for Bread Fortification

The use of novel ingredients, such as flours of microalgae and insects or their derivatives, is the most recent trend in bread fortification with the goal to produce healthier products.

Microalgae are an enormous biological resource; they can be used to enhance the nutritional value of food, due to their well-balanced chemical composition. Moreover, they are cultivated as a source of highly valuable molecules such as essential amino acids; dietary fibre; fatty acids; minerals; phenolic compounds and vitamins as well as α- and β-carotene. The health benefits provided by the incorporation of microalgal oil, microalgae biomass or their metabolites in food is linked to their anti-inflammatory, antioxidant, immunoenhancing activities, and to their role against various diseases such as cardiovascular metabolic, atherosclerosis, and hypertension [90].

Microalgae have been generally introduced into bread formulations to increase the protein content. For example, Achour et al. [91] adding *Spirulina sp.* biomass to white bread formulations, found increased protein content, which varied from 8.18% in the control to 9.90% in the enriched bread with 3% *Spirulina sp.* The authors observed that microalgae incorporation significantly decreased the bread's volume. Ak et al., [92] added 10% *Arthrospira platensis* (*Spirulina*) biomass to conventional bread; it enhanced the proteins and volatile compounds content, also improving the bread shelf life. The sensory assessment was satisfactory even perceiving algal taste. Graça et al. [93] recently studied the impact of introducing *Chlorella vulgaris* on the rheology of wheat flour dough and bread textural properties and concluded that incorporation of the microalgae at concentrations up to 3% produced a positive impact on dough rheology and viscoelastic characteristics. However, higher microalgae concentrations resulted in negative effects, not only on dough rheology but also on bread texture and flavour. Jeong et al., [94], adding *Chlorella* powder to white bread, determined an improvement of its quality characteristics in terms of volume and color while the texture was not significantly affected. The bread with 0.2% of Chlorella powder scored high sensory evaluation. Garzon et al., [95] added up to 3% *Chlorella vulgaris* in breads formulated with either 10% sourdough or chemically acidified doughs. Microalgae incorporation increased the protein and ash content of the breads. Moreover, microalgae enriched breads made with chemically acidified doughs or sourdoughs had higher Total Phenolic Content and antioxidant activity as assessed by FRAP and ABTS methods. One of the main factors limiting the application of microalgae in baked goods is their green colour. In order to avoid the green colour of microalgae in baked products, a good alternative would be to isolate microalgae-derived bioactives and to use them instead of the whole biomass. Fitzgerald et al. [96] studied bread enriched with an enzymatic hydrolysate of the macroalgae *Palmaria palmata* (with antihypertensive properties) at a concentration of 4%. The authors reported a bitter taste of breads containing the algal-derived proteins. Moreover, the results of this study confirmed that the bioactivity of the enzymatic hydrolysate was resistant to the baking process.

Recently, also edible insects have been evaluated as a valuable source of proteins, other than contributing into fat, minerals, vitamins (especially the B group), and fibre (especially chitin) intake. Osimani et al. [97] mixed *Acheta domesticus* (cricket) and wheat flour in various proportions to generate protein-enriched bread. While cricket flour incorporation did enhance the nutritional value, particularly protein content, it also negatively affected the hardness of the loaf. Bread enriched with 10% cricket flour was still globally appreciated according to the sensory evaluation, even if the panel ultimately preferred the bread without insect ingredient, probably because of the characteristic flavour of cricket flour, or because of the unfavourable initial bias of the untrained panelists, which might have negatively influenced them. Haber et al., [98] observed that the addition of grasshopper powder decreased the specific volume of bread and resulted in softer texture. Moreover, the inclusion of 200 g/kg of grasshopper powder to the bread recipe increased the protein content by up to 60%. Sensory panel scored 100 g/kg enriched grasshopper bread similarly to wheat bread, while bread enriched with 200 g/kg received a lower score mainly due to its distinctive odour. de Oliveira et al., [99], added *Nauphoeta cinera* flour to wheat bread dough; the authors observed a linear correlation between the incorporation of insect flour and an increase in loaf volume and firmness. Moreover, bread enriched with 10% insect flour increase of 49.16% in protein compared to standard bread, being however moderately liked by the consumers. Roncolini et al. [100] added mealworm (*Tenebrio molitor*) flour into bread doughs at 5 and 10%

substitution level of soft wheat flour to produce protein fortified breads. All the tested doughs showed the same leavening ability, whereas breads containing 5% mealworm flour showed the highest specific volume and the lowest firmness. An enrichment in protein content and essential aminoacids was observed in the enriched breads especially in breads with 10% flour substitution. By contrast, no differences in lipids were seen between fortified and control breads. Finally, González et al., [101] recently compared the addition of 5% insect flour (*Hermetia illucens*, *Acheta domesticus* and *Tenebrio molitor*) to bread dough. These authors concluded that *Acheta domesticus* was the most suitable flour for bread formulation as it had the best functional properties and flavour of the three insects. The authors also noted that much work is yet to be done to meet consumer expectations and high standards. Moreover, potential spoilage and safety issues need to be further considered.

6 Conclusions

The types of bread in the Mediterranean area are linked to ancient rural traditions that have been innovated over time with the aim to better answer the increasing demands of more conscientious consumers. Nowadays there is a strong tendency to re-evaluate traditional products. Consumers prefer bread with the taste, appearance, and flavour specific to traditional products, but obtained in the safer conditions and with the long shelf-life provided by the new technologies. Making bread a functional food is the new frontier in this sector. The new challenge for research and food industry is to find a compromise between sensory acceptance and healthy properties of the novel formulations.

References

1. Rubel W. Bread: a global history. Islington: Reaktion Books; 2011.
2. Goesaert H, Brijs K, Veraverbeke W, Courtin C, Gebruers K, Delcour JA. Wheat flour constituents: how they impact bread quality, and how to impact their functionality. Trends Food Sci Technol. 2005;16:12–30.
3. Boukid F, Folloni S, Sforza S, Vittadini E, Prandi B. Current trends in ancient grains-based foodstuffs: insights into nutritional aspects and technological applications. Compr Rev Food Sci Food Saf. 2018;17(1):123–36.
4. Cauvain SP, Young LS. Technology of breadmaking: Springer; 2007.
5. Collar C, Angioloni A. Pseudocereals and teff in complex breadmaking matrices: impact on lipid dynamics. J Cereal Sci. 2014;59(2):145–54.
6. Cauvain SP, Young LS. Bakery food manufacture and quality: water control and effects. Hoboken: John Wiley & Sons; 2009.
7. Catzeddu P. Sourdough breads. In: Preedy VR, Watson RR, editors. Flour and breads and their fortification in health and disease prevention. Cambridge: Academic Press; 2019. p. 177–88.
8. Miller RA, Hoseney RC. Role of salt in baking. Cereal Foods World. 2008;53(1):4–6.
9. Mondal A, Datta AK. Bread baking–a review. J Food Eng. 2008;86(4):465–74.

10. Pareyt B, Finnie SM, Putseys JA, Delcour JA. Lipids in bread making: sources, interactions, and impact on bread quality. J Cereal Sci. 2011;54(3):266–79.
11. Tebben L, Shen Y, Li Y. Improvers and functional ingredients in whole wheat bread: a review of their effects on dough properties and bread quality. Trends Food Sci Technol. 2018;81:10–24.
12. Armero E, Collar C. Crumb firming kinetics of wheat breads with anti-staling additives. J Cereal Sci. 1998;28(2):165–74.
13. Davidou S, Le Meste M, Debever E, Bekaert DJFH. A contribution to the study of staling of white bread: effect of water and hydrocolloid. Food Hydrocoll. 1996;10(4):375–83.
14. Guarda A, Rosell CM, Benedito C, Galotto MJ. Different hydrocolloids as bread improvers and antistaling agents. Food Hydrocoll. 2004;18(2):241–7.
15. Boukid F, Carini E, Curti E, Pizzigalli E, Vittadini E. Bread staling: understanding the effects of transglutaminase and vital gluten supplementation on crumb moisture and texture using multivariate analysis. Eur Food Res Technol. 2019;245(6):1337–45.
16. Barcenas ME, Rosell CM. Effect of HPMC addition on the microstructure, quality and aging of wheat bread. Food Hydrocoll. 2005;19(6):1037–43.
17. Magan N, Aldred D, Arroyo M. Mould prevention in bread. In: Couvain SP, editor. Breadmaking, improving quality. Boca Raton: CRC Press-Woodhead Publishing; 2012. p. 597–613.
18. Legan JD. Mould spoilage of bread: the problem and some solutions. Int Biodeterior Biodegradation. 1993;32(1–3):33–53.
19. Giunta F, Pruneddu G, Zuddas M, Motzo R. Bread and durum wheat: intra-and inter-specific variation in grain yield and protein concentration of modern Italian cultivars. Eur J Agron. 2019;105:119–28.
20. Chiavaro E, Vittadini E, Musci M, Bianchi F, Curti E. Shelf-life stability of artisanally and industrially produced durum wheat sourdough bread ("Altamura bread"). LWT-Food Sci Technol. 2008;41(1):58–70.
21. Belderok B, Mesdag J, Mesdag H, Donner DA. Bread-making quality of wheat: a century of breeding in Europe. Berlin: Springer Science & Business Media; 2000.
22. Mefleh M, Conte P, Fadda C, Giunta F, Piga A, Hassoun G, Motzo R. From ancient to old and modern durum wheat varieties: interaction among cultivar traits, management, and technological quality. J Sci Food Agric. 2019;99(5):2059–67.
23. Boukid F, Vittadini E, Prandi B, Mattarozzi M, Marchini M, Sforza S, et al. Insights into a century of breeding of durum wheat in Tunisia: the properties of flours and starches isolated from landraces, old and modern genotypes. LWT- Food Sci Technol. 2018;97:743–51.
24. Boukid F, Dall'Asta M, Bresciani L, Mena P, Del Rio D, Calani L, et al. Phenolic profile and antioxidant capacity of landraces, old and modern Tunisian durum wheat. Eur Food Res Technol. 2019;245(1):73–82.
25. Cappelli A, Cini E, Guerrini L, Masella P, Angeloni G, Parenti A. Predictive models of the rheological properties and optimal water content in doughs: an application to ancient grain flours with different degrees of refining. J Cereal Sci. 2018;83:229–35.
26. Geisslitz S, Wieser H, Scherf KA, Koehler P. Gluten protein composition and aggregation properties as predictors for bread volume of common wheat, spelt, durum wheat, emmer and einkorn. J Cereal Sci. 2018;83:204–12.
27. Hidalgo A, Brandolini A. Heat damage of water biscuits from einkorn, durum and bread wheat flours. Food Chem. 2011;128(2):471–8.
28. Brandolini A, Hidalgo A. Einkorn (*Triticum monococcum*) flour and bread. In: Preedy V, Watson R, Patel V, editors. Flour and breads and their fortification in health and disease prevention. 1st ed. Cambridge: Academic Press; 2011. p. 79–88.
29. Brandolini A, Hidalgo A, Moscaritolo S. Chemical composition and pasting properties of einkorn (*Triticum monococcum* L. subsp. *monococcum*) whole meal flour. J Cereal Sci. 2008;47(3):599–609.
30. Hidalgo A, Brandolini A. Nutritional properties of einkorn wheat (*Triticum monococcum* L.). J Sci Food Agric. 2014;94(4):601–12.

31. Stallknecht GF, Gilbertson KM, Ranney JE. Alternative wheat cereals as food grains: einkorn, emmer, spelt, kamut, and triticale. In: Janick J, editor. Progress in new crops. Alexandria, VA: ASHS Press; 1996. p. 156–70.
32. Pasqualone A, Piergiovanni AR, Caponio F, Paradiso VM, Summo C, Simeone R. Evaluation of the technological characteristics and bread-making quality of alternative wheat cereals in comparison with common and durum wheat. Food Sci Technol Int. 2011;17(2):135–42.
33. Balestra F, Laghi L, Saa DT, Gianotti A, Rocculi P, Pinnavaia G. Physico-chemical and metabolomic characterization of KAMUT® Khorasan and durum wheat fermented dough. Food Chem. 2015;187:451–9.
34. Korczyk-Szabó J, Lacko-Bartošová M. Crumb texture of spelt bread. J Cent Eur Agric. 2013;14(4):1343–52.
35. Callejo MJ, Vargas-Kostiuk ME, Rodríguez-Quijano M. Selection, training and validation process of a sensory panel for bread analysis: influence of cultivar on the quality of breads made from common wheat and spelt wheat. J Cereal Sci. 2015;61:55–62.
36. Bonafaccia G, Galli V, Francisci R, Mair V, Skrabanja V, Kreft I. Characteristics of spelt wheat products and nutritional value of spelt wheat-based bread. Food Chem. 2000;68(4):437–41.
37. Coda R, Nionelli L, Rizzello CG, De Angelis M, Tossut P, Gobbetti M. Spelt and emmer flours: characterization of the lactic acid bacteria microbiota and selection of mixed starters for bread making. J Appl Microbiol. 2010;108(3):925–35.
38. Boukid F, Gentilucci V, Vittadini E, De Montis A, Rosta R, Bosi S, et al. Rediscovering bread quality of "old" Italian wheat (*Triticum aestivum* L. ssp. *aestivum*) through an integrated approach: physicochemical evaluation and consumers' perception. LWT- Food Sci Tech. 2020;122:109043.
39. Sterna V, Zute S, Brunava L. Oat grain composition and its nutrition benefice. Agric Agric Sci Procedia. 2016;8:252–6.
40. Wood PJ. Cereal β-glucans in diet and health. J Cereal Sci. 2007;46(3):230–8.
41. Boita ER, Oro T, Bressiani J, Santetti GS, Bertolin TE, Gutkoski LC. Rheological properties of wheat flour dough and pan bread with wheat bran. J Cereal Sci. 2016;71:177–82.
42. Girgis S, Neal B, Prescott J, Prendergast J, Dumbrell S, Turner C, Woodward M. A one-quarter reduction in the salt content of bread can be made without detection. Eur J Clin Nutr. 2003;57(4):616–20.
43. Liem DG, Miremadi F, Keast RS. Reducing sodium in foods: the effect on flavor. Nutrients. 2011;3(6):694–711.
44. Minervini F, De Angelis M, Di Cagno R, Gobbetti M. Ecological parameters influencing microbial diversity and stability of traditional sourdough. Int J Food Microbiol. 2014;171:136–46.
45. Chin NL, Rahman RA, Hashim DM, Kowng SY. Palm oil shortening effects on baking performance of white bread. J Food Process Eng. 2010;33(3):413–33.
46. Smith PR, Johansson J. Influences of the proportion of solid fat in a shortening on loaf volume and staling of bread. J Food Process Preserv. 2004;28(5):359–67.
47. Azizi MH, Rao GV. Effect of surfactant gel and gum combinations on dough rheological characteristics and quality of bread. J Food Qual. 2004;27(5):320–36.
48. Van Oort M. Enzymes in bread making. In: Whitehurst RJ, van Oort M, editors. Enzymes in food technology. 2nd ed. Hoboken: Wiley-Blackwell; 2010. p. 103–35.
49. Ferrero C. Hydrocolloids in wheat breadmaking: a concise review. Food Hydrocol. 2017;68:15–22.
50. Bardini G, Boukid F, Carini E, Curti E, Pizzigalli E, Vittadini E. Enhancing dough-making rheological performance of wheat flour by transglutaminase and vital gluten supplementation. LWT-Food Sci Technol. 2018;91:467–76.
51. Boukid F, Carini E, Curti E, Bardini G, Pizzigalli E, Vittadini E. Effectiveness of vital gluten and transglutaminase in the improvement of physico-chemical properties of fresh bread. LWT-Food Sci Technol. 2018;92:465–70.
52. Bárcenas ME, De la O-Keller J, Rosell CM. Influence of different hydrocolloids on major wheat dough components (gluten and starch). J Food Eng. 2009;94(3–4):241–7.

53. Rosell CM, Rojas JA, De Barber CB. Influence of hydrocolloids on dough rheology and bread quality. Food Hydrocoll. 2001;15(1):75–81.
54. Gerits LR, Pareyt B, Decamps K, Delcour JA. Lipases and their functionality in the production of wheat-based food systems. Comp Rev Food Sci Food Saf. 2014;13(5):978–89.
55. Gómez M, Del Real S, Rosell CM, Ronda F, Blanco CA, Caballero PA. Functionality of different emulsifiers on the performance of breadmaking and wheat bread quality. Eur Food Res Technol. 2004;219(2):145–50.
56. Cizeikiene, Juodeikiene G, Paskevicius A, Bartkiene E. Antimicrobial activity of lactic acid bacteria against pathogenic and spoilage microorganism isolated from food and their control in wheat bread. Food Control. 2013;31(2):539–45.
57. Guerrini L, Parenti O, Angeloni G, Zanoni B. The bread making process of ancient wheat: a semi-structured interview to bakers. J Cereal Sci. 2019;87:9–17.
58. Ćurić D, Novotri D, Smerdel B. Bread making. In: de Pinho Ferreira Guine R, dos Reis Correia PM, editors. Engineering aspects of cereal and cereal-based products. Boca Raton: CRC Press; 2014. p. 149–74.
59. Pasqualone A. Traditional flat breads spread from the Fertile Crescent: production process and history of baking systems. J Ethnic Foods. 2018;5(1):10–9.
60. Boukid F, Zannini E, Carini E, Vittadini E. Pulses for bread fortification: a necessity or a choice? Trends Food Sci Technol. 2019;88:416–28.
61. Man S, Păucean A, Muste S, Pop A. Effect of the chickpea (Cicer arietinum L.) flour addition on physicochemical properties of wheat bread. Bull UASVM Food Sci Technol. 2015;72(1):41–9.
62. Mohammed I, Ahmed AR, Senge B. Effects of chickpea flour on wheat pasting properties and bread making quality. J Food Sci Technol. 2014;51(9):1902–10.
63. Alasino MC, Osella CA, De La Torre MA, Sanchez HD. Use of sodium stearoyl lactylate and azodicarbonamide in wheat flour breads with added pea flour. Int J Food Sci Nutr. 2011;2(4):385–91.
64. Previtali MA, Mastromatteo M, De Vita P, Ficco DBM, Conte A, Del Nobile MA. Effect of the lentil flour and hydrocolloids on baking characteristics of wholemeal durum wheat bread. Int J Food Sci Technol. 2014;49(11):2382–90.
65. Yorgancilar M, Bilgiçli N. Chemical and nutritional changes in bitter and sweet lupin seeds (Lupinus albus L.) during bulgur production. J Food Sci Technol. 2014;51(7):1384–9.
66. Coda R, Rizzello CG, Gobbetti M. Use of sourdough fermentation and pseudo-cereals and leguminous flours for the making of a functional bread enriched of γ-aminobutyric acid (GABA). Int J Food Microbiol. 2010;137(2–3):236–45.
67. Lin LY, Liu HM, Yu YW, Lin SD, Mau JL. Quality and antioxidant property of buckwheat enhanced wheat bread. Food Chem. 2009;112(4):987–91.
68. Alvarez-Jubete L, Auty M, Arendt EK, Gallagher E. Baking properties and microstructure of pseudocereal flours in gluten-free bread formulations. Eur Food Res Technol. 2010;230(3):437–45.
69. Iglesias-Puig E, Monedero V, Haros M. Bread with whole quinoa flour and bifidobacterial phytases increases dietary mineral intake and bioavailability. LWT-Food Sci Technol. 2015;60(1):71–7.
70. Lin LY, Wang HE, Lin SD, Liu HM, Mau JL. Changes in buckwheat bread during storage. J Food Process Preserv. 2013;37(4):285–90.
71. Sanz-Penella JM, Wronkowska M, Soral-Smietana M, Haros M. Effect of whole amaranth flour on bread properties and nutritive value. LWT-Food Sci Technol. 2013;50(2):679–85.
72. Dall'Asta C, Cirlini M, Morini E, Rinaldi M, Ganino T, Chiavaro E. Effect of chestnut flour supplementation on physico-chemical properties and volatiles in bread making. LWT-Food Sci Technol. 2013;53(1):233–9.
73. Heiniö RL, Noort MWJ, Katina K, Alam SA, Sozer N, De Kock HL, et al. Sensory characteristics of wholegrain and bran-rich cereal foods–a review. Trends Food Sci Technol. 2016;47:25–38.

74. Curti E, Carini E, Bonacini G, Tribuzio G, Vittadini E. Effect of the addition of bran fractions on bread properties. J Cereal Sci. 2013;57(3):325–32.
75. Dziki D, Różyło R, Gawlik-Dziki U, Świeca M. Current trends in the enhancement of antioxidant activity of wheat bread by the addition of plant materials rich in phenolic compounds. Trends Food Sci Technol. 2014;40(1):48–61.
76. Balestra F, Cocci E, Pinnavaia G, Romani S. Evaluation of antioxidant, rheological and sensorial properties of wheat flour dough and bread containing ginger powder. LWT-Food Sci Technol. 2011;44(3):700–5.
77. Das L, Raychaudhuri U, Chakraborty R. Supplementation of common white bread by coriander leaf powder. Food Sci Biotechnol. 2012;21(2):425–33.
78. Peng X, Ma J, Cheng KW, Jiang Y, Chen F, Wang M. The effects of grape seed extract fortification on the antioxidant activity and quality attributes of bread. Food Chem. 2010;119(1):49–53.
79. Sivam AS, Sun-Waterhouse D, Waterhouse GI, Quek S, Perera CO. Physicochemical properties of bread dough and finished bread with added pectin fiber and phenolic antioxidants. J Food Sci. 2011;76(3):H97–H107.
80. Gawlik-Dziki U, Świeca M, Dziki D, Baraniak B, Tomiło J, Czyż J. Quality and antioxidant properties of breads enriched with dry onion (Allium cepa L.) skin. Food Chem. 2013;138(2–3):1621–8.
81. Bhol S, Lanka D, Bosco SJD. Quality characteristics and antioxidant properties of breads incorporated with pomegranate whole fruit bagasse. J Food Sci Technol. 2016;53(3):1717–21.
82. Soares Junior MS, Bassinello PZ, Caliari M, Gebin PFC, Junqueira TDL, Gomes VA, Lacerda DBCL. Quality of breads with toasted rice bran. Food Sci Technol. 2009;29(3):636–41.
83. Fărcaş AC, Socaci SA, Tofană M, Mureşan C, Mudura E, Salanţă L, Scrob S. Nutritional properties and volatile profile of brewer's spent grain supplemented bread. Rom Biotechnol Lett. 2014;19(5):9705–14.
84. Salgado JM, Rodrigues BS, Donado-Pestana CM, dos Santos Dias CT, Morzelle MC. Cupuassu (Theobroma grandiflorum) peel as potential source of dietary fiber and phytochemicals in whole-bread preparations. Plant Foods Hum Nutr. 2011;66(4):384–90.
85. Hayta M, Özuğur G, Etgü H, Şeker İT. Effect of grape (*Vitis Vinifera L.*) pomace on the quality, total phenolic content and anti-radical activity of bread. J Food Process Preserv. 2014;38(3):980–6.
86. Nour V, Ionica ME, Trandafir I. Bread enriched in lycopene and other bioactive compounds by addition of dry tomato waste. J Food Sci Technol. 2015;52(12):8260–7.
87. Mildner-Szkudlarz S, Bajerska J. Protective effect of grape by-product-fortified breads against cholesterol/cholic acid diet-induced hypercholesterolaemia in rats. J Sci Food Agric. 2013;93(13):3271–8.
88. Sudha ML, Dharmesh SM, Pynam H, Bhimangouder SV, Eipson SW, Somasundaram R, Nanjarajurs SM. Antioxidant and cyto/DNA protective properties of apple pomace enriched bakery products. J Food Sci Technol. 2016;53(4):1909–18.
89. Borczak B, Sikora E, Sikora M, Kapusta-Duch J, Kutyła-Kupidura EM, Fołta M. Nutritional properties of wholemeal wheat-flour bread with an addition of selected wild grown fruits. Starch Stärke. 2016;68(7–8):675–82.
90. Sidari R, Tofalo R. A comprehensive overview on microalgal-fortified/based food and beverages. Food Rev Int. 2019;35(8):778–805.
91. Achour HY, Doumandji A, Said S, Saadi S. Evaluation of nutritional and sensory properties of bread enriched with Spirulina. Ann Food Sci Technol. 2014;15:270–5.
92. Ak B, Avsaroglu E, Isik O, Özyurt G, Kafkas E, Etyemez M. Nutritional and physicochemical characteristics of bread enriched with microalgae Spirulina platensis. Int J Eng Res Appl. 2016;6:30–8.
93. Graça C, Fradinho P, Sousa I, Raymundo A. Impact of Chlorella vulgaris on the rheology of wheat flour dough and bread texture. LWT- Food Sci Technol. 2018;89:466–74.
94. Jeong CH, Cho HJ, Shim KH. Quality characteristics of white bread added with chlorella powder. Korean J Food Preserv. 2006;13(4):465–71.

95. Garzon R, Skendi A, Lazo-Velez MA, Papageorgiou M, Rosell CM. Interaction of dough acidity and microalga level on bread quality and antioxidant properties. Food Chem. 2020;95:128710.
96. Fitzgerald C, Gallagher E, Doran L, Auty M, Prieto J, Hayes M. Increasing the health benefits of bread: assessment of the physical and sensory qualities of bread formulated using a renin inhibitory Palmaria palmata protein hydrolysate. LWT-Food Sci Technol. 2014;56(2):398–405.
97. Osimani A, Milanović V, Cardinali F, Roncolini A, Garofalo C, Clementi F, et al. Bread enriched with cricket powder (Acheta domesticus): a technological, microbiological and nutritional evaluation. Innovative Food Sci Emerg Technol. 2018;48:150–63.
98. Haber M, Mishyna M, Martinez JI, Benjamin O. The influence of grasshopper (Schistocerca gregaria) powder enrichment on bread nutritional and sensorial properties. LWT-Food Sci Technol. 2019;115:108395.
99. de Oliveira LM, da Silva Lucas AJ, Cadaval CL, Mellado MS. Bread enriched with flour from cinereous cockroach (Nauphoeta cinerea). Innovative Food Sci Emerg Technol. 2017;44:30–5.
100. Roncolini A, Milanović V, Cardinali F, Osimani A, Garofalo C, Sabbatini R, et al. Protein fortification with mealworm (Tenebrio molitor L.) powder: effect on textural, microbiological, nutritional and sensory features of bread. PLoS One. 2019;14(2):e0211747.
101. González CM, Garzón R, Rosell CM. Insects as ingredients for bakery goods. A comparison study of H. illucens, A. domestica and T. molitor flours. Innovative Food Sci Emerg Technol. 2019;51:205–10.
102. Rinaldi M, Paciulli M, Caligiani A, Sgarbi E, Cirlini M, Dall'Asta C, Chiavaro E. Durum and soft wheat flours in sourdough and straight-dough bread-making. J Food Sci Technol. 2015;52(10):6254–65.
103. Ahrné L, Andersson CG, Floberg P, Rosén J, Lingnert H. Effect of crust temperature and water content on acrylamide formation during baking of white bread: steam and falling temperature baking. LWT-Food Sci Technol. 2007;40(10):1708–15.
104. Primo-Martín C, Van Dalen G, Meinders MBJ, Don A, Hamer RH, Van Vliet T. Bread crispness and morphology can be controlled by proving conditions. Food Res Int. 2010;43(1):207–17.
105. Pagani MA, Lucisano M, Mariotti M. Italian bakery products. In: Zhou W, Hui YH, editors. Bakery products science and technology. 2nd ed. Hoboken: Wiley Blackwell; 2014. p. 685–721.
106. Shittu TA, Raji AO, Sanni LO. Bread from composite cassava-wheat flour: I. effect of baking time and temperature on some physical properties of bread loaf. Food Res Int. 2007;40(2):280–90.
107. Patel BK, Waniska RD, Seetharaman K. Impact of different baking processes on bread firmness and starch properties in breadcrumb. J Cereal Sci. 2005;42(2):173–84.
108. Rinaldi M, Paciulli M, Dall'Asta C, Cirlini M, Chiavaro E. Short-term storage evaluation of quality and antioxidant capacity in chestnut–wheat bread. J Sci Food Agric. 2015;95(1):59–65.
109. Bianchi F, Careri M, Chiavaro E, Musci M, Vittadini E. Gas chromatographic–mass spectrometric characterisation of the Italian protected designation of origin "Altamura" bread volatile profile. Food Chem. 2008;110(3):787–93.
110. Giannou V, Kessoglou V, Tzia C. Quality and safety characteristics of bread made from frozen dough. Trends Food Sci Technol. 2003;14(3):99–108.
111. Council Regulation (EC) 510=2006 of March 20, 2006, on the protection of geographical indications and designations of origin for agricultural products and foodstuff.
112. Quattrini M, Liang N, Fortina MG, Xiang S, Curtis JM, Gänzle M. Exploiting synergies of sourdough and antifungal organic acids to delay fungal spoilage of bread. Int J Food Microbiol. 2019;302:8–14.
113. Paterson A, Piggott JR. Flavour in sourdough breads: a review. Trends Food Sci Technol. 2006;17(10):557–66.

114. Katina K, Arendt E, Liukkonen KH, Autio K, Flander L, Poutanen K. Potential of sourdough for healthier cereal products. Trends Food Sci Technol. 2005;16(1–3):104–12.
115. Arendt EK, Ryan LA, Dal Bello F. Impact of sourdough on the texture of bread. Food Microbiol. 2007;24(2):165–74.
116. Clarke CI, Schober TJ, Arendt EK. Effect of single strain and traditional mixed strain starter cultures on rheological properties of wheat dough and on bread quality. Cereal Chem. 2002;79(5):640–7.
117. Clarke CI, Schober TJ, Dockery P, O'Sullivan K, Arendt EK. Wheat sourdough fermentation: effects of time and acidification on fundamental rheological properties. Cereal Chem. 2004;81(3):409–17.
118. Pétel C, Onno B, Prost C. Sourdough volatile compounds and their contribution to bread: a review. Trends Food Sci Technol. 2017;59:105–23.
119. Gambella F, Paschino F. Produzione di pane" Carasau": comparazione fra un impianto tradizionale e uno semicontinuo. Industrie alimentari. 2004;43(441):1115–20.
120. Moktan K, Ojha P. Quality evaluation of physical properties, antinutritional factors, and antioxidant activity of bread fortified with germinated horse gram (Dolichus uniflorus) flour. Food Sci Nutr. 2016;4(5):766–71.
121. Paciulli M, Mert ID, Rinaldi M, Pugliese A, Chiavaro E. Chestnut and breads: nutritional, functional, and technological qualities. In: Preedy V, Watson R, editors. Flour and breads and their fortification in health and disease prevention. 2nd ed. Cambridge: Academic Press; 2019. p. 237–47.
122. Rinaldi M, Paciulli M, Caligiani A, Scazzina F, Chiavaro E. Sourdough fermentation and chestnut flour in gluten-free bread: a shelf-life evaluation. Food Chem. 2017;224:144–52.
123. Paciulli M, Rinaldi M, Cavazza A, Ganino T, Rodolfi M, Chiancone B, Chiavaro E. Effect of chestnut flour supplementation on physico-chemical properties and oxidative stability of gluten-free biscuits during storage. LWT-Food Sci Tech. 2018;98:451–7.
124. Paciulli M, Rinaldi M, Cirlini M, Scazzina F, Chiavaro E. Chestnut flour addition in commercial gluten-free bread: a shelf-life study. LWT-Food Sci Technol. 2016;70:88–95.
125. Ragaee S, Guzar I, Dhull N, Seetharaman K. Effects of fiber addition on antioxidant capacity and nutritional quality of wheat bread. LWT-Food Sci Technol. 2011;44(10):2147–53.
126. Slavin J. Why whole grains are protective: biological mechanisms. Proc Nutr Soc. 2003;62(1):129–34.
127. Slavin J. Whole grains and human health. Nutr Res Rev. 2004;17(1):99–110.
128. Sogari G, Li J, Lefebvre M, Menozzi D, Pellegrini N, Cirelli M, et al. The influence of health messages in nudging consumption of whole grain pasta. Nutrients. 2019;11(12):2993.
129. Kurek M, Wyrwisz J. The application of dietary fiber in bread products. J Food Process Technol. 2015;6(5):447–50.
130. Žilić SA, Dodig D, Šukalović VHT, Maksimović M, Saratlić G, Škrbić B. Bread and durum wheat compared for antioxidants contents, and lipoxygenase and peroxidase activities. Int J Food Sci Technol. 2010;45(7):1360–7.
131. Gani A, Wani SM, Masoodi FA, Hameed G. Whole-grain cereal bioactive compounds and their health benefits: a review. J Food Process Technol. 2012;3(3):146–56.
132. Mousia Z, Edherly S, Pandiella SS, Webb C. Effect of wheat pearling on flour quality. Food Res Int. 2004;37(5):449–59.

Chapter 4
Italian Dried Pasta: Conventional and Innovative Ingredients and Processing

Paola Conte, Antonio Piga, Alessandra Del Caro, Pietro Paolo Urgeghe, and Costantino Fadda

Abstract Despite the myths and legends that still surround the true origin of pasta, it can be considered a centuries old traditional Italian food nowadays appreciated all over the world for its nutritional value, versatility, palatability, cheapness, and ease of preparation. The widespread popularity of this regularly eaten cereal-based product is due to the way it intercepts cultural and consumer food trends, as well as to its ability to combine tradition and innovation, while maintaining—or improving—productivity, efficiency and high standard quality. Among the wide variety of pasta products currently produced by manufacturers in hundreds of geometrical shapes and size, dried pasta (*pasta secca*) can certainly be considered the most appreciated and consumed kind of pasta in Italy. Starting from these points, this chapter will primarily focus on the evolution of both processing and formulation of the dried pasta, with special emphasis on the use of unconventional ingredients that may enhance healthfulness and functionality of the final product.

Keywords Fresh pasta · Dry pasta · Extrusion · Drying · Fortification

1 Introduction

The term pasta is commonly used to describe one of the oldest cereal-based products, second only to bread. Nowadays, pasta products are widespread and appreciated all over the world not only for their nutritional value, but also for their versatility, palatability, cheapness, long shelf-life, and ease of preparation [1]. Although pasta is traditionally considered one of the symbols of Italy's culinary traditions, the true origin of pasta as well as the time when it was invented are still subject of conjectures. Some believe that pasta was imported to Italy from China in the thirteen century by the Venetian explorer and merchant Marco Polo; others date back the

P. Conte (✉) · A. Piga · A. Del Caro · P. P. Urgeghe · C. Fadda
Dipartimento di Agraria, Università degli Studi di Sassari, Sassari, Italy
e-mail: pconte@uniss.it

F. Boukid (ed.), *Cereal-Based Foodstuffs: The Backbone of Mediterranean Cuisine*, https://doi.org/10.1007/978-3-030-69228-5_4

origin of pasta to periods long before the Middle Ages, arguing that a "fine sheet of dough" called *lagana* made by mixing cereals and water—and very similar to the current *lasagna*—was already known to ancient Romans and Etruscan civilization. Yet others believe that pasta was invented by the nomadic Arabs that, after the invasion of the regions of the Mediterranean basin in the eight century BC, foster the development and the spread of pasta-like products in the southern regions of Italy—mostly Sicily—which became the center of the production of dried pasta [2–4]. Nowadays, however, despite the myths and legends that still surround the origin of this food, it can be said that Italy remains the cradle of the pasta tradition, as well as a reference point for making the best-quality pasta. It is also the country that produces, exports and consumes pasta the most.

In fact, according to the Union of Organizations of Manufactures of Pasta Products of the European Union (UN.A.F.P.A), in 2015 pasta production reached at about 14.3 million tons globally, out of which 4.6 million tons in EU, two million tons in the United States, more than one million tons in Turkey, Brazil and Russia, and four million tons in the remaining countries. Italy, with an output of more than 3.3 million tons (which represents at about 25 and 71% of the world and EU production, respectively) and a consumption of 23.5 kg per capita per year, accounts for the largest pasta producer and consumer in the world [5]. In addition, in 2017 more than half of the Italian pasta production (at about 58%) was exported all over the world (mainly to Germany, France, United Kingdom, United States and Japan), confirming the leadership position of Italy in the global pasta market [6].

The wide variety of existing pasta products, which are currently produced by the manufacturers in hundreds of geometrical shapes and size, are generally divided into two broad categories: fresh (*pasta fresca*) and dried (*pasta secca*) pasta. The latter, which approximately accounts for about 90% of the alimentary pasta consumption, can certainly be considered the most appreciated and consumed kind of pasta in Italy [7].

Despite the smaller market of fresh pasta is gaining popularity among Italian consumers, who always more frequently tend to associate a fresh product with an artisanal product, this chapter will primarily focus on the evolution of both formulation and manufacturing process of dried pasta.

2 Italian Pasta: Legal Requirements, Raw Materials and Classification

In terms of ingredients, pasta can be considered as one of the simplest cereal-based products. In fact, apart from differences in terms of processing requirements, all the produced kinds of pasta are usually prepared using few common ingredients: wheat semolina or flour, water, and sometimes, eggs and other optional ingredients. The history, tradition and quality requirements of the "Italian style" pasta, however, make durum wheat semolina the most suitable, as well as the only legally allowed raw material for the production of high-quality pasta [8]. The essential characteristics of durum wheat (*Triticum turgidum* var. *durum)* include hardness, intense yellow color, unique flavor, and cooking quality. In particular, a high protein content and a strong gluten, which influence the rheological properties of semolina, are of fundamental importance in predicting the overall quality of the cooked products [9, 10]. As strictly specified by the Italian regulation, in fact, "durum wheat semolina pasta", "low grade durum wheat semolina pasta", and "durum wheat whole-meal semolina pasta" has to be obtained by extruding, laminating, and drying a dough exclusively prepared with durum wheat semolina, low grade durum wheat semolina, durum wheat whole-meal semolina and water [11, 12] (Figs. 4.1 and 4.2). As reported in Table 4.1, the Italian law also specifies the legal requirements for dry pasta products, including moisture, ash and protein content, and acidity degree. If the importance of moisture content goes beyond the dictates of the law, as it is guarantee of a good product preservation under normal environmental conditions, protein content continue to be the primary quality factor for pasta production [13].

Fresh pasta, which is not subjected to drying, but must comply with the same legal requirements of dried pasta except for humidity and acidity (at least 24% and ≤ 7 degrees, respectively) (Table 4.1), may be manufactured by adding soft wheat flour in the formulation and must be stored at temperature of 4 ± 2 °C. The artisanal products, which are usually handmade, are locally commercialized as unpackaged foods whit a short shelf life (no longer than 5 days) [12]. The industrial packaged fresh pasta, instead, thanks to the application of appropriate thermal treatments (at least equivalent to pasteurization) and packaging techniques (such as modified atmosphere MAP) able to guarantee the microbial stability of the products, could be sold with sell-by dates up to 90 days [12, 14]. Prepackaged fresh pasta must have a moisture content of not less than 24% and a water activity (a_w) ranging between 0.92 and 0.97. Lower values of both humidity (at least 20%) and a_w (lower that 0.92), which are required, by law, to guarantee transport and storage of fresh pasta at room

Fig. 4.1 Different kinds of durum wheat semolina short pasta currently sold in Italy

Fig. 4.2 Different kinds of long-cut pasta currently sold in Italy. From the top to the bottom: durum wheat semolina square spaghetti; durum wheat whole semolina spaghetti; durum wheat semolina spaghetti (classic n.5)

Table 4.1 Legal characteristics of dry pasta products in Italy (DPR n. 187/2001)

	Characteristics				
		Ash (% dm)		Protein (% dm) (Nitrogen × 5.7)	Acidity (degrees)[a]
Types of pasta	Moisture (%) maximum	Minimum	Maximum	minimum	maximum
Durum wheat semolina pasta	12.5	–	0.9	10.5	4
Durum wheat low-grade semolina pasta	12.5	0.9	1.35	11.5	5
Durum wheat whole-meal semolina	12.5	1.4	1.8	11.5	6
Egg Pasta	12.5	–	1.1	12.5	–

dm dry matter
[a]The level of acidity is the number of cubic centimeters of normal alkaline solution required to neutralize 100 g of dry substance

temperature, can be achieved by applying stabilization treatments to the fresh products.

The production of dry pasta prepared using soft wheat flour is forbidden, except in the case of pasta manufactured in Italy that is intended for export outside the country, as well as in the case of pasta made in another country that will be imported and sold in Italy (only if appropriately labeled) [11, 12, 15]. The use of soft wheat flour is also permitted in the production of other two categories of pasta recognized by the Italian law, namely egg and special pasta, as long as its amount does not exceed 3%. Special pasta is the generic term used to describe all the types of pasta that, although must be labeled as "durum wheat" and/or "durum wheat low grade" and/or "durum wheat whole-meal semolina pasta", can be produced with the addition of ingredients (such as spinach, tomato, spices, mushrooms, and so on) other than soft wheat flour, provided that they are to be mentioned in the list of ingredients. Instead, the term egg pasta is referred to those products that must be produced by adding to the durum wheat semolina at least four hen's eggs (corresponding to a minimum of 200 g) per kg of semolina, in either liquid or dry form [12, 15].

Even if Italy is without doubt the country with the most specific legislation for pasta production, other countries, such as France and Greece, have decreed that pasta must be produced only from durum wheat [16]. The United Kingdom does not have specific regulations regarding the formulation of pasta, while other Europeans countries, such as Spain, Germany, Austria, and Belgium allow the use of soft wheat flour in the production of pasta products [17, 18]. The regulations of the United States, on the other hand, specify that pasta can be produced from durum flour, semolina, flour, farina, or any combination of two or more of these, with water and with or without optional ingredients [19]. Apart from the above-mentioned differences existing in terms of legal requirements among countries, what stands out strongly is that it is not possible to mention pasta without also mentioning wheat, or rather, that it is not possible to produce high quality Italian dry pasta without using durum wheat.

3 The Manufacture of Dried Pasta

3.1 An Overview

The manufacture of pasta in its earliest form was a simple process in which semolina and water were manually kneaded by the *pastaio* (Italian term used to describe the pasta maker) to form a dough, which could be sheeted, extruded by hand press, and finally sun-dried. Since that moment, the process has continuously evolved, with a growing focus on the technological innovation.

The first step towards the mechanization of pasta production was the invention of the wooden extrusion presses in the early 1700s. The efficiency of the process continued to increase in the late 1800s, with the development of new equipment, such as mixers, kneaders, hydraulic presses and drying cabinets, but it was only in about 1930 that, with the invention of the continuous press, this batch method was replaced by a continuous and automatic process [20]. Most of the latest improvements that have been introduced in the pasta manufacturing industry—starting from 1970s—were primarily directed to: (a) minimize the oxidative degradation of the natural pigment during pasta processing by applying vacuum during mixing and extrusion stages; (b) optimize the drying process through the use of higher temperature able to produce pasta with superior cooking quality; (c) shorten processing times and improve affordability and sustainability [13, 21]. In the last 50 years, however, the technological innovation of this process was very limited, so that pasta

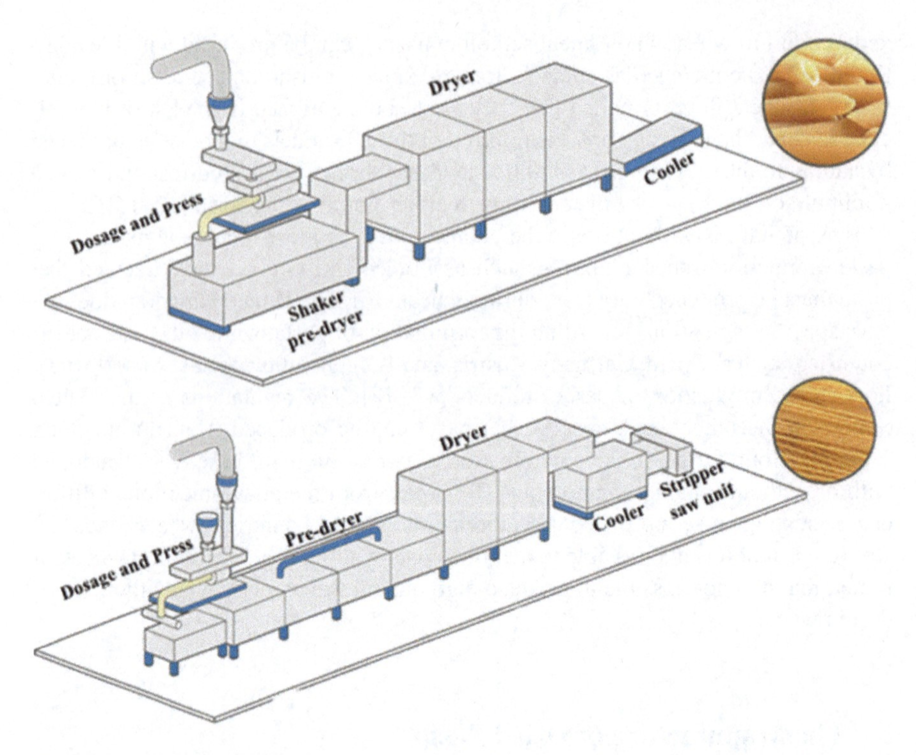

Fig. 4.3 Designs for automated short-cut (top) and long-cut (bottom) pasta lines (Courtesy of Pavan, Italy)

manufacturing can be considered a technologically advanced industry (mature technology) that can produce countless types of pasta, maximizing capacity, productivity and efficiency at no cost to the quality [1].

Pasta production process, which is currently performed *via* automation, involves the following subsequent basic operations: dosing, hydration of semolina, mixing-kneading, forming (by extrusion for dried pasta or lamination for fresh pasta), and drying (Fig. 4.3).

3.2 *Dosing, Hydration and Mixing-Kneading*

During the first stages of the manufacturing process, solid raw materials and liquid ingredients are combined under certain moisture and temperature conditions to obtain a homogeneous mass.

As an optimal water level is required to obtain a cohesive and viscoelastic dough, the amount of water added to semolina has a major impact in determining the dough properties. The optimum water level generally ranges from 25 to 34 kg per 100 kg of semolina, depending on both the initial moisture content of semolina (~14–15%)

and the kind and shape of pasta to be produced [22, 23]. Instead, optimum water temperature, which also has a significant influence on the development of the dough, varies depending on the temperature of semolina and any other ingredients, on the type of pasta, on both the speed and type of the mixer, as well as on the type of drying used in the process [13]. To ensure the use of the right amounts and proportions of each component, both semolina and water are dispensed using different types of feeders. Semolina feeders can be either volumetric or gravimetric, while water dosing is generally carried out by using either a mechanically or a volumetrically based system [24]. Volumetric screw feeders work by discharging a certain volume of the solid ingredient per unit of time with a feed rate determined on the basis of calibration curves and by altering the screw speed. On the contrary, gravimetric feeders measure directly the weight of the material and maintain a predetermined feed rate, which is measured in units of weight per time. Centrifugal pumps, on the other hand, are generally used in water dosing, while peristaltic pumps are preferred when liquid eggs or any other high-viscosity fluids are included in the pasta formulation [1]. During this preliminary dosing step, it is important to keep both raw materials and water feed rates as constant as possible to ensure an even distribution of water among semolina particles [22, 24]. In fact, if the hydration of semolina is not performed in the right way, it can lead to the formation of lumps and small spots of un-hydrated particles, which result in the presence of white specks or streaks in the finished dried product [20]. The appearance of such defects, being also influenced by semolina particle-size distribution, can be reduced by using semolina with a uniform and fine granulation (common range between 150 and 350 μm). In fact, fine particles, which absorb water more rapidly than coarse ones, are ideally suited for obtaining a homogeneous dough within a short time, as required by the mixing step [13, 20]. Furthermore, to ensure a proper semolina hydration, all the ingredients may undergo to a high-speed premixing stage, before entering in the main mixing chamber. Pre-mixers, which are usually installed ahead of the mixer, are equipped with a single or double high-speed rotational shaft that, by means of centrifugal forces, can assure the optimum hydration of semolina particles in less than two seconds [24]. After this operation, a further mixing occurs in the main mixer, before driving the mass to the extrusion stage. At the initial mixing time and in conjunction with the hydration of semolina particles, the added water triggers a complex series of chemical reactions with the semolina components, mostly protein and starch, which start to hydrate. As the mixing stage proceed, gliadin and glutenin, the two storage proteins of wheat, become hydrated and begin to unfold and interact by forming intra-molecular and inter-molecular bonds, thus leading to the formation of an only partially developed protein network due to the limited hydration level and low energy input [13]. In fact, the moistened mass obtained at the end of the mixing process does not resemble a homogeneous dough, but a mixture of agglomerated lumps of semolina, which can be transformed in a cohesive dough in the subsequent kneading stage [25, 26].

The modern pasta mixers usually work under vacuum to prevent the formation of air bubbles in the developing dough, thus minimizing the appearance of quality defects in the finished product. In fact, a de-aerated product, in addition to a better translucent appearance and brightness, has a more compact structure with a lower tendency to breakage. The reduction of oxygen also inhibits the oxidation of the desirable yellow pigments, thus avoiding changes in the final color of pasta [13, 27].

3.3 *Extrusion and Shaping*

After hydration and mixing stages, the wetted mixture of semolina and water are conveyed into the extrusion unit, where pasta dough development largely occurs [28]. The extrusion system is composed of a stainless-steel screw rotating in a cylindrical barrel, which pushes the dough towards a head press on which is attached a die, which may have different geometries depending on the shape of pasta to be produced. The extruder is then able to simultaneously perform two essential functions: (a) the completion of the development of the continuous network of gluten protein, which was already but incompletely formed during mixing, as well as (b) the shaping of the dough into a specific form (Fig. 4.4) [29–31].

Once the moistened mass is transferred inside the barrel, the turning of the screw causes a rapid increase of the pressure (in the range of 5–10 MPa in a single-screw extruder), allowing the transition from a hydrated granular mass to a fully compacted dough [23]. In fact, during the homogenization of the dough under pressure, gluten proteins are stretched and aligned on the basis of the rotational movement of the screw [32]. Furthermore, as the dough is pushed towards the end of the screw, its flow-out is restricted by the presence of the die that, by creating a back pressure along the screw, further stimulates dough development. Concomitantly, friction produced between the dough and the surface of both barrel and screw—that is necessary for both the development and the progress of the dough into the barrel—must remain at low levels to avoid an undesirable increase of the dough temperature, which should not exceed 45–50 °C [28]. In fact, higher temperature generated by the extrusion process could cause negative effects on gluten network quality and, in turn, on the behavior of the final pasta during cooking. In particular, temperatures above 55 °C stimulate abnormal swelling of the starch granules and the coagulation of proteins into an irreversible gel, thus leading to the formation of weaker pasta strands with less resistance to overcooking and to a soft and sticky product. Therefore, in order to dissipate the surplus heat generated inside the extrusion unit, but also to keep the dough temperature below the set limit (45–50 °C), extruders are equipped with a cooling system consisting of an external jacket in which cold water (20–30 °C) is circulated [32, 33]. As previously mentioned for the mixing stage, another element of fundamental importance is the possibility of extruding the dough under vacuum. This extrusion technique can significantly improve the appearance of the pasta products—especially in terms of color—as well as their structural properties.

As the extrusion proceeds, the unshaped but completely developed dough is forced under pressure through small holes of a die, from which it will come out in the desired shape, before being cut. In order to guarantee a uniform distribution of

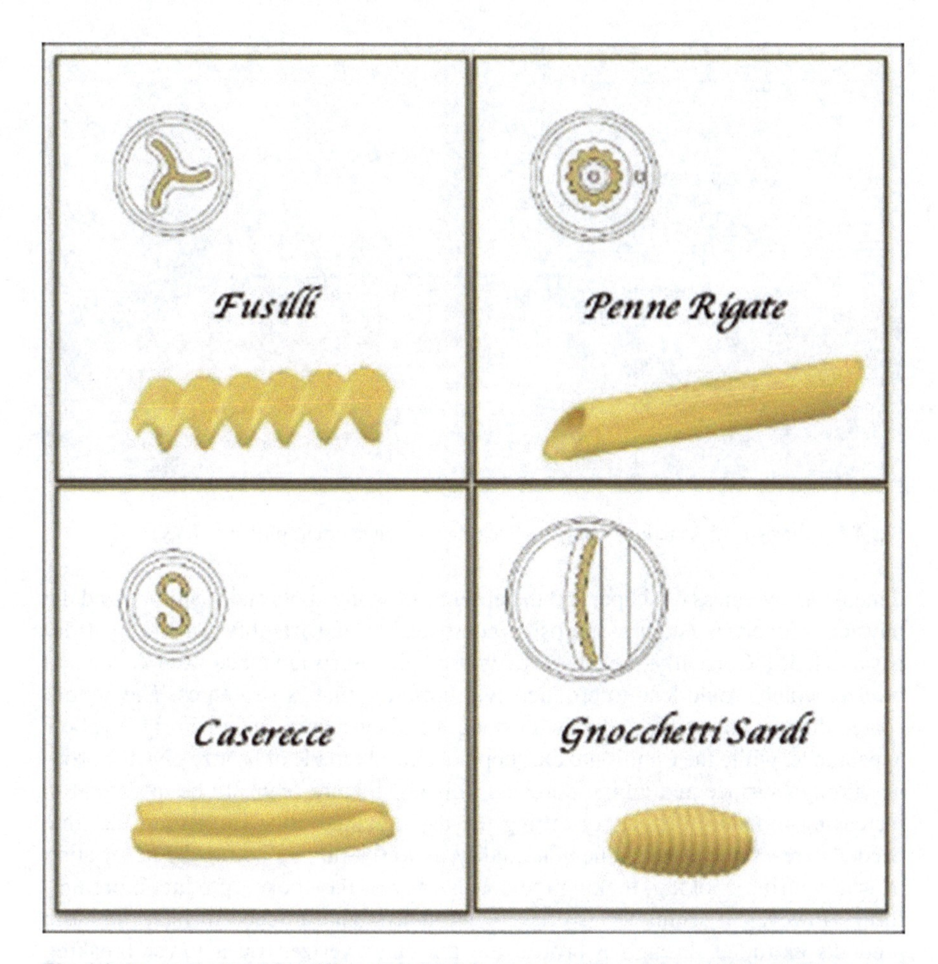

Fig. 4.4 Inserts designed for different kinds of short-cut pasta currently sold in Italy (Courtesy of Pavan, Italy)

the pressure throughout the dough, as well as to ensure an even flow of the dough through the orifices of the die, the whole system has to be carefully designed [30]. Rectangular straight die supports are generally used to extrude long pasta into strands that are then spread on metal sticks (Fig. 4.5) and cut from the die to a uniform and desired length. Circular die supports are instead used for short pasta products that, once extruded, are cut in pieces by using a rotary cutter equipped with one or more blades. An irregular rate of discharge throughout the die, however, might lead to the production of pieces of different length, which can vary depending on both extrusion rate and cutter speed. In both long and short pasta, to keep the strands/pieces from sticking together, the cut products are subjected to a blast of hot air, before being sent to the dryer [32]. During pasta shaping, a major role is also played by the materials used to make the different dies supports. In fact, in addition to forming dough, dies might deeply affect the surface characteristics (such as color,

Fig. 4.5 Conveyor sticks in long-cut pasta line (Photograph courtesy of Pavan, Italy)

dimension, evenness of shape, and the absence of white spots and fractures) and the physical properties (such as porosity, compactness and fragility) of the final dry pasta [34, 35]. Currently, the most widely used die inserts are made from Teflon and bronze, which would lead to products with different matrix structures. The Teflon-coated die inserts produce pasta with smooth and even surface and a bright-yellow appearance, while the traditional die supports entirely made of bronze give the product a rough surface and a high porosity. This last feature seems to be important in increasing moisture diffusivity during the drying stage, thus shortening the time needed to reach the target value of humidity, as well as in facilitating the penetration of water during cooking, making the consistency of the cooked product more uniform. However, it seems to also have some detrimental effects. In fact, the pasta products extruded through a bronze die are characterized by a lower breaking strength and a greater starch release in the cooking water in comparison to those extruded with a Teflon die. Moreover, the bronze dies have a lower resistance to wear and a lower production yield [34, 35]. However, despite the numerous disadvantages, pasta extruded through bronze die is without doubt the most appreciated by Italian consumers. Such preference is probably due to its rough surface that not only helps the sauce to easily adhere to each piece of pasta, but also allows pasta to absorb water rapidly during boiling for obtaining an *al dente* texture. *Al dente* is a typical Italian expression used to describe the texture that pasta should have after being cooked for a short time in abundant boiling water: it should be firm to the bite, with a chewy texture, but not sticky.

3.4 Drying

On leaving the die, the fresh pasta is progressively dehydrated from an initial value of moisture content of about 30% to a final value of 12.5%, which is low enough to ensure the microbiological stability of the product (a_w lower than 0.6 and close to

0.5) under normal storage conditions and to reduce numerous enzymatic activities. Basically, the unit operation of drying requires two simultaneous and important process-controlling factors: (a) the transfer of heat (which is generated in the drying chamber by an heat exchanger) from air to pasta, and (b) the transfer of water—or water vapor—from the inner core of pasta to its surface and, then, from pasta to air. Humidity, temperature, and speed of air, which depend on pasta temperature, moisture, shape and dimension, are the most significant technological parameters that define the drying conditions. A proper management of the drying process must attain a uniform and homogeneous removal of water and, then, an even moisture distribution across the stable product, while maintaining (or improving) the overall quality of the final dried pasta (such as high firmness and flexibility, low stickiness and cooking loss, and the absence of physical defects). Therefore, mechanical damage and stresses, such as surface and internal structural tensions between the pasta components, due to the formation of an excessive moisture content gradient between the surface and the internal core of the product, as well as to the different affinity of protein and starch with water, must be avoided [30, 36].

Drying, which is undoubtedly the most crucial and critical stage of the entire pasta manufacturing process, consists of three subsequent and closely interconnected phases, namely pre-drying, final drying, and cooling (stabilizing). Sometimes, before being conveyed to the pre-drying section of the dryer, pasta products might be subjected to another preliminary phase, which in Italian is called *incartamento*. It is not—as very often erroneously believed—a pre-drying step, but rather a preliminary superficial dehydration of the product, which consists of passing a current of dry and warm air over the pasta exiting the die to achieve a dual purpose: forming a thin crust capable of preventing pieces from sticking together, before being placed on rods (long pasta) (Fig. 4.5) or vibrating mesh trays (short pasta) and conveyed to the pre-dryer; and avoiding that pasta becomes misshapen, by strengthen its structure. This phase, however, is not necessary if, immediately after shaping, pasta is directly conveyed to the pre-drying unit in which this preliminary step is conducted under controlled thermo-hygrometric conditions [31, 32].The pre-drying phase is responsible for the majority loss of pasta moisture that decreases from the initial 30% to around 18 to 20%, corresponding to about one-third of the total water amount of pasta and to about 10% of the total drying time [36]. During this phase the fresh pasta is still in its plastic state, which means that it may undergo severe and fast drying, while supporting deformation without causing any internal tensions and, in turn, any cracks formation. Moreover, the rapid dehydration of pasta shortens the overall drying time and inhibits superficial enzymatic browning [30]. However, as drying proceeds, the water removal is sufficient to determine the transition from a plastic to an elastic state, which depends not only on the variation of the moisture content of the product, but also on the temperature of the dryer. In general, the higher the temperature, the lower the moisture content at which the change of the physical state of pasta occurs. In fact, at higher temperature (>70 °C) the plastic state is maintained for a long time, even when the humidity of the product drops to relatively low values (19–18%). On the contrary, at low temperature, (<50 °C) the physical state of pasta changes when moisture content decreases below 23–22% [31]. In any case, during the actual drying phase, in which the moisture content of pasta is reduced from 18–20% to approximately 12.5%, the water removal is conditional upon satisfying a correct and constant balance between the quantity and concentration levels of water in the various layers of the product, as typical of the elastic state. It follows that, if the drying diagram is not designed in the right way and the ventilation of the product is too fast, stresses generated between areas with different moisture content become higher than the mechanical resistance of the pasta matrix that, being not able to redistribute such stresses and to relax, leads to the formation of points of fracture (checking) [36]. In order to guarantee an even redistribution of the residual moisture inside the product, thus avoiding stress fractures and cracking, the drying process is currently performed by alternating short periods of hot air circulation with long resting times. The pasta drying process is then completed by the cooling phase during which pasta is slowly brought to thermo-hygrometric equilibrium with the surrounding environment. Then, the properly stabilized and cooled final pasta can be immediately packaged in cellophane or polyethylene bags, without any inconvenient.

3.4.1 Drying: Innovations and Implications for Pasta Quality

Until the 1970s, the drying process was considered a laborious and expensive process that took place at very low temperatures (up to 60 °C) and very long times (between 14 and 20 h). Since those days, the need to shorten processing time, increase microbiological safety and plant productivity, while maintaining or improving the quality of the product, have resulted in many technological innovations primarily related to the increase in drying temperatures. Therefore, in the last few decades, low-temperature (LT) pasta drying has been replaced firstly with high-temperature (HT) drying, in which the applied temperature ranges from 60 to 80 °C, and, successively, with ultrahigh-temperature (UHT) drying, where the temperature applied is between 80 and 110 °C [16, 37]. One of the first benefit associated with the use of both HT and UHT drying is certainly the drastic shortening of the processing times that, according to the shape and type of pasta, have been reduced to about 2–5 h. Higher hygiene and greater cost saving are other improvements that can be attributed to the use of high-temperature drying cycles. In addition, this technology has enabled the production of pasta with acceptable or even superior quality in terms of both appearance and cooking behavior, even if raw materials of poor quality are used [26, 38]. Apart from the above-mentioned improvements, however, the use of high-temperature drying cycles can have detrimental effects on pasta quality characteristics, as in the case of pasta color. In fact, although some authors reported that the use of both HT and UHT drying improve the yellowness and brightness of pasta—probably as a consequence of lipoxygenases inactivation—high temperature can represents one of the main cause of pasta color alteration due to the potential occurrence of the Maillard reaction (i.e. non-enzymatic browning) and the consequent formation of undesirable brown or reddish color. While in the case of LT drying the non-enzymatic browning is not an issue, this reaction is of substantial concern when higher temperature cycles are applied. In fact, the concomitant occurrence of factors, such as a_w, moisture content, and drying time—besides the presence of reducing sugars and free amino groups—might create the optimum conditions for the Maillard reaction to occur. As it was demonstrated, a moderate HT drying conditions (~80 °C) could slightly, but significantly, enhance pasta yellowness in comparison with the conventional drying cycle (LT), but the application of high-temperature drying (~100 °C) at low product moisture content (~15%), by promoting advanced Maillard reaction, might led to the formation of red-brown melanoidins in the final pasta. [38–41]. Besides color, also the cooking properties of pasta are affected by HT drying. Such influence is strongly related to the temperatures applied in the various phases of the drying process—and, therefore, to different moisture content of the products—that may impact on the physical state of both protein and starch. Notwithstanding the great influence that protein content and composition could have. When pasta is obtained by applying LT drying cycles it exhibits a low organized gluten network and no modifications in the structure of starch granules are observed. Therefore, its cooking properties are defined by the protein content and gluten strength and, then, by the quality of semolina used [42, 43]. If high temperatures are employed at the start of the drying process with high product moisture (at about 30%), their application does not enhance the pasta cooking quality, due to a premature denaturation of proteins and some starch gelatinization. On the contrary, high temperatures applied during the final phase of drying at low products moisture (15%), and after a pre-drying phase at low temperature, are more effective in improving the pasta cooking quality. The reason of such better results is due to the formation of a continuous network of almost completely coagulated gluten proteins that, trapping the starch molecules, limits their swelling and solubilization into the cooking water, thus leading to pasta with improved cooking quality [39, 42–45]. These changes in pasta textural properties and cooking behavior are even more pronounced when the drying temperature is increased from 80 to 100 °C. It follows that, when using high temperatures—especially at the final drying—the cooking quality is governed by protein content only, lowering the role of semolina quality in pasta cooking behavior [41, 43, 46, 47]. In addition, the application of high temperatures also affect pasta texture, by enhancing firmness and reducing stickiness of the cooked pasta, especially when they are applied at a late stage of the drying process [39, 41].

4 Innovative Ingredients in Dried Pasta Production

Nowadays, the demand for foods able to provide health benefits and to prevent nutrition-related diseases (such as diabetes and obesity), beyond the basic nutritional functions, is on the rise. In fact, over the last decades, the always more aware health and wellness consumers are increasingly influential in redefining the concept of food, which has changed substantially. In this context a regularly eaten and

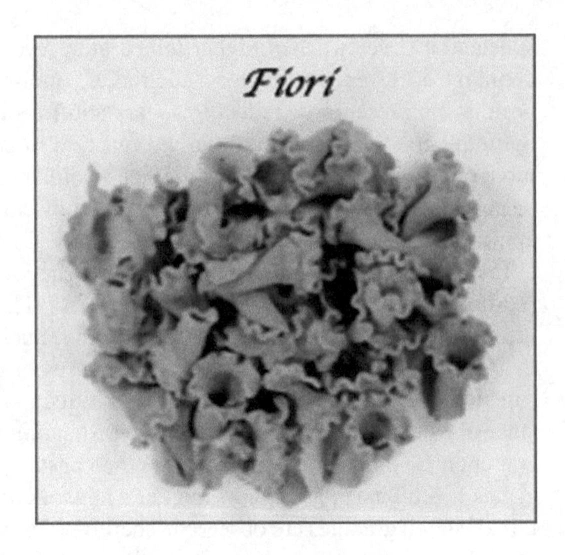

Fig. 4.6 Organic durum wheat semolina pasta enriched with hemp flour (*Cannabis sativa* spp.) currently sold in Italy

universal food like pasta, due to the way it intercepts cultural and consumer food trends, is considered a good career for the addition of different nutritional and health-promoting substances to the diet (Fig. 4.6). This trend was also endorsed by governments and health-promoting organizations, such as the Federal Drug and Administration (FDA) and the World Health Organization (WHO). In this regard, it should be noted that, as far back as in 1949, it was the same FDA to include pasta among the first foods authorized for enrichment with vitamins and iron. In fact, although pasta is recognized as low in sodium and lipids with no cholesterol, a rich source of complex carbohydrates (~70%), and a food with moderate to low post-prandial glycemic response, it is deficient in terms of protein and essential amino acids, such as lysine and threonine, as well as in vitamins and iron [48, 49]. Therefore, with the aim of meeting the need of consumers, while enhancing the nutritional quality, healthfulness, and functionality of pasta, a wide range of non-traditional ingredients have been added to the conventional pasta formulation. However, due to the detrimental effects that such non-traditional ingredients could have on the rheological properties of the dough and, therefore, on the overall quality of the final pasta products (mostly appearance and cooking quality), the partial or total replacement of semolina with non-traditional ingredients in pasta formulation still represents a critical factor.

Starting to the enrichment of pasta with dietary fiber and protein that, up to now, is the most common, a brief overview of the main innovative ingredients used and their effects on pasta structure and quality is reported in the following sections.

4.1 Dietary Fiber

With an average value of 3% reported in commercial pasta products, which is still too far from the recommended daily intake of 28 g/day (referred to a daily intake of 2000 calories), durum wheat semolina pasta is usually characterized by an intermediate to low dietary fiber content [50–52] (Table 4.2).

Dietary fibers consist of non-digestible carbohydrates, lignin, and other substances that are resistant to digestion and absorption in the early sections of the digestive tract with a partial or complete fermentation in the colon by gut microbiota [53]. The widely recognized health benefits associated with the inclusion of dietary fibers as part of a balanced diet vary from the ability to lower serum cholesterol and to normalize blood glucose and insulin levels, as in the case of the incompletely or slowly fermented soluble fraction, to the capacity to prevent intestinal constipation and to increase postprandial satiety, as in the case of the incompletely or slowly fermented insoluble fraction. Therefore, in an attempt to produce high-fiber pasta products for the health food market, soluble fiber, such as inulin-type fructans, β-glucans from barley and oat, and psyllium [54–57], as well as insoluble fiber, such as wheat, barley, oat and rice brans [58–61] have been included—among others—in conventional dry pasta formulations. In general, several authors agreed in considering that the incorporation of different kinds of dietary fibers in traditional pasta formulation compromises the formation of a continuous and cohesive gluten-starch matrix, thus interfering with other pasta quality parameters such as water absorption, optimal cooking time, and cooking loss [54, 56, 58–60]. A decreased water absorption and a lower cooking time are probably due to the changes in pasta microstructure. For instance, modifications observed in the gluten-starch matrix due to a dilution effect of gluten by replacement of semolina with inulin individually added or in combination with psyllium and barley balance, allowing a faster transfer of water in the inner core of pasta strands, led to the reduction of the cooking time of the enriched-pasta and a significant increase in the cooking loss [54]. Similar results were obtained in non-traditional pasta products obtained by using two types of inulin with differing degrees of polymerization and crystallinity, being the effects more pronounced at increasing percentages (up to 20%) of semolina substitution [56]. Instead, in comparison to the traditional product, no changes in both cooking time and cooking loss were observed in pasta enriched with only barley balance (up to 20%) that, in addition, showed increased firmness values probably due to a more compact structure and a reduced starch swelling [54, 55]. Yet, others demonstrated that the pasta enrichment with brans from various cereals could cause a physical disruption of the gluten matrix, promoting water uptake and reducing cooking times, thus facilitating granule swelling and rupture and a consequent increase in cooking loss, which was worsened by higher levels of semolina substitution [59, 60]. Moreover, bran-enriched pasta, in addition to decreased firmness values, showed an undesirable increase of the reddish-brown color and a lower overall acceptability of the final product [58–60]. A lower ability of the starch-protein matrix to retain its physical integrity during cooking was also observed in pasta

Table 4.2 Nutrition values of some kinds of traditional and enriched short and long-cut pasta currently sold in Italy (https://www.barilla.com/it-it/)

Types of pasta	Nutrition values						
	Protein	Dietary fiber	Total fat	Saturated fat	Carbohydrates	Sugars	Calories (kcal)
Short pasta							
Durum wheat semolina Fusilli	12.5	3	2	0.5	71.2	3.5	359
Durum wheat whole semolina Fusilli	13	8	2.5	0.5	64	3.5	347
Red Lentil Penne	25	12	2.4	0.5	47.4	1.8	335
Five Cereals[a] Fusilli	13	6.8	2.2	0.5	66.6	3	352
Chikpea Casarecce	21	14	6.2	1.1	45.1	2.9	348
Long-cut pasta							
Durum wheat semolina Spaghetti	12.8	3	2	0.5	70.9	3.5	359
Durum wheat whole semolina Spaghetti	13	8	2.5	0.5	64	3.5	347
Red Lentil Spaghetti	25	12	2.4	0.5	47.4	1.8	335

[a]Five cereals: durum wheat, barley, spelt, corn, rye

products enriched with barley flour containing ~11% β-glucans (60%) [57]. Therefore, it seems well established that the presence of fibers could affects—positively or negatively—the quality of the enriched-pasta by interfering with the starch-gluten matrix, but what stands out strongly is that not all types of fibers behave in the same way and that the incorporation of fiber blends may influence their functionality, leading to the development of fiber-enriched pasta with high quality [54].

4.2 Legume Flours

It is well known that the incorporation of legume flours of various origin in cereal-based products can improve their nutritional value, not only in terms of protein content, but also in terms of protein biological value [62]. In fact, legumes, besides being a rich source of protein (18–25%), contain high levels of the essential amino acid lysine, which is usually deficient in cereal grains that, in turn, are rich in sulfur-containing amino acids—namely methionine and cysteine—usually lacking in legumes. Hence, when legumes are consumed in combination with cereals, they can complement the protein quality and provide a more nutritionally balanced final product [53]. For these reasons, the main raw materials of vegetable origins used in the development of protein-enriched pasta are legumes (Table 4.2). Among these, the most studied are undoubtedly chickpea and soybean, which are added to the traditional pasta formulation either as flour or isolates and concentrates [62]. While the enrichment of pasta with various kinds of legumes allows deficiencies in lysine and protein content to be compensated, the effects of their incorporation on pasta quality (appearance, texture, cooking quality and sensory properties) strongly depends on both the kinds and the substitution levels of legumes used [63–66]. It has been demonstrated that the replacement of durum wheat semolina with chick-pea flour, when levels of substitution are up to 15%, led to the development of enriched pasta with increased protein and lysine content, more red color, low amount of amylose leached into the cooking water, reduced stickiness values, as well as similar firmness value in comparison to semolina pasta, without detrimental effects in the overall acceptance. Addition of increasing percentages of chickpea flour, instead, led to a weakening of the gluten matrix—probably due to a dilution effect by the added protein—and, therefore, to a lower firmness [65]. Contrasting results were previously obtained by other authors, who observed an increase in firmness and color intensity, but a decrease in the overall quality of dried pasta enriched with chickpea flour up to 30% [63]. However, such contrasting behavior, especially in terms of increased cooking loss, may be due to the different type of chickpeas and/or to the different method of analysis used in the two research studies (i.e. cooking loss measured by solid loss *versus* cooking loss measured as amount of amylose leached into the cooking water) [65]. Yet, increased protein content, color intensity and firmness values were obtained by other authors when assessing the effects of semolina replacement with chickpea flour at different levels of substitution (up to 50%). In this case, the authors stated that the combination of chickpea flour and wheat semolina led to the development of strength and extensible dough, allowing pasta to retain its firmness and elasticity, as long as chickpea incorporation level does not exceed 10%. However, they also reported that for increasing amounts of chickpea flour, due to the dilution of gluten, the pasta processing, handling, and cooking characteristics deteriorated remarkably with a consequent drop in consumers acceptance [66].

The replacement of durum wheat semolina with soy flour in the production of non-traditional pasta was also explored [67–69]. It was observed that the incorporation of increasing amount of toasted and partially defatted soy flours (up to 25% and 50%, respectively) did not negatively affect the cooking properties and the consumers acceptability of the resulting pasta products, which, in turn were characterized by an increased protein content [68]. The authors reported that soy incorporation led to the formation of a weak gluten matrix, due not only to a gluten dilution effect, but also to the competition for water between the added non-gluten proteins and the gluten, as well as to the disruption of starch-protein complexes, being the weakening effect more pronounced when toasted soy flour was used. In addition, at the optimal cooking time all the samples enriched with soy flour showed a cooking loss higher than that of semolina pasta. However, owing to the competition for water of soy protein with both starch and gluten, all the samples showed a good behavior in terms of resistance to cooking and overcooking, thus counterbalancing the negative effect derived by the weakening of the dough. Furthermore, no significant differences in texture properties among semolina and soy-enriched samples were observed from a sensorial point of view [68]. Similar results were obtained by other authors when analyzing flavor and texture sensory attributes of pasta enriched by adding up to 35% soy flour [67].

It has been also demonstrated that the incorporation of legumes in pasta conventional formulations could affect different steps of the pasta-making process. In fact, the substitution levels of chickpea flour above 30% make the extrusion phase difficult, due to the formation of particle agglomerates during mixing [65]. Similar findings were also obtained in the production of legume-enriched pasta by using split bean and fava bean flours (35%). The authors reported that lower hydration level and higher mixing speed are required to limit the formation of such aggregates, thus avoiding the formation of a sticky dough that would make the subsequent extrusion phase difficult [70].

4.3 *Insect Flours*

In the last decades, the potential use of edible insects has been proposed as a novel and alternative protein source to the development of cereal-based foods with added value. Although edible insects are considered part of the traditional diet in over 100 countries of Asia, Africa and South America, it was only in 2013 that the Food and Agriculture Organization of the United Nations (FAO) identified insects as a food resource for both human and animal consumption [71, 72]. More than 2000 species of insects have been currently considered as human foods and, among them, the most commonly consumed are beetles (31%), caterpillars (18%), ants, bees, and wasps (14%), closely followed by crickets, grasshoppers, and locusts (13%) [71, 73]. The insect species currently reared in the European Union, which fulfil the safety conditions for insect production for feed—as identified by the European Food Safety Authority (EFSA) in 2015—are reported in Table 4.3. Insects are generally considered highly nutritious foods with a favorable nutritional profile for humans. The major constituents of insects are proteins (from 13 to 81% dry matter)

Table 4.3 Insect species currently reared in the European Union for insect production for feed use as identified by the European Food Safety Authority (EFSA) in 2015

Order	Family	Species	Common name
Orthoptera	Gryllidae	*Acheta domesticus* L.	House cricket
Orthoptera	Gryllidae	*Gryllodes sigillatus*	Banded cricket
Orthoptera	Gryllidae	*Gryllus assimilis*	Field cricket
Coleoptera	Tenebrionidae	*Tenebrio molitor* L.	Yellow mealworm
Coleoptera	Tenebrionidae	*Alphitobius diaperinus*	Lesser mealworm
Diptera	Muscidae	*Musca domestica* L.	Common housefly
Diptera	Stratiomydae	*Hermetia illucens* L.	Black soldier fly

Commission Regulation (EU) 2017/893

with high biological value, lipids (from 10 to 60%) with high levels of monounsaturated and/or polyunsaturated fatty acids, and dietary fibers (from 2 to 29%), in proportions the range depending on both the origin and the insects' diet. Moreover, insect proteins have good amino acids profiles—in some cases similar or even superior to soy proteins—and digestibility levels higher than those of plant proteins. Insects also contain minor components, such as vitamins (B-group and vitamin E) and minerals (copper, iron, magnesium, manganese, phosphorous, selenium, and zinc) [73, 74]. Furthermore, if compared with many other animal farms, insects require less water and land, have lower emissions of greenhouse gases and ammonia, as well as a more favorable feed conversion efficiency, thus representing a major benefit also for the environment [75]. In the European Union, from a regulatory perspective, edible insects are currently considered as "novel foods" and their commercialization is subject to the granting of a specific authorization released under Regulation [76–79]. Despite this, however, insect's consumption for food purposes is almost negligible. In fact, the vast majority of Europeans still have a poor propensity to accept insects as edible foods, since they are generally considered a source of contaminant, vehicles of allergies, dirty, and disgusting [73]. In an attempt to overcome this consumer unwillingness to use insects as food, it was thought to process them into non-visible forms, such as flour, to be added in different food matrices [74]. To this end, in the last few years, the number of research papers reporting the incorporation of insect flours, defatted flours and protein concentrates into different food matrices, including meat [80, 81] and cereal-based products (mostly bread and snacks) [82, 83], has growing faster [84]. To the best of our knowledge, however, there are only three research focusing on the development of dried pasta by using insect flour, and only one of them concerns the partial substitution of durum wheat semolina in traditional pasta formulation [85]. The other two, in fact, are related to the incorporation of insect flour in buckwheat- and millet-based dried pasta [74, 86]. Regarding the durum wheat semolina pasta, the authors evaluated the effects of the incorporation of cricket powder (up to 15%) on the appearance, cooking quality, nutritional value and sensory properties of the obtained insects-enriched pasta. They found that, in addition to an enhanced nutritional value in terms of protein, lipids, and mineral content, the resulting supplemented pasta showed an increase of the optimal cooking time and firmness values, and a decrease in cooking weight and loss at all supplementation levels. They also observed a clear darkening of the pasta color with more red color values in all the enriched samples that, however, were judged by consumers similar to that of the whole semolina pasta, provided that the cricket powder amount does not exceed 5%. In general, sensory analysis showed that the insects-enriched pasta met the consumer's expectation, even if at 10% and 15% of cricket powder supplementation the distinctive flavor and dark color were poorly judged. This is obviously a first attempt in a new field that is only beginning to evolve. Further studies are needed to better understand how the addition of edible insects can affect the characteristics of the durum semolina dry pasta.

4.4 Microalgae

Algae belong to a large and heterogeneous group of photosynthetic multicellular or unicellular forms of living organisms which, depending on their size, can be divided into two categories: macroalgae (or marine algae) and microalgae (or filamentous algae). In particular, microalgae, which constitute at about 75% of the total algal species, has been regarded as another attractive source for the development of highly valuable healthy products. Among them, the most well-known and produced are undoubtedly *Spirulina* (*Arthrospira*) and *Chlorella* species [87, 88].

The increasing interest in the inclusion of microalgae in traditional food formulations is mainly connected to their excellent protein profile—in terms of both quantity and biological value—but also to their richer composition, especially regarding minor components, such as vitamins (biotin, folic acid, pantothenic acid, niacin, thiamin, riboflavin, vitamin A, C, and E), minerals (iodine, potassium, iron, and magnesium), and bioactive compounds (lycopene, chlorophyll, β-carotene, phycobiliproteins), in proportions that vary depending on the microalgae species, as well as on the composition, temperature and light regime of the growth environment. Furthermore, microalgae consumption has been associated with a decreased risk of developing cardiovascular diseases, certain forms of cancer, and degenerative chronic diseases [87, 88].

It has been demonstrated that the fortification of durum wheat semolina pasta by using microalgae biomass from *Chlorella vulgaris* (green and orange) and *Spirulina maxima* (0.5–2%) species led to a final product in which almost all the pasta quality characteristics are maintained, without detrimental effects on pasta sensory acceptance. Overall, the cooking quality of pasta—mostly in terms of cooking loss—was not affected by the incorporation of the three experimental microalgae species, even if *Spirulina maxima* seemed to be more effective than both orange and green *Chlorella vulgaris* in preserving the pasta quality characteristics, especially when it is added at intermediate supplementation level (1%). As expected, the most significant change concerned the color of the enriched pasta (green and orange) that was, however, widely preferred by the sensory panel [89]. Similar results were obtained by the same authors when studying other two species of microalgae, namely *Isochrysis galbana* and *Diacronema vlkianum,* which were included in durum wheat semolina fresh pasta (0.5–2%) as a source of polyunsaturated fatty acids. In this case, the two microalgae species, while preserving (or slightly improving) the pasta cooking behavior, significantly enhanced the nutritional value of the enriched pasta samples, in terms of protein and mineral content, as well as in terms of fatty acids composition, despite a skewed ratio of ω-3 and ω-6 fatty acids. On the basis of the obtained results, the authors underlined the importance of microalgae incorporation in conventional durum wheat semolina pasta as a means to improve the nutritional value of the enriched products [90]. Interesting results in the development of durum wheat semolina dried pasta with enhanced nutritional value and high sensory acceptability by using other marine ingredients have been also reported [91–93].

4.5 By-products

In the last few years, a great deal of attention have been paid to the use of food industry by-products in the development of non-traditional enriched pasta products, pursuing a dual objective: (a) re-using products that are wasted during industrial processing and transformation, thus preventing and reducing food waste, which have negative environmental and economic impacts, and (b) transforming the waste products, which often are good sources of valuable compounds, in useful ingredients for the production of highly valuable healthy products. Olive oil industry waste, brewers' spent grain, maize bran, grape marc, apple peel powder, gac fruit, fraction from the debranning of different cereal sources, and fish waste flour are just some of the by-products more recently added to durum wheat dried pasta formulation in an attempt to improve both nutritional value and bio-functionality of this widely consumed cereal-based product, while maintaining its structure and overall quality [94–102]. Different olive oil industry by-products, such as olive pomace, olive paste powder, and olive mill waste-water, are recently used as additional sources of phenolic compounds and dietary fibers in durum wheat semolina pasta [94–96]. Fortification of pasta with increasing levels of olive pomace (up to 10%) significantly affected appearance, texture, cooking quality and nutritional value of the final product [94]. The presence of high amount of dietary fibers interfered with the formation of a proper structure of pasta, promoting water absorption and reducing the ability of the weakened gluten network to retain gelatinized starch, thus increasing cooking loss. Increasing values of firmness, adhesiveness and a more intense red color were also observed in all the enriched pasta samples [94]. Increasing levels of olive pomace incorporation significantly improved the total polyphenols content and the antioxidant activity of both uncooked and cooked pasta, even if a 70% drop in polyphenols content—and a concomitant decrease in antioxidant activity—was observed in all the cooked samples, probably due to their thermal degradation during boiling or solubilization into the cooking water. Nutritional starch fractions were also improved, but without significantly affecting the predicted glycemic index [94]. The loss of polyphenols and other bioactive compounds during the pasta-making process was also reported in durum wheat pasta enriched with nine different debranning fractions of soft wheat (with progressively minor portions of outer layers and endosperm of the wheat grain) [100]. The authors, when evaluating the effect of both processing and cooking on the phenolic acids profile of the enriched pasta samples observed that the presence of water and oxygen during the kneading phase caused a reduction in the more reactive fraction of the free phenolic acids, while maintaining unchanged the bound fraction; on the contrary, during the cooking phase, the bound fraction increased (mostly bound ferulic acid), while the free phenolic acids did not vary. This increase was explained by the fact that, probably, the boiling water, by promoting the thermal extraction of the bound ferulic acid previously linked to the cells wall, increased both bound phenolic acids concentration and *in vitro* antioxidant activity of the cooked pasta [100]. The loss of total polyphenols content and anthocyanins were also reported in semolina-based pasta enriched with an amount (25%) of debranning fraction from purple wheat (corresponding to an abrasion level of 3.7%) suitable for improving both phenolics and dietary fiber content, while avoiding excessive changes in the structural integrity of the pasta product. The significant losses observed in the uncooked samples were further increased during the cooking of pasta, leading to a reduction of almost 50 and 65% for the total polyphenols and anthocyanins content, respectively. Despite this strong reduction, which was probably due to the thermal treatment during both drying and cooking of pasta, the final content of both anthocyanins and total phenolics was found to be much higher in comparison to that reported for other pasta products obtained from whole pigmented grains [101].

The effects of the incorporation of by-products generated by the fruit processing industry in the overall quality of semolina dried pasta were also reported [98]. The authors found that the addition of 15% apple peel powder was effective in increasing not only the total polyphenols content and antioxidant activity of the fortified products in comparison to the reference semolina pasta, but also to enhance the pasta yellowness. However, they also found that apple peel powder supplementation slightly deteriorated the quality characteristics of the enriched pasta, such as water absorption, optimum cooking time, cooking loss, and firmness, with a consequent drop in the overall sensory acceptance of the final products [98].

The recent studies reported above demonstrated that by-products can be considered as promising functional ingredients to be incorporated in durum wheat semolina pasta. However, further studies are needed to evaluate and optimize the production process conditions in such a way to prevent both losses of bioactive compounds and detrimental effects on pasta quality parameters.

5 Conclusions

In this chapter, the production of dried pasta has been presented in terms of both processing and ingredients characteristics. What stand out strongly is that the production of high-quality pasta depends not only on the quality of the raw material used, but also on the complex phenomena that involved their components under mechanical and thermal energy input during processing. So far, in a world that is continuously evolving, a simple cereal-based product, as pasta is, has demonstrated to be able to combine tradition and innovation, without losing its peculiar attributes deeply rooted in the Italian culinary tradition. However, although the modern and completely automated pasta plants allow to transform a high-quality semolina into pasta of superior quality through an excellent control of the processing conditions, the new trends more and more oriented towards the development of non-traditional, healthy and, at the same time, sustainable products, require further evolution of this ancient product and its process.

References

1. Carini E, Curti E, Minucciani M, Antoniazzi F, Vittadini E. Pasta. In: Ferreira Guine R, Correia P, editors. Engineering aspects of cereal and cereal-based products. Boca Raton: CRC Press; 2013. p. 211–38.
2. Agnesi E. The history of pasta. In: Kruger JE, Matsuo RB, Dick JW, editors. Pasta and noodle technology. St. Paul: American Association Cereal Chemists; 1996. p. 1–12.
3. Serventi S, Sabban F. Pasta: the story of a universal food. New York: Columbia University Press; 2000.
4. Shelke K. Pasta and noodles. A global history. London: Reaktion Books Ltd; 2016.
5. UNAFPA Union of Organizations of Manufactures of Pasta Products of the European Union. 2015. http://www.pasta-unafpa.org/ingstatistics1.htm. Accessed 20 May 2020.
6. AIDEPI Associazione delle Industrie del Dolce e della Pasta Italiane (Association of Italian manufacturing industries of sweetes and pasta). 2017. http://www.aidepi.it/pasta/73-in-cifre.html. Accessed 20 May 2020.
7. Di Monaco R, Cavella S, Di Marzo S, Masi P. The effect of expectations generated by brand name on the acceptability of dried semolina pasta. Food Qual Prefer. 2004;15:429–37.
8. Feillet P, Dexter JE. Quality requirements of durum wheat for semolina milling and pasta production. In: Kruger JE, Matsuo RB, Dick JW, editors. Pasta and noodle technology. St. Paul: American Association Cereal Chemists; 1996. p. 95–131.
9. D'Egidio MG, Mariani BM, Nardi S, Novaro P, Cubadda R. Chemical and technological variables and their relationships. A predictive equation for pasta cooking quality. Cereal Chem. 1990;67:275–81.
10. Padalino L, Mastromatteo M, Lecce L, Spinelli S, Contò F, Del Nobile MA. Effect of durum wheat cultivars on physico-chemical and sensory properties of spaghetti. J Sci Food Agric. 2014;94:2196–204.
11. Italian Legislation. Law n. 580, July 4. Disciplina per la lavorazione e commercio dei cereali, degli sfarinati, del pane e delle paste alimentari. 1967. Official Journal n.189, July 29.
12. Presidential Decree (DPR) n. 187, February 9. Regolamento per la revisione della normativa sulla produzione e commercializzazione di sfarinati e paste alimentari, a norma dell'articolo 50 della legge 22 febbraio 1994, n. 146. 2001. Official Journal n. 117, May 22.
13. Milatovich L, Mondelli G. La Tecnologia della Pasta Alimentare. Pinerolo: Chiriotti Editore; 1990.
14. Alamprese C, Casiraghi E, Rossi M. Structural and cooking properties of fresh egg pasta as a function of pasteurization treatment intensity. J Food Eng. 2008;89:1–7.
15. Presidential Decree (DPR) n. 41, March 5. Regolamento recante modifiche al decreto del Presidente della Repubblica 9 febbraio 2001, n. 187, concernente la revisione della normativa sulla produzione e commercializzazione di sfarinati e paste alimentari. 2013. Official Journal n. 95, Apr 23.

16. Mefleh M, Conte P, Fadda C, Giunta F, Piga A, Hassoun G, et al. From ancient to old and modern durum wheat varieties: interaction among cultivar traits, management, and technological quality. J Sci Food Agric. 2019;99:2059–67.
17. Kill RC, Turnbull K. Pasta and semolina technology. Oxford: Blackwell Science Ltd; 2001.
18. UNAFPA Union of Organizations of Manufactures of Pasta Products of the European Union. 2020. http://www.pasta-unafpa.org/ing-documents1.htm. Accessed 25 May 2020.
19. FDA Food and Drug Administration. Department of Health and Human Services. Code of Federal Regulations Title. 2019;21(139):110.
20. Sissons M. Pasta. In: Wrigley C, Corke H, Seetharaman K, Faubion J, editors. Encyclopedia of food grains. 2nd ed. Oxford: Academic; 2015. p. 79–89.
21. Rosentrater KA, Evers AD. Kent'S technology of cereals an introduction for students of food science and agriculture. Cambridge: Woodhead Publishing; 2017.
22. Dalbon G, Drivon D, Pagani MA. Pasta: continuous manufacturing process. In: Kruger JE, Matsuo RB, Dick JW, editors. Pasta and noodle technology. St. Paul: American Association Cereal Chemists; 1996. p. 13–58.
23. Petitot M, Abecassis J, Micard V. Structuring of pasta components during processing: impact on starch and protein digestibility and allergenicity. Trends Food Sci Technol. 2009;20:521–32.
24. Dawe PR. Pasta mixing and extrusion. In: Kill RC, Turnbull K, editors. Pasta and semolina technology. Oxford: Blackwell Science Ltd; 2001. p. 86–118.
25. Icard-Vernière C, Feillet P. Effects of mixing conditions on pasta dough development and biochemical changes. Cereal Chem. 1999;76:558–65.
26. Marchylo BA, Dexter JE. Pasta production. In: Owens G, editor. Cereal processing technology. Oxford: Academic; 2001. p. 109–30.
27. Finnie S, Atwell WA. Wheat flour. 2nd ed. Cambridge: Woodhead Publishing; 2016. p. 131–211.
28. De Cindio B, Baldino N. Pasta: manufacture and composition. In: Caballero B, Finglas PM, Toldrá F, editors. Encyclopedia of food and health. Oxford: Academic; 2015. p. 235–41.
29. Matsuo RR, Dexter JE, Dronzek BL. Scanning electron microscopy study of spaghetti processing. Cereal Chem. 1978;55:744–53.
30. Antognelli C. The manufacture and applications of pasta as a food and as a food ingredients: a review. J Food Technol. 2016;15:125–45.
31. Mondelli G. Essicazione Statica della Pasta. Tecnologia e Pratica Operativa. Milano-Bologna: Avenue Media; 2016. p. 1–57.
32. Manthey FA, Twombly W. Extruding and drying of pasta. In: Hui Y, Sherkat F, editors. Handbook of food science, technology, and engineering, vol. 4. Boca Raton: CRC Press; 2005. p. 158.1–15.
33. Sissons M. Role of durum wheat composition on the quality of pasta and bread. FoodReview. 2008;2:75–90.
34. Lucisano M, Pagani MA, Mariotti M, Locatelli DP. Influence of die material on pasta characteristics. Food Res Int. 2008;41:646–52.
35. Mercier S, Des Marchais L, Villeneuve S, Foisy M. Effect of die material on engineering properties of dried pasta. Procedia Food Sci. 2011;1:557–62.
36. Pagani MA, Lucisano M, Mariotti M. Section IV. Bakery products: traditional italian products from wheat and other starchy flours. In: Huy IH, Chandan RC, Clark S, Cross NA, Dobbs JC, Hurst WJ, Nollet LML, Shimoni E, Sinha N, Smith EB, Surapat S, Toldrá F, Tichenal A, editors. Handbook of food product manufacturing. Hoboken: John Wiley & Sons; 2007. p. 327–88.
37. Pollini CM, Dexter JE. THT technology in the modern industrial pasta making process. In: Kruger JE, Matsuo RB, Dick JW, editors. Pasta and noodle technology. St. Paul: American Association Cereal Chemists; 1996. p. 59–74.
38. Sensidoni A, Peressini D, Pollini CM. Study of the Maillard reaction in model systems under conditions related to the industrial process of pasta thermal VHT treatment. J Sci Food Agric. 1999;79:317–22.

39. Dexter JE, Matsuo RR, Morgan BC. High temperature drying: effects on spaghetti properties. J Food Sci. 1981;46:1741–6.
40. Acquistucci R. Influence of Maillard reaction on protein modification and colour development in pasta. Comparison of different drying conditions. Lebens Wiss Technol. 2000;33:48–52.
41. Zweifel C, Handschin S, Escher F, Conde-Petit B. Influence of high-temperature drying on structural and textural properties of durum wheat pasta. Cereal Chem. 2003;80:159–67.
42. de Noni I, Pagani MA. Cooking properties and heat damage of dried pasta as influenced by raw material characteristics and processing conditions. Crit Rev Food Sci Nutr. 2010;50:465–72.
43. Marti A, Seetharaman K, Pagani MA. Rheological approaches suitable for investigating starch and protein properties related to cooking quality of durum wheat pasta. J Food Qual. 2013;36:133–8.
44. Malcolmson L, Matsuo R, Balshaw R. Textural optimization of spaghetti using response surface methodology: effects of drying temperature and durum protein level. Cereal Chem. 1993;70:417–23.
45. Bruneel C, Pareyt B, Brijs K, Delcour JA. The impact of the protein network on the pasting and cooking properties of dry pasta products. Food Chem. 2010;120:371–8.
46. Güler S, Köksel H, Ng PKW. Effects of industrial pasta drying temperatures on starch properties and pasta quality. Food Res Int. 2002;35:421–7.
47. Cubadda RE, Carcea M, Marconi E, Trivisonno MC. Influence of gluten proteins and drying temperature on the cooking quality of durum wheat pasta. Cereal Chem. 2007;84:48–55.
48. Chillo S, Laverse J, Falcone PM, Del Nobile MA. Quality of spaghetti in base amaranthus wholemeal flour added with quinoa, broad bean and chickpea. J Food Eng 2008; 84:101–107.
49. Fuad T, Prabhasankar P. Role of ingredients in pasta product quality: a review on recent developments. Crit Rev Food Sci Nutr. 2010;50:787–98.
50. Barilla: Pasta by shape. 2020. https://www.barilla.com/en-us/product-results/pasta/shape/long?sort=alpha. Accessed 15 July 2020.
51. Garofalo: Durum Wheat Semolina Pasta. 2020. https://www.pasta-garofalo.com/us/products/durum-wheat-semolina-pasta/. Accessed 15 July 2020.
52. FDA Food and Drug Administration. Daily value on the new nutrition and supplement facts labels. 2020. https://www.fda.gov/media/135301/download. Accessed 15 July 2020.
53. Conte P, Fadda C, Drabińska N, Krupa-Kozak U. Technological and nutritional challenges, and novelty in gluten-free breadmaking: a review. Polish J Food Nutr Sci. 2019;69:5–21.
54. Peressini D, Cavarape A, Brennan MA, Gao J, Brennan CS. Viscoelastic properties of durum wheat doughs enriched with soluble dietary fibres in relation to pasta-making performance and glycaemic response of spaghetti. Food Hydrocoll. 2020;102:10561.
55. Aravind N, Sissons M, Egan N, Fellows CM, Blazek J, Gilbert EP. Effect of β-glucan on technological, sensory, and structural properties of durum wheat pasta. Cereal Chem. 2012;89:84–93.
56. Aravind N, Sissons MJ, Fellows CM, Blazek J, Gilbert EP. Effect of inulin soluble dietary fibre addition on technological, sensory, and structural properties of durum wheat spaghetti. Food Chem. 2012;132:993–1002.
57. Montalbano A, Tesoriere L, Diana P, Barraja P, Carbone A, Spanò V, et al. Quality characteristics and in vitro digestibility study of barley flour enriched ditalini pasta. LWT - Food Sci Technol. 2016;72:223–8.
58. Ciccoritti R, Nocente F, Sgrulletta D, Gazza L. Cooking quality, biochemical and technological characteristics of bran-enriched pasta obtained by a novel pasta-making process. LWT. 2019;101:10–6.
59. Kaur G, Sharma S, Nagi HPS, Dar BN. Functional properties of pasta enriched with variable cereal brans. J Food Sci Technol. 2012;49:467–74.
60. Aravind N, Sissons M, Egan N, Fellows C. Effect of insoluble dietary fibre addition on technological, sensory, and structural properties of durum wheat spaghetti. Food Chem. 2012;130:299–309.
61. Manno D, Filippo E, Serra A, Negro C, De Bellis L, Miceli A. The influence of inulin addition on the morphological and structural properties of durum wheat pasta. Int J Food Sci Technol. 2009;44:2218–24.

62. Marconi E, Messia MC. Pasta made from nontraditional raw materials: technological and nutritional aspects. In: Sissons M, Marchylo B, Abecassis J, Carcea M, editors. Durum wheat chemistry and technology. St. Paul: American Association Cereal Chemists; 2012. p. 201–11.
63. Zhao YH, Manthey FA, Chang SKC, Hou HJ, Yuan SH. Quality characteristics of spaghetti as affected by green and yellow pea, lentil, and chickpea flours. J Food Sci. 2005;70:S371–6.
64. Mercier S, Villeneuve S, Mondor M, Des Marchais LP. Evolution of porosity, shrinkage and density of pasta fortified with pea protein concentrate during drying. LWT - Food Sci Technol. 2011;44:883–90.
65. Wood JA. Texture, processing and organoleptic properties of chickpea-fortified spaghetti with insights to the underlying mechanisms of traditional durum pasta quality. J Cereal Sci. 2009;49:128–33.
66. Sabanis D, Makri E, Doxastakis G. Effect of durum flour enrichment with chickpea flour on the characteristics of dough and lasagne. J Sci Food Agric. 2006;86:1938–44.
67. Shogren RL, Hareland GA, Wu YV. Sensory evaluation and composition of spaghetti fortified with soy flour. J Food Sci. 2006;71:S428–32.
68. Baiano A, Lamacchia C, Fares C, Terracone C, La Notte E. Cooking behaviour and acceptability of composite pasta made of semolina and toasted or partially defatted soy flour. LWT - Food Sci Technol. 2011;44:1226–32.
69. Kaur G, Sharma S, Nagi HPS, Ranote PS. Enrichment of pasta with different plant proteins. J Food Sci Technol. 2013;50:1000–5.
70. Petitot M, Boyer L, Minier C, Micard V. Fortification of pasta with split pea and faba bean flours: pasta processing and quality evaluation. Food Res Int. 2010;43:634–41.
71. Baiano A. Edible insects: An overview on nutritional characteristics, safety, farming, production technologies, regulatory framework, and socio-economic and ethical implications. Trends Food Sci Technol. 2020;100:35–50.
72. FAO Food and Agriculture Organization of the United Nations: Edible insects. Future prospects for food and feed security. 2013. http://www.fao.org/3/i3253e/i3253e.pdf. Accessed 15 Jul 2020.
73. Rumpold BA, Schlüter OK. Nutritional composition and safety aspects of edible insects. Mol Nutr Food Res. 2013;57:802–23.
74. Biró B, Fodor R, Szedljak I, Pásztor-Huszár K, Gere A. Buckwheat-pasta enriched with silkworm powder: technological analysis and sensory evaluation. LWT - Food Sci Technol. 2019;116:108542.
75. Veldkamp T, Bosch G. Insects: a protein-rich feed ingredient in pig and poultry diets. Anim Front. 2015;5:45–50.
76. European Commission. Regulation (Eu) 2015/2283 of the European Parliament and of the Council of 25 November 2015 on novel foods, amending Regulation (EU) No 1169/2011 of the European Parliament and of the Council and repealing Regulation (EC) No 258/97 of the European Parliam. Off J Eur Union. 2015;327:1–22.
77. European Commission. Commission Implementing Regulation (EU) 2017/2468 of 20 December 2017 laying down administrative and scientific requirements concerning traditional foods from third countries in accordance with Regulation (EU) 2015/2283 of the European Parliament and of t. Off J Eur Union. 2017;351:55–63.
78. European Commission. Commission Implementing Regulation (EU) 2017/2469 of 20 December 2017 laying down administrative and scientific requirements for applications referred to in Article 10 of Regulation (EU) 2015/2283 of the European Parliament and of the Council on novel foo. Off J Eur Union. 2017;351:64–71.
79. European Commission. Commission Regulation (EU) 2017/893 of 24 May 2017 amending Annexes I and IV to Regulation (EC) No 999/2001 of the European Parliament and of the Council and Annexes X, XIV and XV to Commission Regulation (EU) No 142/2011 as regards the provisions on proc. Off J Eur Union. 2017;138:92–115.
80. Kim HW, Setyabrata D, Lee YJ, Jones OG, Kim YHB. Effect of house cricket (Acheta domesticus) flour addition on physicochemical and textural properties of meat emulsion under various formulations. J Food Sci. 2017;82:2787–93.

81. Smetana S, Ashtari Larki N, Pernutz C, Franke K, Bindrich U, Toepfl S, et al. Structure design of insect-based meat analogs with high-moisture extrusion. J Food Eng. 2018;229:83–5.
82. Azzollini D, Derossi A, Fogliano V, Lakemond CMM, Severini C. Effects of formulation and process conditions on microstructure, texture and digestibility of extruded insect-riched snacks. Innov Food Sci Emerg Technol. 2018;45:344–53.
83. Roncolini A, Milanović V, Cardinali F, Osimani A, Garofalo C, Sabbatini R, et al. Protein fortification with mealworm (Tenebrio molitor L.) powder: Effect on textural, microbiological, nutritional and sensory features of bread. PLoS One. 2019;14:e0211747.
84. Gravel A, Doyen A. The use of edible insect proteins in food: challenges and issues related to their functional properties. Innov Food Sci Emerg Technol. 2020;59:102272.
85. Duda A, Adamczak J, Chelminska P, Juszkiewicz J, Kowalczewski P. Quality and nutritional/textural properties of durum wheat pasta enriched with cricket powder. Foods. 2019; https://doi.org/10.3390/foods8020046.
86. Jakab I, Tormási J, Dhaygude V, Mednyánszky ZS, Sipos L, Szedljak I. Cricket flour-laden millet flour blends' physical and chemical composition and adaptation in dried pasta products. Acta Aliment. 2020;49:4–12.
87. Sathasivam R, Radhakrishnan R, Hashem A, Abd Allah EF. Microalgae metabolites: a rich source for food and medicine. Saudi J Biol Sci. 2019;26:709–22.
88. Torres-Tiji Y, Fields FJ, Mayfield SP. Microalgae as a future food source. Biotechnol Adv. 2020;41:107536.
89. Ḿonica F, Batista AP, Nunes MC, Gouveia L, Bandarra NM, Raymundo A. Incorporation of Chlorella vulgaris and Spirulina maxima biomass in pasta products. Part 1: preparation and evaluation. J Sci Food Agric. 2010;90:1656–64.
90. Fradique M, Batista AP, Nunes MC, Gouveia L, Bandarra NM, Raymundo A. Isochrysis galbana and Diacronema vlkianum biomass incorporation in pasta products as PUFA's source. LWT - Food Sci Technol. 2013;50:312–9.
91. Iafelice G, Caboni MF, Cubadda R, Criscio T Di, Trivisonno MC, Marconi E. Development of functional spaghetti enriched with long chain omega-3 fatty acids. Cereal Chem 2008; 85: 146–151.
92. Prabhasankar P, Ganesan P, Bhaskar N. Influence of Indian brown seaweed (Sargassum marginatum) as an ingredient on quality, biofunctional, and microstructure characteristics of pasta. Food Sci Technol Int. 2009;15:471–9.
93. Verardo V, Ferioli F, Riciputi Y, Iafelice G, Marconi E, Caboni MF. Evaluation of lipid oxidation in spaghetti pasta enriched with long chain n-3 polyunsaturated fatty acids under different storage conditions. Food Chem. 2009;114:472–7.
94. Simonato B, Trevisan S, Tolve R, Favati F, Pasini G. Pasta fortification with olive pomace: Effects on the technological characteristics and nutritional properties. LWT - Food Sci Technol. 2019; https://doi.org/10.1016/j.lwt.2019.108368.
95. Padalino L, D'Antuono I, Durante M, Conte A, Cardinali A, Linsalata V, et al. Use of olive oil industrial by-product for pasta enrichment. Antioxidants. 2018;7:59. https://doi.org/10.3390/antiox7040059.
96. Cedola A, Cardinali A, D'Antuono I, Conte A, Del Nobile MA. Cereal foods fortified with by-products from the olive oil industry. Food Biosci. 2020;33:100490. https://doi.org/10.1016/j.fbio.2019.100490.
97. Spinelli S, Padalino L, Costa C, Del Nobile MA, Conte A. Food by-products to fortified pasta: a new approach for optimization. J Clean Prod. 2019;215:985–91.
98. Lončarić A, Kosović I, Jukić M, Ugarčić Z, Piližota V. Effect of apple by-product as a supplement on antioxidant activity and quality parameters of pasta. Croat J Food Sci Technol. 2014;6:97–103.
99. Chusak C, Chanbunyawat P, Chumnumduang P, Chantarasinlapin P, Suantawee T, Adisakwattana S. Effect of gac fruit (Momordica cochinchinensis) powder on in vitro starch digestibility, nutritional quality, textural and sensory characteristics of pasta. LWT. 2020;118:108856.

100. Fares C, Platani C, Baiano A, Menga V. Effect of processing and cooking on phenolic acid profile and antioxidant capacity of durum wheat pasta enriched with debranning fractions of wheat. Food Chem. 2010;119:1023–9.
101. Parizad PA, Marengo M, Bonomi F, Scarafoni A, Cecchini C, Pagani MA, et al. Bio-functional and structural properties of pasta enriched with a debranning fraction from purple wheat. Foods. 2020;9:163. https://doi.org/10.3390/foods9020163.
102. Monteiro MLG, Mársico ET, Deliza R, Castro VS, Mutz YS, Soares Junior MS, et al. Physicochemical and sensory characteristics of pasta enriched with fish (Oreochromis niloticus) waste flour. LWT - Food Sci Technol. 2019;111:751–8.

Chapter 5
Cereal-Derived Foodstuffs from North African-Mediterranean: From Tradition to Innovation

Fatma Boukid

Abstract North African traditional cereals-based foodstuffs (i.e. couscous, burghul and freekeh) are trending upward given the diversity of ingredients, forms and shapes, and versatility (staple or side). Such versatile foods are consumed worldwide and commercialized using different technologies. The concept of their industrialization might be viewed as shifting away from their authenticity or, conversely, as renewing/ rediscovering nutritious foods in response to specific health trends or convenience. Understanding the know-how of the traditional manners can provide appropriate knowledge to effectively produce industrial-scale, and at the same time authentic foodstuffs.

Keywords Couscous · Burghul · Freekeh · Cereals · Tradition · Industrialization

1 Glossary

1.1 Traditional Foods

To effectively define traditional foods, several attempts have been made from different perspectives (i.e. sociocultural, manufactural and consumers-driven definitions; detailed in Table 5.1. Relying on food regulation was also required to provide a full understanding of the situation. European regulations have improved in the sector of traditional foods during the last years (Table 5.1).

F. Boukid (✉)
Food Safety and Functionality Programme, Food Industry Area, Institute of Agriculture and Food Research and Technology (IRTA), Catalonia, Spain
e-mail: fatma.boukid@irta.cat

F. Boukid (ed.), *Cereal-Based Foodstuffs: The Backbone of Mediterranean Cuisine*, https://doi.org/10.1007/978-3-030-69228-5_5

Table 5.1 Glossary: multi-driven definitions of traditional food

Viewpoint	Definition	References
Sociocultural	Traditional foods can be defined as foods that have been used in local societal groups and whose recipes have been perpetuated from generation to generation by word of mouth or in writing	[1]
Manufactural	Traditional foods are authentic in their recipe, origin of raw material and or production process; which are commercially available for the last 50 years as a part of the gastronomic heritage	[2]
Consumers	Traditional foods are frequently consumed or associated with specific celebrations and/or seasons, transmitted from one generation to another, made in a specific way according to the gastronomic heritage, distinguished and known thanks to its sensory properties and associated with a certain local area, region or country	[3]
EU definition	Traditional foods were defined as foods consumed outside the EU, which includes foods made from plants, microorganisms, fungi, algae and animals (e.g. chia seeds, baobab fruit, insects, water chestnuts).	[4]

Table 5.2 Common traditional foods-based on cereals in north Africa

Food item	Cereals used	Foodways
Flat bread	Common wheat, durum wheat, sorghum	Flour, water, salt and leavening agent (optional) were used to make the dough. Then, the dough is flattened to make bread.
Bazin	Starch from durum wheat, barley, soft wheat	An unleavened bread based on a cooked dough made with flour, water and salt. Bazin can be baked or steamed.
Barkoukech	Durum wheat	Flour was agglomerated until obtaining large granules (around 1 mm), and then steamed and dried. The obtained product is generally served as a soup.
Mhammes	Durum wheat	Flour is hand rolled to form granules to be sun-dried. The obtained product is generally served as a soup.
Bsisa	Durum wheat and chickpea, or millet	Grains, legumes or pseudocereals were roasted and grinded. The obtained flour is generally used as beverage with water and sugar or a dense cream when mixed with oil
Malthouth	Barley	Is cracked grain based on boiling, drying and cracking the grain.
Assida	Durum wheat, common wheat, sorghum	Flour and water were cooked until obtaining unitary dough. The obtained product is served either as a sweet with sugar and olive oil or with a sauce (tomatoes, olive oil and spices)

1.2 *Traditional Cereal-Derived Foodstuffs: The Case of North Africa*

Cereals are widely cultivated in north African countries, and their derived foods are widely consumed as primary staple foods for centuries. Table 5.2 summarized the most common traditional foods consumed in north Africa. These foods might present certain similarities or dissimilarities in their ingredients or making process

Table 5.3 The most popular traditional north African cereals-based foods

Food item	Common names	Definition
Couscous	Kuskus, couscous, maftoul, moghrabieh, seksu, kusksi, keskesu, kouskousaki	Couscous is prepared from durum wheat semolina (Triticum durum) the elements of which are bound by adding potable water and which has undergone physical treatment such as cooking and drying [5] Couscous, a granular paste product made from mixing semolina with water, is considered one of the major food staples in Egypt, Libya, Tunisia, Algeria and Morocco. It is mainly served with fish, lamb, or poultry and vegetables with broth or spicy sauce [6]
	Short-cut pasta type couscous	Couscous, a paste product made from mixing semolina with water to form unitary dough. Once the dough goes through extrusion machine, a cutting machine allows to obtain the couscous granules.
Burghul	*Bulghur*, bourghol, boulgour, borghol, bulgur	Burghul is coarse or fine cracked product of several different wheat species, most often from durum wheat. Burghul is traditionally consumed in Central Asia, Turkey, the Middle East (Lebanon, Palestine, Syria) and north African cuisines, where it is used in salads such as tabouleh, soups and baked products such as kibbeh, soups or boiled and consumed like [7]
Freekeh	Frikeh, farik, farīkah	Freekeh, a dried green durum wheat product, is a staple food in North Africa and the Middle East especially Syria. Freekeh is either boiled or steamed and served with lamb or poultry [8]

among countries. Couscous, burghul and freekeh (Table 5.3) are among the most economically and culturally important staple foods in the Mediterranean region, mainly North African countries (e.g. Libya, Tunisia, Algeria and Morocco) and Middle Est countries (e.g. Syria, Lebanon, Palestine, Jordan, and Egypt). Couscous, burghul and freekeh are typically durum wheat-based products, which older women used to prepare at home during the summer to have the family's stock for the following year.

2 Preface to the Topic

Cereals and their derived food products are the backbone of the Mediterranean diet, in which they are by far the main source of energy, protein, and minerals. Traditional foods have been consumed locally or regionally for a long time as a way of expression (e.g. culture, history, geography, climate, agriculture, and lifestyle). Until now, north African countries still retain traditional habits, where women annually prepare traditional cereals-based products as a part of the folklore. Given the rising global exchange in foodstuffs, the culinary borders are becoming increasingly blurred. In this frame, the consumption of north African traditional cereals-based foodstuffs has been trending upward not as ethnic food but more as a global food.

Couscous, burghul and freekeh are particularly among the food trends worldwide as a staple or side. The interest in these cereals-based foods can be explained by the increasing demand of international consumers for better quality, more diversity and highly beneficial products. Therefore, their industrial production was driven towards ingredients diversity or/and process innovations to be steps ahead of the next trend (e.g. gluten-free, fortified using legumes, pseudo-cereals and minor cereals, organic, refined, wholegrain, pre-packaged and ready-to eat-meals). Industrialization took two pathways, either small- or large-scale. Small-scale manufacturers were engaged to preserve traditional processes. Large-scale manufactures are, however, in front of a more challenging situation to develop effective mechanization and to produce machine-made traditional products with high-quality, affordable price and at the same time safe. To this end, the present review is not to merely summarize what we know until now, but also to explore the main north African cereals-based products (i.e. couscous, burghul and freekeh); to extend the current knowledge by critically investigating the advantages and the limitations of their processing; and to assess their nutritional profile and safety challenges.

3 The Most Popular-Traditional Cereal-Derived Foodstuffs: Why Everything Old Is New Again?

Traditional cereal-based products (i.e. couscous, burghul, and freekeh) are marketed worldwide as "the new old trend". The reasons of interest to look favorably to couscous, burghul, and freekeh are explained as follows:

3.1 Socio-Cultural Importance

For centuries, couscous, burghul, and freekeh were and still a part of the patrimony of Mediterranean area and Middle east countries. These traditional products are symbols of socio-historical accounts and stand as cultural artifacts that link communities. Such foods are vehicles of expression of personal, social, and cultural identity, which contributed in vanishing geographical and psychosocial borders among Mediterranean populations [9]. In the framework of globalization, traditional foods were transferred as ethnic food to provide culinary variety and a sense of adventure [10, 11]. These foods have been standardized and revisited by manufacturers to fit the culinary and gastronomic habits of European countries [12]. But it was up to the consumers to make the decision of either accepting or rejecting food with foreign origin. North African cereals-based food went beyond their classic uses through innovative applications to suit the lifestyle of international consumers. More specifically, the immigration (1960s and 1980s) from North Africa to Europe and North America contributed to the international widespread of couscous [9].

Currently, couscous is not anymore considered as migrating dish for Maghrebi immigrants [12]. European-Mediterranean countries (e.g. France and Italy) are now familiar with couscous and they considered an everyday food [13]. For instance, in Cecily, international festival of couscous is an annual celebration since 1998. Likewise, burghul, the man's first processed food, is popular in Central Asia, Turkey, the Middle East, Canada, Europe and USA cuisines and it is used in more than 250 meals [6, 14]. Freekeh is mainly produced in Turkey, North Africa, and the Middle East [15], but recently it is gaining more popularity. There products in their traditional version are staple food, but within the process of their "evolution" (i.e. inclusion in new culinary habits), they swished to side dishes in response to convenience.

3.2 Technological Importance

Given their convenient and cooking options (easy and quick), traditional products (i.e. couscous, burghul and freekeh) are well-appreciated by the international food market. Such products might be classified as products with "portfolio" given their high diversity (e.g. several types of cereals, forms and shapes) and versatility (ready- or semi-ready-to-eat; soups, meat balls, tabbouleh and salad) [15–17]. They have pellicular flavor profile and desirable taste [16]. These products have also a long shelf-life given their low water activity, high resistance to insect, mites and microorganisms [18–20]. As well, no special conditions are required for storage, commercialization, transport and packaging [21].

3.3 Nutritional Importance

The nutritional value is closely associated with the raw ingredient and processing and consequently their impact on human health depends on the nutritional value, which differ from refined to wholegrain flour. Wholegrain cereals are rich in bioactive components associated with benefits for health, including dietary fiber, phenolic compounds, lignans, vitamins and minerals [22, 23]. Couscous, burghul and freekeh are traditionally made with durum wheat wholegrain. Wholegrain intake is associated with several health benefits [22, 24]. Bulgur is also stored for military and human nutrition purposes in some countries because it does not absorb radiation (today, bulgur is one of the important wheat products in the USA and it is included in the special list of food rations in nuclear fallout shelters) [18, 25]. Furthermore, bulgur is used for food aid in famine regions by the World Food Program, because of its nutritional value and long-term storage ability, even in improper conditions, owing to resistance to insect and mold activity [17]. Freekeh is a rich source of amino acids, fibers and minerals and reduced phytic [15]. Currently, these products are prepared using different cereals (gluten-containing or gluten-free), pseudocereals and legumes to respond to consumer's needs, and consequently a new spectrum of health-beneficial foods can be obtained through their innovation.

3.4 Economic Importance

Given their high nutritional value, long shelf life and affordable price, the demand of the market toward traditional cereal-based product increased all over the world. Couscous worldwide annual production reached 420,000 tons, where the European production (mainly France, Italy and Spain) is around 145,000 tons/year, and USA and Canada reach 20,000 tons/year [26]. Burghul production has mostly been known in Central Asia, the Middle East, Balkans and Turkey [27]. In Turkey, approximately one million tons are annually produced [28]. In the Middle East, its annual consumption is around 25–35 kg/person higher than that in Turkey (15 kg/person) [28, 29]. Since burghul was declared as a wholegrain by Whole Grain Council, it became more available in the USA and European countries [17]. Freekeh is mostly consumed in the Middle East, where the annual consumption exceeded 500,000 tons/year [30]. For consumers, the price of freekeh is high (two or three times higher than that of bulgur and pasta) [31], probably because this product is seasonable. Otherwise, freekeh provides an economic advantage for local farmers, since it is commonly sold at a price that is three times as high as wheat specially in Jordan [32]. Commercialization and industrialization of freekeh is still limited compared to couscous and burghul. Recently, Greenwheat Freekeh™, an Australian brand for freekeh production, is marketed worldwide and gaining strong foothold in the international food markets (Europe and USA). Freekeh and burghul are also registered in the database of whole grain content-based foods in Australia.

4 Couscous

4.1 Raw Ingredient

According to the *codex Alimentarius* (standard CODEX STAN 202–1995), couscous might be prepared from a mixture of coarse (diameter between 475 and 700 μm) and fine (diameter between 130 and 183 μm) semolina or "coarse medium" (diameter between 183 and 700 μm) semolina of durum wheat (*Triticum durum*). No food additives shall be added during the industrial processing of couscous [5]. Because of consumers' quest for health-beneficial food, several ingredients have been inserted to upgrade the nutritional value of couscous. Durum wheat has been partially or totally replaced by other cereals (e.g. sorghum and barley), pseudocereals (e.g. quinoa) or legumes [33–35]. Such products are also labelled "couscous" followed by a specification of the type of the additional ingredient used [5].

4.2 *Processing*

4.2.1 Traditional Processing

Traditional couscous is made with durum wheat semolina (refined or whole) and water following the diagram illustrated in Fig. 5.1. The main operations for making couscous are agglomeration, rolling, sifting, steam-cooking, and drying [36, 37].

Hydrating and mixing: generally medium semolina is used to make homemade couscous. Before starting, a salted solution was prepared to be added gradually to semolina. Women keep mixing with their hands the watered semolina to homogenize the mixture. If the quantity of water exceeded the desired level (formation of dough pieces), more semolina can be added.

Wet granulation consists in agglomerating semolina particles to increase their size and/or to shape to form larger assemblies, called agglomerates [38, 39]. The wet agglomeration requires the addition of water to generate capillary bridges which adds cohesion between particles [40]. The agglomerate size is dependent on the source and physical characteristics of the semolina (e.g., particle size and particle shape) and the amount of water added to a given amount of dry material [41].

Rolling: the agglomerates are rolled by circular movement with hands to make small fine particles. Flour or salt water was alternately added until the grains were completely coated, and the granules acquired a soft surface [42].

Sifting is performed to obtain agglomerates with uniform particle size. The first sieve (with the highest diameter around 3 mm) to calibrate the couscous, in forcing the grains by hand pressure to pass through the sieve. This operation is performed twice to enable the uniformity of the granules. Then, the second sieving was performed with the medium sieve (diameter around 2 mm) to separate the couscous and

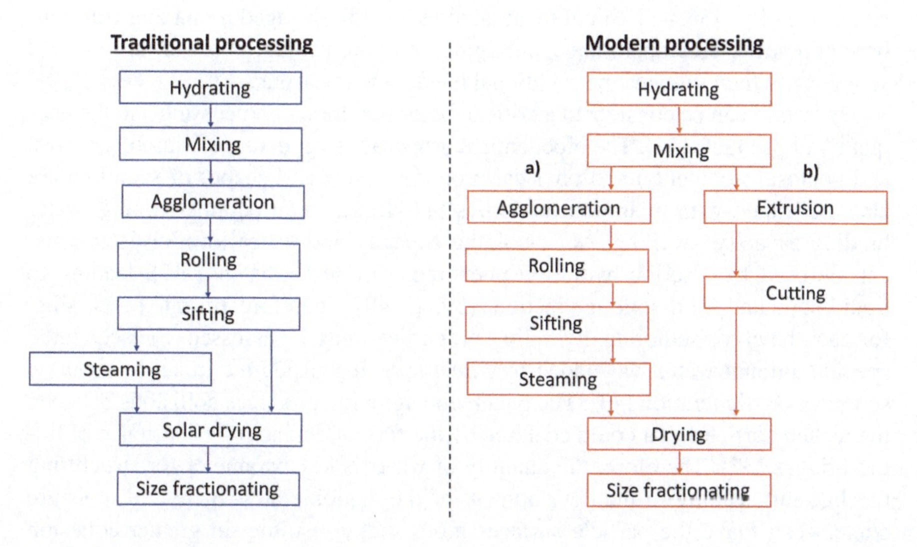

Fig. 5.1 Couscous-making: handmade versus machine-made

non-aggregated particles. Couscous was spread on a cheesecloth to avoid their agglomeration. The non-aggregated particles must undergo a new cycle of agglomeration [42].

Steaming is an optional step depending on the type of couscous. Couscous made with refined flour is steamed in a "couscoussier" (steamer), while when it is made with wholegrain, this step can be skipped, and the couscous can be directly sun-dried.

Sun-drying must be performed properly to enable a long shelf life of the couscous. The particles were spread on a cheesecloth and sun-dried with mixing. This process can take from 3 to 6 days during the summer season.

Size fractionating is grading the resultant particles into different size particle: fine, medium and coarse couscous. Sizing was performed using a third type of sieve (diameter around 1 mm). The resulted couscous grains were firm and completely dry [42].

4.2.2 Industrial Processing

In the industrial context, couscous can be made by either granulation or extrusion technique. Small-scale manufactures produce couscous by agglomeration process using mechanical low shear mixers [36, 38, 43]. Large-scale manufacturers developed a new processing method, extrusion, to promote acceptance and consumption of couscous in other regions.

Agglomeration Technique (Fig. 5.1a)

The agglomeration is a key operation in the elaboration of couscous grains of durum wheat semolina [36, 43]. Small manufactures are still engaged in making couscous based on successive operations: wet agglomeration, rolling, cooking, drying, and screening, which imitates the traditional handmade couscous.

Hydrating can be considered a critical factor for storage, processing and the end-quality of products [44]. The biochemical properties (e.g. damaged starch, proteins, and pentosans contents) and physical properties (size and shape) of semolina are also associated with hydration properties [44–46]. Several parameters (e.g. water binding capacity, swelling, water-holding capacity and water sorption isotherms) can be used to describe hydration phenomena of semolina [44, 47], leading to liquid-solid and solid-solid interactions [39, 43, 48]. Therefore, prior to processing, for each batch of semolina, hydration properties must be assessed. Subsequently, specific amount water was added by direct spraying inside the mixer to promote semolina agglomeration [43]. The pneumatic agitation produces collisions between the wetted particles that could coalesce by the formation and the persistence of liquid bridges [37]. Therefore, the quantity of water is key parameter for structuring mechanisms leading to the development of the agglomerates. Increased moisture content can make the particle surfaces more sticky resulting in greater adhesion between the particles and against wall surface, that leads to cohesion and difficult flow [46]. Controlling the wetting/mixing process is still largely dependent on the technical know-how of the operators [43].

Agglomeration processes is made up of three phases: wetting and nucleation, coalescence and growth and attrition and breakage [39, 43, 49]. Agglomeration is a complex industrial process to control, depending on the particles properties, the interaction water-particles system and the operating condition of the mixing [47, 50]. Such a process is a critical stage that determines the final quality of couscous (i.e. size distribution, density and shape), production yield (i.e. efficiency of the production line to generate lower amounts of recycling) and functionalities of the couscous grains [36, 38, 43, 48, 51].

In practice, industrial agglomeration is carried out using mechanical or pneumatic energy (which ensures particle agitation and water homogenization) to increase the size of wetted particles to reach agglomerates phase [36, 37, 43, 52]. The suitable conditions applied are low shear rather than high shear to avoid forming unitary doughy mass instead of watering semolina [43, 49]. As for contact time, manufactures keep blending until the mix reaches the optimum agglomeration using a mechanical mixer (e.g. paddle mixer) for a short time (3–5 min) to provide granules of a moisture content of about 30–35%. In this case, high heterogenicity was obtained in the size of the wet agglomerates [with up to 60% particles out of range (too large or too small)] [43, 49]. The management of these fluctuations is still challenging and is still basically depend on technical know-how of operators, mainly by visual monitoring or sampling procedures [43, 53]. Agglomeration mechanisms are not fully understood and only few technical and scientific works are available on the internal structure of agglomerates and how to manage the process of agglomeration [37, 38].

Recently, Rondet et al., (2016) developed a novel reverse-phase wet granulation process, that was previously applied in the pharmaceutical field [54, 55]. The novel process is conducted as two main successive phases. During the first step, water was added to semolina to produce a dough under continuous mixing. Then, agglomerates were generated by adding the dry powder to the dough, during the mixing. In this case, the fragmentation mechanisms induce dough disruption and agglomerates production (Rondet et al., 2016). This novel approach is supposed to have the potential advantages of eliminating the traditional variations associated to the stage of granule nucleation, and of reducing the potential risk of uncontrolled growth of the granules [54, 55].

Rolling/wet-sifting is carried out for dual objective: (1) to select the expected range of agglomerates based on the size of their diameters (between 1 and 2 mm) and (2) to strengthen the formed agglomerates using an inclined rotary drum [52]. This equipment consisted of a cylindrical perforated drum that rotated to allow smaller materials to drop out during the rotation process [52]. Due to the inclination of the drum, the remaining particles travel onward to the subsequent screening rings (with increasing meshes diameters) to be separated. Continuous rolling drum is used by industry given it relatively simple, low expensive, requiring little operating and maintenance costs compared to other separation systems [52]. Rolling greatly contributes to the final sensory attributes of the couscous grains [45, 56]. Indeed, couscous resulting from rotary drum are more spherical and less porous, than those that are sieved on traditional horizontal vibrating sieves (Abecassis et al., 2012) or on fluidized bed, thanks to the low shear stresses [57–59].

Steaming: couscous agglomerates are steam-cooked (8 min, 180 °C) to reach a final content of 37%, and approximately 55 or 60% of the starch is gelatinized. As the relative humidity of the surrounding air is increased, wheat flours tend to absorb water molecules, which may form liquid bridges between powder particles and result in greater powder cohesion and reduced followability [46].

Drying is carried in a rotary dryer out until the product moisture is reduced below 13%, preferably to 10–12%. When vibrating sieves were used, couscous layer is carefully separated through a lump breaker to avoid granules agglomeration.

Size fractionation or dry-sifting: after cooling, dried couscous was sifted into fine and medium-sized couscous grains. The vibrating screen is one of the most general pieces of equipment in the separation of granular materials based on size. Over-sized agglomerates are milled and then resifted, while undersized particles are recycled [36].

Extrusion Processing: Pasta Type-Couscous

Pasta type-couscous production is a patented technique (Patent US5334407A) based extrusion technology for large-scale industrial production (see diagram in Fig. 5.1b).

Hydration: semolina or middlings of semolina was hydrated using water to form unitary dough mass. Optimal proportions must be optimized for each batch [60].

Extrusion barrel is made of three zones, in which the dough can be cooked, hydrated and dehydrated and then extruded through an extrusion die [size openings = 7 ± 0.5 mm^2].

After hydration, the mixture was advanced through the extruder barrel by rotatable screws through:

1. **cooking zone:** in which the temperature and moisture content of the dough increased [around 90–130 °C via air steam, around 35–50%, respectively]. The dough was moderately sheared to improve its unity and integrity. However, excessive shearing leads to dough structural breakdown thereby increasing the amount of undesirable soluble starch in the final product;
2. **venting zone:** where the temperature is reduced (around 80–100 °C). Subsequently, the cooked dough is dehydrated to reach a moisture content around 30–40%;
3. **forming zone:** where the temperature and moisture content of the dough is maintained stable (80–100 °C and 25–35%, respectively) as in the cooking zone. Pressure was first increased (55–90 bar) to compress and compact the dough and enhance its development and contribute to the structural integrity or unity of the end-product. Then, pressure decreased (13–41 bar) to increase the retention time, which then dough was extruded through the extrusion die.

Cutting: a blade flush with the die face of the extruder contacts the extrudate cutting from the extrudate stub-like particles.

Drying can be achieved using different techniques: belting or tossing in a rotary screen or free-falling or suspending in a fluidized bed or floating in drying atmosphere. Drying involves also vibrations of the cut particles to maintain the uniformity of the temperature (around 70–85 °C) and to avoid the agglomeration of particles within the drying atmosphere. The final moisture content is around 10–12%.

Size fractionating is not required because particle size was previously adjusted at the end of the extrusion die, and consequently couscous particles were uniform and ready for packaging.

4.3 Nutritional Profile

The nutritional value of couscous is influenced by several factors, including raw ingredient, formulation, process of production and cooking method.

Different commercial types of couscous [durum wheat semolina (whole ad refined), barley and buckwheat] showed no relevant difference in their proteins content [from 11.4% (buckwheat) to 12.6% (wholegrain durum wheat)] [61]. However, lipid content was slightly higher in buckwheat-based couscous (3%) than that of whole wheat (2.2%), barley (1.9%) and refined semolina (1.5%). Carbohydrates had the highest value (70.2) in refined durum wheat. Barley-based couscous showed the highest fiber content (11%). Fiber also was higher in wholegrain-based couscous (8.8%) compared to that refined (3.8%).

In response to consumers pursuit to health-beneficial foodstuffs, nutritional improvement of cereal-based foodstuffs (e.g. bread, pasta and biscuits) was driven by fortification (e.g. legumes minor cereals and pseudocereals) [62–64]. In this frame, several attempts were made to enhance the nutritional profile of couscous. For instance, couscous (refined wheat semolina) fortified with increasing levels of chickpea flour (up to 100%) had increased minerals (Ca, Mg, K, P, Fe and Zn), protein and cellulose contents [34]. Nevertheless, technological properties (color, firmness and overall acceptability) were negatively influenced, and consequently the appropriate substitution was 50% [34]. Couscous was enriched with 20% buckwheat or legumes (soybean flour, chickpea flour, common bean flour, lentil flour and lupine flour) and 10% wheat germ. The resulting product had increased mineral, protein contents. More especially, buckwheat addition enabled the highest ash content [65]. Soy and oat flours-based couscous had higher protein content and minerals (Ca, K and Fe) that that conventional. Likewise, gluten-free couscous (rice-based) was nutritionally enhanced particularly an increase in protein and minerals contents by legumes addition (chickpea, proteaginous pea and field beans; at 50%) [33]. The fortification of maize and sorghum flours-based couscous with 10% soybean or moringa leaves powder increased protein (1.5–2 times) and minerals contents (2 times) as compared to the traditional formulation [66].

Traditional formulation of couscous in Algeria is mainly based on fermented wheat as a main ingredient together with semolina, eggs and fat [42]. Results showed that couscous production reduced protein content [fermented wheat: 10.099% ± 0.011; couscous: 9.610% ± 0.021]. However, starch, lipid and ash contents increased given the addition of semolina, fat and eggs. In Turkey, undersized burghul is commonly inserted in extruded couscous [67]. Regardless of the formulation, the addition of burghul flour (25, 50, 75 and 100%) increased protein, ash, polyphenols and fiber as compared to that made with 100% semolina. Such a result might be attributed to the fact that grains used for burghul production are partially debaranned, and subsequently a part of the nutritional features of wholegrain were preserved unlike refined semolina. On the other hand, increasing the quantity of burghul flour had negative effect on the sensory attributes, bulk density, cooking loss and functional properties. Therefore, 50% addition enabled the best compromise between nutritional and technological properties.

Rahmani and Muller (1996) investigated the effect of traditional and modern dying on phytochemicals composition of couscous. Thiamin content was not influenced by the drying method (traditional: 0.26 ± 0.07 mg/100 g; modern: 0.22 ± 0.02 mg/100 g). However, riboflavin content of sun-dried couscous was lower than that commercial (0.037 ± 0.006 g/100 g and 0.066 ± 0.01 μg/100 g) [68]. Similarly, Yuksel et al., (2017) reported a reduction in total phenols and flavonoid contents after drying (bed dryer or microwave). Bioactive components present in couscous are also influenced by cooking method [69]. Total polyphenols showed the highest value in raw couscous made with buckwheat (1117.7 mg GAE/100 g (d.w.)), followed by barley (624.8), whole grain (574.5) and refined durum wheat semolina (470). After cooking in boiling water, polyphenols were reduced by around 16% due to their solubilization in cooking water [69]. Likewise, steam-cooking reduced thiamin and riboflavin contents (15 and 36%, respectively) [68].

5 Burghul

5.1 *Raw Ingredient*

Burghul is traditionally produced from *Triticum durum* wheat, but other type of cereals (e.g. bread wheat, durum, barley, rye, oat, triticale, einkorn and maize) [18, 27] or legumes (e.g. chickpea, lupin and soybean) [70–73] can be used.

5.2 *Processing*

5.2.1 Traditional Processing

The traditional processing for making burghul (illustrated in Fig. 5.2a) consists of dry-cleaning whole-wheat grain for eliminating external material, and then carefully washing with water to remove the dust and clod particles. The cleaned kernels

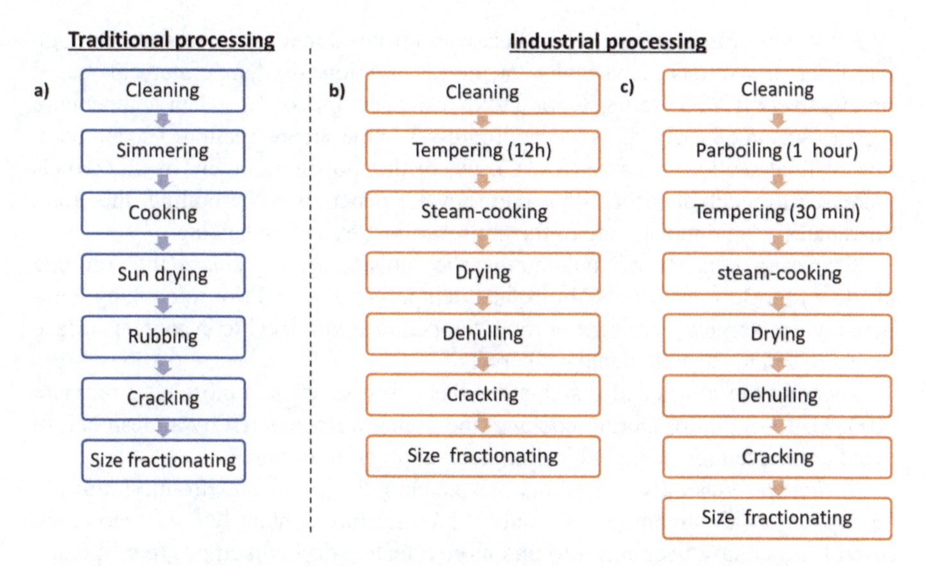

Fig. 5.2 Burghul-making: (**a**) Traditional handmade burghul (**b**) Conventional industrial processing (US2884327); (**c**) Modern industrial processing (US3132948A)

were left overnight for simmering in water until the grains become tender. The simmered particles were cooked in boiling water to gelatinize the starch and then spread in the sun to dry. Dried grains are dehulled to remove the outer bran layers and cracked with a stone or coarse mill. Particles were sorted to different particle sizes (fine, medium and coarse) [19, 74].

5.2.2 Industrial Processing

Conventional Method

The patent US2884327 (see Fig. 5.2b) described a continuous basis processing to produce burghul at industrial scale. Following cleaning, wheat grains was tempered until at least 40% moisture content. Moistened grains were cooked with steam under super-atmospheric pressure. Then, grains were dried in drying towers until about 12% moisture content, dehulled to remove some of the bran, cracked, and sorted. Then, kernels were milled and sorted. Although this process is a continuous operation, several limitations are present including the longtime of tempering (12 hours).

Modern Processing

According to the patent US3132948A (see Fig. 5.2b), modern processing of burghul can be as follows:

Cleaning and washing wheat kernels are subjected to conventional cleaning operations (dry and wet).

Parboiling (Fig 5.2c): cleaned wheat was advanced in a rotatable screw through the different sections of parboiler. In the first section, the temperature increased progressively from 55 to 68 °C for a total residence time of 1 h. Then, temperature increases from 68 to 85 °C as wheat progresses in the second section. The temperature of final section is stable (85 °C), where the moisture content of the kernels increased to reach about 40–45% due to water penetration. Parboiling, therefore, guarantees a rapid moistening of the grain, without excessive heating.

Tempering consists of maintaining the moistness of grains stable (around 40–45%) for 30 min at 74 °C in isolated tempering device. Prior to cooking, tempered wheat kernels were kept in room temperature for 30 °C to evaporate surface moisture and to separate damaged kernels.

Cooking: the moistened grains are subjected to steam at atmospheric pressure (100 °C; 15–20 min). During cooking, the grains were covered by a mesh belt to avoid excessive moistening which might result in grain rupture.

Drying was generally carried out by spreading the grains on wire-mesh trays to be exposed to air stream (82 °C) until 9.1% moisture content [29, 72]. However, several works have been made to find more effective drying methods (e.g. spouted bed, microwave, tray and infrared) with additional benefits (e.g. shorter drying time, higher product quality and lower cost) [75–77]. Based on pilot-scale trials, the effect of drying methods (solar, microwave and tray drying) was determined on physical properties of burghul (fine and pilaf) [77]. The optimum drying was achieved faster with microwave drying, followed by solar, tray, and sun drying. Solar drying allowed the highest yield of pilaf burghul (followed by microwave-, tray-, and sun-dried), while drying method did not influence that of fine burghul. As for color, solar gave the highest lightness and yellowness fine and pilaf burghul. Protein extractability did not depend on drying method, while water and oil absorption were slightly higher after solar and tray-drying, respectively. The drying method had no impact on sensory properties (e.g. flavor, mouthfeel, and appearance) [77]. Kahyaoglu et al. (2010) combined spouted bed drying [different temperatures (50, 70, and 90 °C)] with microwave [different powers (288 and 624 W)]. The results showed that microwave assisted spouted bed drying shortened the time of processing (around 60–85%), and the yield was not affected as compared to spouted bed drying [75]. The best result was obtained by combining the lowest microwave power (288 W) with the lowest temperature of spouted bed (50 °C) by saving energy and preserving an acceptable physical quality in the final product [75]. Therefore, microwave combined with hot air-drying treatment could be favorable for bulgur production. Infrared drying allowed an important reduction in drying time [sun-dried for 72 h or dried in oven at 60 °C for 18 h] without side effects on burghul quality (i.e. color, particle size, cooking loss, ash, dietary fiber and water absorption) [76]. Despite their effectiveness, the main shortcoming of large-scale application of microwave or infrared is their high equipment and operation expenses [75, 76].

Dehulling: dried grains pass through abrasion and friction machinery to remove the outer-bran allowing nutritious parts, such as the aleurone layer to remain within the intact kernels [78].

Cracking is generally carried out using a single breakage mill (stone, hammer, disc and roller). The yield of single roller was the highest (99.19%) as compared to double disc (94.84%) and vertical disc mills (97.60%). As for particles uniformity, the resulting burghul had irregular shape (long shape) using single roller [15, 79], while double disc mill enabled a better uniformity of particle size (oval shape) [18, 27]. However, the quantity of by-product (flour) increased due to abrasion effect using double mill [79].

Polishing (optional): is carried out with tempering followed by mechanical kneading (Balcı et al., 2017). In brief, water is directly sprayed onto the surface of burghul and left to rest (25–30 min) for water penetration [18, 27]. Then, tempered kernels were squeezed by fiction to smooth their sharp corners (oval shape) [80]. Such operation can indeed, improve the yellowness of burghul, which is an important quality trait for consumer acceptability and business competitiveness [80, 81]. But additional production cost due high energy consumption is the main limitation [80].

Size fractionating: burghul is classified as coarse (>3.5 mm), pilaf (3.5–2.0 mm), medium (2.0–1.0 mm), fine (1.0–0.5) and undersize burghul (<0.5 mm) [15]. Undersized burghul is around 15% of total produced burghul, which mainly used for animal feed [35].

5.3 Nutritional Profile

Bulghur can be considered a health-beneficial food given its highly nutritional composition [18, 27]. Particularly, dietary fiber in bulghur was higher than wheat flour (6.8 times) followed by pasta (4.3 times) and rice (3.5 times) [15]. A rich mineral content (phosphorus, zinc, magnesium and selenium) was also reported to be involved in the prevention from some gastrointestinal diseases such as constipation and colon cancer [17]. Moreover, bulghur had a low glycemic index and high resistant starch because during cooking starch gelatinized, and then retrograded during drying [18, 27].

Depending on drying method, loss of vitamins (thiamin, niacin, pantothenic acid, pyridoxine, riboflavin) was more pronounced in sun-dried bulghur followed by hot-air oven dried (at 80 °C) [82]. Bioactive compounds of einkorn and durum burghul was also investigated as function of different cooking methods (traditional, microwave, autoclave) and drying techniques (microwave and hot air) [67]. Results showed that yellow pigments, total phenolic content, antioxidant capacity and individual phenolic acids were reduced regardless of the processing. More specially, microwave cooking enabled minimum losses in polyphenols, probably due the short time needed for the treatment (6–7 min) [75]. Furthermore, the debranning caused relevant reduction in total protein, ash and ß-carotene contents as compared to the raw wheat [83]. To avoid such loss, debranning machine can be adjusted to remove only the external part of bran and keep aleurone layers (rich in bioactive compounds) [75].

6 Freekeh

6.1 Raw Ingredient

Freekeh is a cereal-based food made from green (under-mature) wheat (*Triticum turgidum* var. durum) that is roasted and rubbed to create its unique flavor. Freekeh was made chiefly from wheat, but freekeh from green barley, green triticale and other green grains is currently commercialized.

6.2 Processing

6.2.1 Traditional Processing

Traditionally, freekeh is produced from green wheat, which is early harvest durum wheat at the milky stage. After harvest, green wheat piles left to dry in open air (24 h). After that, they were carefully scorched on the ground to burn the husks, the chaff and straw. Because the grains are still green and moisturized, they are fire resistant. When the kernel color become darker, the burning is stopped. Roasted grains are, therefore, sun dried, manually threshed, and cracked using stone mill. The final product is characterized by a greenish color and a smoky flavor [31].

6.2.2 Industrial Processing

Conventional Processing: Small-Scale Production

In the conventional method (Fig. 5.3), green wheat is cut, collected in piles and left to dry for 2–4 h in an open place [8]. Then, farmers set piles on fire using the dried remains of other crops. When the grains are slightly blackened, burning is stopped and the spikes are left to cool down. Afterward, the grains are threshed using a threshing machine. Freshly roasted green wheat is characterized by sweet and chewy taste, and roasted flavor. Green wheat can be sold immediately fresh or after drying to reduce its moisture content from 40 to 12%. Drying is critical step, where grains should be dried in less than 1 day to avoid spoilage in the shadow, and not directly under the sun to avoid bleaching. Roasted freekeh production provides an economic advantage for local farmers since it is commonly sold at a price higher than wheat or burghul. Such production, however, is limited because the processing is non-mechanized and time-consuming (Fig. 5.3) [8].

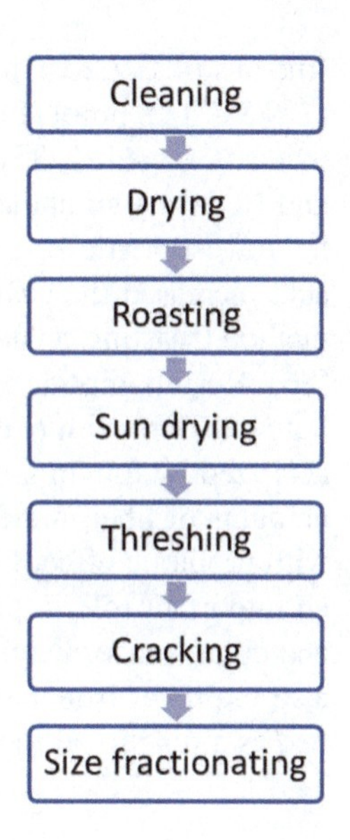

Fig. 5.3 Standard diagram for freekeh-making

Modern Processing: Large-Scale

Mechanization of freekeh production is a continuous process involving four key steps: (1) roasting the collected sheaves by direct exposing to fire in a metal conveyor belt, (2) threshing the roasted sheaves to separate green wheat kernels from other residues, (3) drying the threshed kernels to around 12% moisture content, and (4) cracking the dried kernels using roll machine [32]. Recently, a south Australian company Greenwheat Freekeh™ developed a technology to harvest and treat green wheat to automatically produce freekeh.

6.3 Nutritional Profile

Roasted freekeh had superior nutritional quality compared to rice and wheat in terms of proteins, dietary fiber, minerals (potassium, magnesium, and calcium), and vitamins (A, B1, B2, C, and E) [84]. Nutritional profile may depend on several factors such environmental conditions (e.g. fertilization, climate and location), type or/ and variety of the cereal used, date of harvest, maturation period and production method [8]. Selecting the appropriate harvesting date is the major parameter controlling the nutritional quality [8, 31]. Ozbou et al., (2001) found that fiber, protein and ash contents decreased with increasing maturation time (13, 16, 19, 22, 25 days post anthesis) [85]. Likewise, it was found that immature wheats had rich nutrient composition at early maturity stages such as total dietary fiber (17.3–20.4%), fructans (4.1–7.2%), total phenolics (4602.5–4838.3 mgGAE/kg) and antioxidants (729.2–782.8 μmolTE/100 g) [86]. Moreover, phytic acid increased with increasing maturity stage [31, 85]. When grains are harvested at early stages of maturation (13 and 16 days after anthesis), grains are characterized by the highest fiber content and the lowest phytic acid content [85, 86]. As an antinutrient, the reduction of phytic acid increased the availability of dietary minerals (Ca, Zn, and Fe) [83]. Cooking method (roasting on flames and boiling at atmospheric pressure) was also evaluated [86]. Results reported a reduction in thiamin and riboflavin contents occurred in both cultivars with maturation regardless of cooking method [83]. Freekeh is characterized by low in glycemic index and cholesterol probably attributed to the high amounts of fiber, minerals, antioxidant components, zeaxanthin and lutein as well with probiotic effect [31, 86]. Therefore, the consumption of such a food can play an important role in preventing cardiovascular diseases, diabetic diseases, obesity and colon cancer thanks to its highly nutritious composition [31, 86]. High fructan also improves iron and calcium availability [86]. Therefore, freekeh can be classified as a functional food with highly-beneficial health attributes [85, 87, 88].

7 Safety Challenges

Although traditional foods are becoming mainstream in several European countries, safety is yet to be enlightened. Potential hazards have been reported in cereals derived-foodstuffs, which if inappropriately managed, can seriously threaten consumer healthiness [89–91]. The safety concept can include several aspects such as biological, physical or chemical contamination [92]. A successful hazard identification and safety assessment must cover the potential risks from crop growth, harvesting, preprocessing, processing (e.g. home and restaurants and industrial), postprocessing and home preparation. Undoubtedly, the great development in analytical methods provided valuable tools for safety controlling to stay steps ahead in the public health research.

7.1 Couscous

7.1.1 Microbiological Risks

The occurrence of microbial contamination in cereals can be mainly associated with the presence of *Pseudomonadaceae, Micrococcaceae, Lactobacillaceae* and *Bacillaceae* [93]. Several pathogenic bacteria have been found in durum wheat such as *Salmonella, Escherichia coli* and *Bacillus* [94–97]. More specifically, *Bacillus* are known to be involved in foodborne illnesses due to their spore-forming ability and toxin production (thermolabile and thermostable) (e.g. lichenysin by A.

licheniformis and cereulide toxin by *B. cereus)* [98–100]. Considering these potential risks, microbiological investigations of couscous is necessary for public health or food safety reasons. Indeed, it was identified that couscous is among the most frequent food vehicles of *Bacillus cereus*-induced food-borne in France from 2007 to 2014 [101]. Furthermore, Ziane et al. (2014) isolated aerobic spore-forming bacteria (20 CFU/g) from three Algerian commercial couscous after heat treatment (80 °C, 10 min). The species identified were *Bacillus licheniformis*, *Bacillus cereus* and *Bacillus subtilis*. *Bacillus cereus* is well recognized as causative of food poisoning, with the ability to survive heat treatment during couscous-making [102]. As well, these bacteria can survive cooking temperature and produce toxins in couscous, especially during long waiting time after preparation [102].

Fungal contamination can be divided into types: 'field fungi', which invade and produce their toxins before harvest; and 'storage' fungi [103]. The occurrence of *Aspergillus* and *Penicillium* was mainly observed in grains during storage due to high relative humidity and temperature. Such contamination can severely damage the quality of the grain (e.g. unacceptable odors, grain discoloration, production of free fatty acid and mycotoxin) [103–106]. The presence of *A. flavus* (aflatoxins), *Aspergillus niger, Aspergillus carbonbarious, Penicillium chrysogenum and Penicillium verrucosum* (ochratoxin), *Fusarium oxysporum and Fusarium chlamydosporum* (fumonisins and trichothecenes) can be due to lack of good agriculture and food manufacturing practices throughout the food chain [107].

7.1.2 Mycotoxin

Mycotoxins are secondary metabolites of *Aspergillus, Penicillium* and *Fusarium* species [108], which can severely threaten the health of humans and animals [109]. These metabolites can be carcinogenic, mutagenic, teratogenic, and immunosuppressive fungal metabolites [110–113]. Cereals are susceptible to be contaminated with mycotoxins [99, 114–116] such as durum wheat [117–119].

Mycotoxins may not be completely removed during food processing [106, 120], which explains their natural occurrence in cereals derivates (e.g. bread, pasta, lasagna, biscuits and noodles) [108, 121–123]. Nevertheless, the incidence of mycotoxin in couscous was much less documented.

Careful evaluation throughout adequate sampling and valid analytical methods are required to prevent the report of false positive results [124]. The presence of ochratoxin A was investigated in a small set of cereal-based foods [$n = 11$ (national and imported); couscous ($n = 2$)] available in the Libyan market [125]. The results revealed that ochratoxin A content was within European specifications (<3 μg/kg) in the locally produced couscous (0.59 μg/kg). However, ochratoxin A was found to exceed the limits in thee imported couscous (15.50 μg/kg), macaroni (national production; 10.70 μg/kg), rice (imported; 3.25 μg/kg) and wheat flour [4.80 μg/kg for national production; 1.89 μg/kg for imported]. In another study, a larger set of products ($n = 51$) was considered for possible contamination with mycotoxin [126]. The incidence of *Fusarium* mycotoxins [enniatins (EN A, EN A_1, EN B and EN B_1), beauvericin and fusaproliferin) was determined in cereals (wheat, barley, maize and sorghum) and derivates (pasta and couscous). Enniatins showed high frequency of contamination (96%), while beauvericin and fusaproliferin were not detected. The highest value of enniatins was found in sorghum (683.9 mg/kg). More recently, Zinedine et al. (2017) conducted a multi-mycotoxin evaluation ($n = 22$) of couscous ($n = 96$) consumed in Morocco. Results showed that 98% of total couscous samples were contaminated by at least one mycotoxin. Enniatin B (ENB), Enniatin B1 (ENB1), Enniatin A1 (ENA1) and zearalenone (ZEA) showed the highest incidences in contaminated samples. Total aflatoxins (B1 + B2 + G1 + G2) exceeded the European limits (4 ng/g) in 3.6% of the samples (Zinedine et al., 2017). NIV, ZEA and STG were detected [52.4–462.2 µg/kg; 22.0–132.1 µg/kg; 32.6–273.4 µg/kg, respectively]. DON was also detected with a maximum value of 106.6 µg/kg, which was within the European limits (1750 µg/kg). As for the estimated exposure risk to mycotoxin in couscous, the average Moroccan consumer intakes were sometimes above the tolerable daily intake [26]. Keeping in mind the above mentioned, investigating the fate mycotoxin during cooking couscous will be of great interest to enable a reliable assessment of the exposure.

7.1.3 Heavy Metals

Accumulation of heavy metals in the soils is a serious environmental concern for their properties as non-biodegradable and tendency to accumulation in all media (liquid and solid) [127–130] and a potential risk for public health due to its adverse effect on human health (e.g. cytotoxic, mutagenic and carcinogenic effects) [131–133]. For the human body, traces of heavy metals (e.g. Zn, Cu, and Ni) are essential for biological systems as structural and catalytic components of proteins but may be toxic when safe concentrations are exceeded [134–136]. Other metals (e.g. Cd, Pb, and Cr) are classified toxic independently of their concentrations [134–137]. In the case of cereals, several studies imparted a significant accumulation of heavy metals [130, 134, 138, 139]. Wheat grains acted as accumulator of heavy metals with different bioconcentration factor: Pb (100%) > Cd (89%) > Cr (67%) > Zn (56%) > Cu (44%) > Co (22%) [140]. However, Bermudez et al. (2011) found that the mean concentrations of As, Cd, Ni and Pb were below the tolerance limits in wheat grains, while the other minerals (Cr, Cu, Fe, Mn and Zn) exceeded the limits stated by FAO/WHO. Mathebula et al. (2017) also reported that Cd and Cr were detected in whole grain used for bread production (57–156 µg/kg and 19–63 µg/kg, respectively) [141]. Although mineral accumulation is closely related to genetic factors, soil type and agronomic practices, investigating residual heavy metal in foodstuffs was required due to the high magnitude of cereal based-products consumption [142]. Nevertheless, a scarcity in scientific documentation was found in couscous contamination with heavy metals. Fiłon et al. (2012) evaluated Pb and Cd contamination in a range of cereal products (flour, bread, pasta, rice, bran and couscous) available in the polish market. The highest Cd level was found in pastas (58 ± 33.0 μg/kg) and the highest Pb level was in couscous (120 ± 89.9 μg/kg), which are within the permitted limits by the FAO/WHO (100 μg/kg and 85 μg/kg, respectively) [143].

7.2 *Burghul*

7.2.1 Microbiological Risks

Numerous microbiological methods have been used for the identification of the type and distribution level of fungal pathogens in burghul. Based on conventional method (cultures technique), enumeration of molds and aflatoxigenic (*Aspergillus)* of raw wheat (durum, aestivum and triticale) and the final product (burghul) remained within limits during storage [18]. Correspondingly, sensory quality data revealed that burghul quality was acceptable after 6 months of storage at 37 °C. Stable quality during storage can be due the inactivation of enzymes in burghul [144]. However, the counting values of molds and aflatoxigenic *Aspergillus* in commercial burghul in Brazil was positive (273.3 CFU/g and 133.3 CFU/g, respectively) [145]. Molecular methods such as polymerase chain reaction showed that commercial burghul was heavily contaminated with mycotoxigenic fungi (*F. graminearum)* [146]. This method also enabled the discrimination of 42 monosporic isolates as *Aspergillus flavus* with aflatoxigenic potential in burghul [145], and therefore the occurrence of microbial and fungal in burghul is a serious concern to keep monitoring for preventive measures.

7.2.2 Mycotoxin

The occurrence of deoxynivalenol (DON) in Lebanese burghul was evaluated using high-performance liquid chromatography with ultraviolet light detection [147]. The results revealed that burghul contamination (8.75%) was higher than wheat (7.5%), but without surpassing the permitted limits [1250 μg/kg, Commission Regulation No. 1126/2007]. In this case, although the tolerable daily intake is respected, potential exposition must be monitored to avoid high levels of DON contamination specially for staple food such as burghul. A collection of Turkish products (210 wheat flour and 210 burghul) was screened for aflatoxins contamination using thin layer chromatography scanner methods [148]. Results showed that none of the wheat flour and bulgur samples were found to be positive for aflatoxin contamination. However, Faria et al. (2017) confirmed the presence of aflatoxin B1 and aflatoxin B2 using high-performance liquid chromatography method in burghul purchased from Brazilian market. Furthermore, Turhan et al., (2017) assessed aflatoxin B1, total aflatoxin and ochratoxin A by enzyme-linked immunosorbent assay technique in commercial burghul (*n* = 113). The results showed that aflatoxin B1 contamination was within the limits (≤3 ppb), while total aflatoxin and ochratoxin A surpassed the limits (by 7.08% and 2.65%, respectively) [149]. Therefore, monitoring exposure levels as well as raising consumer and producer's awareness are necessary [148, 149].

7.2.3 Heavy Metals

Rice is reported as one of the main sources of exposure to toxic heavy metals [130, 139, 150], while wheat was more often reported to be less contaminated [134, 151]. Heavy metals accumulation in burghul was, therefore, evaluated as potential alternative to rice. As was detected in rice and burghul ($n = 50$), but average As concentration in bulgur was 10 times lower than rice [152]. Such result might be attributed to the fat that rice is particularly susceptible to As accumulation compared to other cereals as it is generally grown under flooded conditions where As mobility is high [153, 154]. In fact, As concentration in white rice (0.14 mg/kg) was almost 3 times higher than in wheat (0.05 mg/kg) and 14 times higher than in maize (0.01 mg/kg) ($n = 549$; [155]). Consequently, estimated carcinogenic risks using in-effect of burghul (0%) was significantly than that of (53%). Sofuoglu and Sofuoglu (2018) determined a larger number of potentially toxic elements ($n = 9$; Cd, Co, Cr, Cu, Mn, Ni, Pb, Sr, and Zn) in burghul and rice ($n = 50$). The estimations of chronic-toxic risks for Cd and Co, and carcinogenic risks for Pb associated with burghul consumption was bellow respective threshold, whereas all were below those for rice consumption. Although Cu, Mn, and Zn levels were higher in burghul than rice, all associated chronic-toxic risks were below the threshold level. In another study, metals amounts ($n = 12$; Al, As, Cd, Cr, Cu, Fe, Hg, Mn, Ni, Pb, Sn, and Zn) were assessed in burghul [156]. Toxic metal concentrations were low and suitable for daily intake and essentials minerals were within the dietary reference intake (Erdogan et al., 2015). Overall, these studies promoted burghul consumption as a "safer" and "healthier" food than rice based-foods [152, 157].

7.3 Freekeh

Until now, no studies were found covering the contamination of freekeh, probably because commercial production is lunched recently. Roasting is key step in freekeh process diagram, where high temperatures are applied thereby Millard reactions products can be formed. Such process can be associated with processing contamination such as acrylamide. Acrylamide is found in several foods such as potato, coffee and cereal products like bread, cereals, and coffee after processing [158]. Acrylamide, a processing contaminant formed during the cooking or high-temperature processing, is currently among the most pressing problems facing the food industry [159, 160]. Acrylamide is generated during the Maillard reaction involving free reducing sugars and asparagine, at high temperatures above 120 °C [160–162]. The European Commission indicative value of acrylamide levels is between 600 and 1000 mg/kg for potato-based products and 50–500 mg/kg in cereal-based foods (European Commission, 2013). Many preprocessing factors can affect acrylamide content, including variety, crop management and storage conditions [158, 162, 163]. The major factors to control are those associated with the processing (e.g. pH, temperature, reaction time and moisture content) [158, 160]. Therefore, controlling processing contaminants is required beside chemical and microbiological risks.

8 Concluding Remarks and Future Outlooks

Even though north-African traditional cereal-based products (i.e. couscous, burghul and freekeh) were handmade for centuries, nowadays they are commercialized worldwide. Indeed, given their high versatility and suitability to modern lifestyle, they went through the process of industrialization. The shifting from tradition to innovation had several advantages and limitations, which are summarized in Table 5.4. Despite the great technological advancement in equipment's and analytical tools, the management of several operations still rely on the know-how of operators, probably due to the lack of fundamental knowledge in some processing (e.g. granulation, polishing and extrusion) and the behavior of the particles during the processing.

To track the trend, manufacturers are facing challenges to identify the optimal formulation and process to enhance the nutritional profile. To go beyond the trend, the key challenge for innovation are safety and convenience to maintain and expand the market [3]. In the case traditional foods, safety and risk assessment are merely reported in scientific evidence. Probably, their inclusion within the novel food legislation could be an efficient way to highlight the need for further investigation. Noteworthy, the road from tradition to innovation may increase or decrease the potential risks to public health. Beside the common microbiological and chemical contaminants, processing contaminants are currently one of the most critical matters that the food industry is facing. Therefore, hazard identification and exposure assessment are deemed necessary.

Acknowledgments The author would like to acknowledge the important role of Mediterranean women in preserving food tradition and transmitting the foodway knowledge through generations.

Table 5.4 North African traditional cereals-based foods: from tradition to innovation

Processing	Advantages	Disadvantages
Couscous		
Traditional	– Traditional mouthfeel – Economic	Size heterogeneity → non-uniform cooking
Industrial	*Agglomeration technique:* – Traditional mouthfeel	– High recycling rates [43] → fractionating and regrinding are required after drying to reduce oversized particles → extra expenses. – Size heterogeneity → non-uniformity of down-stream processes [50] -non-uniform cooking → sticky or gummy couscous [49]
	Extrusion technique: – Intense yellow color – High degree of starch gelatinization – No sifting step is needed – Uniform particle size → uniform cooking	In case of excessive shearing, particles might be sticky
Burghul		
Traditional	– Traditional mouthfeel – Sun drying is gentle, simple, effective, and the least expensive [8]	– Sun-drying may cause quality degradation and contamination problem + loss of vitamins – Favor mold development and mycotoxin secretions – Consistent good quality cannot be guaranteed [8]
Industrial	*Conventional*: Traditional mouthfeel	– Longtime of tempering (12 h) + size heterogenicity – Batch method and not automated: highly labor-intensive and of low capacity – Cooking under super-atmospheric pressure → high economic costs.
	Modern: Continuous operation +Paraboiling enable quick moistening + size uniformity	– Paraboiling: heat temperature might induce nutrients loss. – Excess of water might cause the rupture of grains. – Polishing: supplementary costs are needed
Freekeh		
Traditional	– Traditional mouthfeel – Sun drying is gentle, simple and the least expensive [8]	– Non-mechanized and slow → time consuming – Sun-drying cannot be guaranteed to provide a consistent good -quality product [8]
Industrial	Size uniformity + roasting is controlled and enable stable product	– Not a completely mechanized process – Adequate for small manufacturers and not large-scale production.

References

1. Dilis V, Vasilopoulou E, Alexieva I, Boyko N, Bondrea A, Fedosov S, et al. Definition and documentation of traditional foods of the Black Sea Area Countries: potential nutrition claims. J Sci Food Agric [Internet]. 2013;93:3473–7. John Wiley & Sons, Ltd. [cited 2020 Apr 10]. Available from: http://doi.wiley.com/10.1002/jsfa.6238
2. Kuhnlein HV, Chan HM. Environment and contaminants in traditional food systems of northern indigenous peoples. Annu Rev Nutr [Internet]. Ann Rev. 2000;20:595–626 [cited 2020 Apr 10]. Available from: http://www.ncbi.nlm.nih.gov/pubmed/10940347.
3. Guerrero L, Guàrdia MD, Xicola J, Verbeke W, Vanhonacker F, Zakowska-Biemans S, et al. Consumer-driven definition of traditional food products and innovation in traditional foods. A qualitative cross-cultural study. Appetite. 2009;52:345–54.
4. EFSA. Novel and traditional food: guidance finalised. Eur Food Saf [Internet]. 2016. [cited 2020 Apr 10]. Available from: https://www.efsa.europa.eu/en/press/news/161110
5. Alimentarius C. Codex Stan 202–1995 Codex Alimentarius Volume 7: Codex Standard for Couscous - IHS, Inc [Internet]. 1995 [cited 2019 Jul 11]. Available from: http://products.ihs.com/Ohsis-SEO/363722.html
6. Coskun F. Production of couscous using the traditional method in Turkey and couscous in the world. Afn J Agric Res. 2013;8:2609–15 [cited 2018 Feb 10]. Available from: http://www.academicjournals.org/AJAR
7. Toufeili I, Olabi A, Shadarevian S, Antoun MA, Zurayk R, Baalbaki I. Relationships of selected wheat parameters to burghul-making quality. J Food Qual [Internet]. 1997;20:211–24 [cited 2018 Nov 9]. Available from: http://doi.wiley.com/10.1111/j.1745-4557.1997.tb00465.x
8. Al-Mahasneh MA, Rababah TM, Bani-Amer MM, Al-Omari NM, Mahasneh MK. Fuzzy and conventional modeling of open sun drying kinetics for roasted green wheat. Int J Food Prop [Internet]. 2013;16:70–80. Taylor & Francis Group; [cited 2019 Jul 11]. Available from: http://www.tandfonline.com/doi/abs/10.1080/10942912.2010.528108
9. Béji-Bécheur A, Ourahmoune N, Özçağlar-Toulouse N. The polysemic meanings of couscous consumption in France. J Consum Behav [Internet]. 2014;13:196–203 [cited 2018 Nov 9]; Available from: http://doi.wiley.com/10.1002/cb.1478
10. Barrena R, García T, Sánchez M. Analysis of personal and cultural values as key determinants of novel food acceptance. Application to an ethnic product. Appetite. 2015;87:205–14.
11. Tey YS, Arsil P, Brindal M, Liew SY, Teoh CT, Terano R. Personal values underlying ethnic food choice: means-end evidence for Japanese food. J Ethn Foods [Internet]. 2018;5:33–9. No longer published by Elsevier; [cited 2019 Jul 1]. Available from: https://www.sciencedirect.com/science/article/pii/S2352618117301877
12. Durmelat S. Introduction: colonial culinary encounters and imperial leftovers. French Cult Stud [Internet]. 2015;26:115–29. SAGE PublicationsSage UK: London, England; [cited 2019 Jul 1]; Available from: http://journals.sagepub.com/doi/10.1177/0957155815572572
13. Huybregts L, Roberfroid D, Lachat C, Van Camp J, Kolsteren P. Validity of photographs for food portion estimation in a rural West African setting. Public Health Nutr [Internet]. 2008;11:581–7 [cited 2019 Jul 1]. Available from: https://www.cambridge.org/core/product/identifier/S1368980007000870/type/journal_article
14. Singh S, Singh N. Effect of debranning on the physico-chemical, cooking, pasting and textural properties of common and durum wheat varieties. Food Res Int. 2010;43:2277–83.
15. Yıldırım A, Bayram M, Öner MD. Bulgur milling using a helical disc mill. J Food Eng. 2008;87:564–70.
16. Celik I, Isik F, Gursoy O. Couscous, a traditional Turkish food product: production method and some applications for enrichment of nutritional value. Int J Food Sci Technol [Internet]. 2004;39:263–9 [cited 2018 Nov 9]. Available from: http://doi.wiley.com/10.1111/j.1365-2621.2004.00780.x
17. Erbaş M, Aykın E, Arslan S, Durak AN. Adsorption behaviour of bulgur. Food Chem. 2016;195:87–90.

18. Bayram M. Modelling of cooking of wheat to produce bulgur. J Food Eng. 2005;71:179–86.
19. Bayram M, Öner MD. Bulgur milling using roller, double disc and vertical disc mills. J Food Eng. 2007;79:181–7.
20. Odabaş Hİ, Koca I. Application of response surface methodology for optimizing the recovery of phenolic compounds from hazelnut skin using different extraction methods. Ind Crop Prod. 2016;91:114–24.
21. Ziane M, Desriac N, Le Chevalier P, Couvert O, Moussa-Boudjemaa B, Leguerinel I. Identification, heat resistance and growth potential of mesophilic spore-forming bacteria isolated from Algerian retail packaged couscous. Food Control [Internet]. 2014;45:16–21. Elsevier; [cited 2018 Feb 15]. Available from: https://www.sciencedirect.com/science/article/pii/S0956713514001868
22. Adil İH, Yener ME, Bayındırlı A. Extraction of total phenolics of sour cherry pomace by high pressure solvent and subcritical fluid and determination of the antioxidant activities of the extracts. Sep Sci Technol [Internet]. 2008;43:1091–110 [cited 2018 Mar 4]. Available from: http://www.tandfonline.com/doi/abs/10.1080/01496390801888243
23. Gil A, Ortega RM, Maldonado J. Wholegrain cereals and bread: a duet of the Mediterranean diet for the prevention of chronic diseases. Public Health Nutr [Internet]. 2011;14:2316–22 [cited 2018 Nov 7]. Available from: http://www.journals.cambridge.org/abstract_S1368980011002576
24. Abdel-Aal E-SM, Rabalski I. Effect of baking on free and bound phenolic acids in wholegrain bakery products. J Cereal Sci. 2013;57:312–8.
25. Yıldırım A, Bayram M, Öner MD. Ternary milling of bulgur with four rollers. J Food Eng. 2008;84:394–9.
26. Zinedine A, Fernández-Franzón M, Mañes J, Manyes L. Multi-mycotoxin contamination of couscous semolina commercialized in Morocco. Food Chem [Internet]. 2017;214:440–6. Elsevier; [cited 2018 Nov 9]. Available from: https://www.sciencedirect.com/science/article/pii/S0308814616311207
27. Bayram M, Öner MD. Determination of applicability and effects of colour sorting system in bulgur production line. J Food Eng. 2006;74:232–9.
28. Tacer Caba Z, Boyacioglu MH, Boyacioglu D. Bioactive healthy components of bulgur. Int J Food Sci Nutr [Internet]. 2012;63:250–6 [cited 2019 Jul 11]. Available from: http://www.ncbi.nlm.nih.gov/pubmed/22136100.
29. Kahyaoglu LN, Sahin S, Sumnu G. Spouted bed and microwave-assisted spouted bed drying of parboiled wheat. Food Bioprod Process. 2012;90:301–8.
30. Corporation GR and D. Modern world market looms large for ancient wheat product - Grains Research and Development Corporation. Grains Research and Development Corporation; 2018 [cited 2019 Jul 11]; Available from: https://grdc.com.au/resources-and-publications/groundcover/ground-cover-issue-101/modern-world-market-looms-large-for-ancient-wheat-product
31. Bayram. An analysis of scorched immature wheat: frekeh. Cereal Foods World [Internet]. 2008 [cited 2019 Jul 11]; Available from: http://www.aaccnet.org/publications/plexus/cfw/pastissues/2008/abstracts/CFW-53-3-0134.html
32. Al-Mahasneh MA, Rababah TM. Effect of moisture content on some physical properties of green wheat. J Food Eng. 2007;79:1467–73.
33. Boucheham N, Benatallah L, Zidoune MN. P102 Couscous à base de riz et de féverole pour malades cœliaques. Cah Nutr Diététique [Internet]. 2011;46:S102. Elsevier Masson; [cited 2018 Feb 15]. Available from: https://www.sciencedirect.com/science/article/pii/S0007996011701869
34. Demir B, Bilgiçli N, Elgün A, Demir MK. The effect of partial substitution of wheat flour with chickpea flour on the technological, nutritional and sensory properties of couscous. J Food Qual [Internet]. 2010;33:728–41. Blackwell Publishing Inc; [cited 2018 Feb 16]. Available from: http://doi.wiley.com/10.1111/j.1745-4557.2010.00359.x

35. Yüksel AN, Öner MD, Bayram M. Usage of undersize bulgur flour in production of short-cut pasta-like couscous. J Cereal Sci. 2017;77:102–9.
36. Barkouti A, Delalonde M, Rondet E, Ruiz T. Structuration of wheat powder by wet agglomeration: case of size association mechanism. Powder Technol. 2014;252:8–13.
37. Bellocq B, Cuq B, Ruiz T, Duri A, Cronin K, Ring D. Impact of fluidized bed granulation on structure and functional properties of the agglomerates based on the durum wheat semolina. Innov Food Sci Emerg Technol. 2018;45:73–83.
38. Hafsa I, Mandato S, Ruiz T, Schuck P, Jeantet R, Mejean S, et al. Impact of the agglomeration process on structure and functional properties of the agglomerates based on the durum wheat semolina. J Food Eng. 2015;145:25–36.
39. Rondet E, Delalonde M, Ruiz T, Desfours JP. Fractal formation description of agglomeration in low shear mixer. Chem Eng J. 2010;164:376–82.
40. Rondet E, Cuq B, Cassan D, Ruiz T. Agglomeration of wheat powders by a novel reverse wet agglomeration process. J Food Eng [Internet]. 2016;173:92–105. Elsevier; [cited 2019 Jul 11]; Available from: https://www.sciencedirect.com/science/article/pii/S0260877415300479
41. Coba de la Peña T, Pueyo JJ. Legumes in the reclamation of marginal soils, from cultivar and inoculant selection to transgenic approaches. Agron Sustain Dev [Internet]. 2012;32:65–91. Springer-Verlag; [cited 2018 Mar 1]. Available from: http://link.springer.com/10.1007/s13593-011-0024-2
42. Bekhouche F, Merabti R, Bailly J-D. "Lemzeiet": Traditional couscous manufacture from fermented wheat (Algeria): process and technological and nutritional quality "Lemzeiet": Traditional couscous manufacture from fermented wheat (Algeria); investigation of the process and estimation of the technological and nutritional quality. J Food Sci Technol Int Res J [Internet]. 2013;4:167–75 [cited 2019 Jul 11]. Available from: https://hal.archives-ouvertes.fr/hal-01431455
43. Saad MM, Barkouti A, Rondet E, Ruiz T, Cuq B. Study of agglomeration mechanisms of food powders: application to durum wheat semolina. Powder Technol. 2011;208:399–408.
44. Roman-Gutierrez AD, Guilbert S, Cuq B. Description of microstructural changes in wheat flour and flour components during hydration by using environmental scanning electron microscopy. LWT - Food Sci Technol [Internet]. 2002; 35:730–40. Academic Press [cited 2019 Jul 11]. Available from: https://www.sciencedirect.com/science/article/pii/S0023643802909321
45. Hébrard A, Oulahna D, Galet L, Cuq B, Abecassis J, Fages J. Hydration properties of durum wheat semolina: influence of particle size and temperature. Powder Technol. 2003;130:211–8.
46. Landillon V, Cassan D, Morel M-H, Cuq B. Flowability, cohesive, and granulation properties of wheat powders. J Food Eng. 2008;86:178–93.
47. Rondet E, Ruiz T, Cuq B. Rheological and mechanical characterization of wet agglomerates processed in low shear mixer. J Food Eng. 2013;117:67–73.
48. Oulahna D, Hebrard A, Cuq B, Abecassis J, Fages J. Agglomeration of durum wheat semolina: thermodynamic approaches for hydration properties measurements. J Food Eng. 2012;109:619–26.
49. Barkouti A, Rondet E, Delalonde M, Ruiz T. Influence of physicochemical binder properties on agglomeration of wheat powder in couscous grain. J Food Eng [Internet]. 2012;111:234–40. Elsevier; [cited 2018 Nov 9]. Available from: https://www.sciencedirect.com/science/article/pii/S0260877412000982
50. Iveson SM, Litster JD, Hapgood K, Ennis BJ. Nucleation, growth and breakage phenomena in agitated wet granulation processes: a review. Powder Technol. 2001;117:3–39.
51. Cuq B, Mandato S, Jeantet R, Saleh K, Ruiz T. 7 – Agglomeration/granulation in food powder production. Handb Food Powders [Internet]. 2013. Elsevier; [cited 2018 Nov 9]. pp. 150–77. Available from: https://linkinghub.elsevier.com/retrieve/pii/B9780857095138500078
52. Bellocq B, Ruiz T, Delaplace G, Duri A, Cuq B. Screening efficiency and rolling effects of a rotating screen drum used to process wet soft agglomerates. J Food Eng. 2017;195:235–46.
53. Mandato S, Rondet E, Delaplace G, Barkouti A, Galet L, Accart P, et al. Liquids' atomization with two different nozzles: modeling of the effects of some processing and formulation conditions by dimensional analysis. Powder Technol. 2012;224:323–30.

54. Wade JB, Martin GP, Long DF. Controlling granule size through breakage in a novel reverse-phase wet granulation process; the effect of impeller speed and binder liquid viscosity. Int J Pharm. 2015;478:439–46.
55. Wade JB, Martin GP, Long DF. Feasibility assessment for a novel reverse-phase wet granulation process: the effect of liquid saturation and binder liquid viscosity. Int J Pharm. 2014;475:450–61.
56. Ahmadian-Kouchaksaraie Z, Niazmand R. Supercritical carbon dioxide extraction of antioxidants from Crocus sativus petals of saffron industry residues: optimization using response surface methodology. J Supercrit Fluids. 2017;121:19–31.
57. Jacob M. Chapter 9 granulation equipment. Handb Powder Technol. 2007;11:417–76.
58. Ji J, Fitzpatrick J, Cronin K, Fenelon MA, Miao S. The effects of fluidised bed and high shear mixer granulation processes on water adsorption and flow properties of milk protein isolate powder. J Food Eng. 2017;192:19–27.
59. Morin G, Briens L. A comparison of granules produced by high-shear and fluidized-bed granulation methods. AAPS PharmSciTech [Internet]. 2014;15:1039–48 [cited 2018 Nov 9]. Available from: http://link.springer.com/10.1208/s12249-014-0134-7
60. Ruiz T, Delalonde M, Bataille B, Baylac G, de Crescenzo CD. Texturing unsaturated granular media submitted to compaction and kneading processes. Powder Technol. 2005;154:43–53.
61. Turfani V, Narducci V, Durazzo A, Galli V, Carcea M. Technological, nutritional and functional properties of wheat bread enriched with lentil or carob flours. LWT - Food Sci Technol [Internet]. 2017;78:361–6. Academic Press; [cited 2018 Mar 1]. Available from: https://www.sciencedirect.com/science/article/pii/S0023643816308076
62. Aguilar N, Albanell E, Miñarro B, Capellas M. Chickpea and tiger nut flours as alternatives to emulsifier and shortening in gluten-free bread. LWT - Food Sci Technol [Internet]. 2015;62:225–32 [cited 2018 Mar 1]. Available from: http://linkinghub.elsevier.com/retrieve/pii/S0023643814008342
63. Ferreira SMR, de Mello AP, de Caldas Rosa dos Anjos M, Krüger CCH, Azoubel PM, de Oliveira Alves MA. Utilization of sorghum, rice, corn flours with potato starch for the preparation of gluten-free pasta. Food Chem. 2016;191:147–51.
64. Foschia M, Horstmann SW, Arendt EK, Zannini E. Legumes as functional ingredients in gluten-free bakery and Pasta products. Annu Rev Food Sci Technol [Internet]. 2017;8:75–96. Available from: http://www.annualreviews.org/doi/10.1146/annurev-food-030216-030045
65. Demir MK, Demir B. Utilisation of buckwheat (Fagopyrum esculentum M.) and different legume flours in traditional couscous production in Turkey. Qual Assur Saf Crop Foods. 2016;7:321–326.
66. Hama-Ba F, Silga P, Diawara B. Evaluation de la qualité et de l'acceptabilité de couscous à base de trois formulations de farines composites enrichies au soja (*Glycine max*) et au moringa (*Moringa oleifera*). Int J Biol Chem Sci. 2017;10:2497.
67. Yüksel AN, Öner MD, Bayram M. Usage of undersize bulgur flour in production of short-cut pasta-like couscous. J Cereal Sci [Internet]. 2017;77:102–9. Academic Press; [cited 2018 Feb 15]. Available from: https://www.sciencedirect.com/science/article/pii/S0733521017301364
68. Rahmani N, Muller HG. The fate of thiamin and riboflavin during the preparation of couscous. Food Chem [Internet]. 1996;55:23–7. Elsevier [cited 2018 Feb 15]. Available from: https://www.sciencedirect.com/science/article/pii/0308814695000658
69. Carcea M, Narducci V, Turfani V, Giannini V. Polyphenols in raw and cooked cereals/pseudocereals/legume pasta and couscous. Foods [Internet]. 2017;6:80 [cited 2019 Jul 11]. Available from: http://www.ncbi.nlm.nih.gov/pubmed/28892013
70. Alp H, Bilgiçli N. Effect of transglutaminase on some properties of cake enriched with various protein sources. J Food Sci [Internet]. 2008;73:S209–14 [cited 2018 Mar 4]. Available from: http://doi.wiley.com/10.1111/j.1750-3841.2008.00760.x
71. Bayram M, Öner MD, Kaya A. Influence of soaking on the dimensions and colour of soybean for bulgur production. J Food Eng. 2004;61:331–9.

72. Hayta M, Alpaslan M, Cakmakli U. Physicochemical and Sensory Properties of Soymilk-incorporated Bulgur. J Food Sci [Internet]. 2003;68:2800–3 [cited 2018 Nov 9]. Available from: http://doi.wiley.com/10.1111/j.1365-2621.2003.tb05808.x
73. Yorgancilar M, Bilgiçli N. Chemical and nutritional changes in bitter and sweet lupin seeds (Lupinus albus L.) during bulgur production. J Food Sci Technol [Internet]. 2014;51:1384–9 [cited 2019 Feb 4]. Available from: http://link.springer.com/10.1007/s13197-012-0640-0
74. Bayram M, Öner MD. Stone, disc and hammer milling of bulgur. J Cereal Sci. 2005;41:291–6.
75. Kahyaoglu LN, Sahin S, Sumnu G. Physical properties of parboiled wheat and bulgur produced using spouted bed and microwave assisted spouted bed drying. J Food Eng. 2010;98:159–69.
76. Savas K, Basman A. Infrared drying: a promising technique for bulgur production. J Cereal Sci. 2016;68:31–7.
77. Hayta M. Bulgur quality as affected by drying methods. J Food Sci [Internet]. 2002;67:2241–4. John Wiley & Sons, Ltd (10.1111); [cited 2019 Jul 11]. Available from: http://doi.wiley.com/10.1111/j.1365-2621.2002.tb09534.x
78. Mousia Z, Edherly S, Pandiella SS, Webb C. Effect of wheat pearling on flour quality. Food Res Int [Internet]. 2004;37:449–59. Elsevier; [cited 2019 Jul 11]. Available from: https://www.sciencedirect.com/science/article/pii/S0963996904000481
79. Bayram M, Öner MD. Bulgur milling using roller, double disc and vertical disc mills. J Food Eng [Internet]. 2007;79:181–7 [cited 2018 Nov 9]. Available from: http://linkinghub.elsevier.com/retrieve/pii/S026087740600121X
80. Balcı F, Bayram M. Modification of mechanical polishing operation using preheating systems to improve the bulgur color. J Cereal Sci. 2017;75:108–15.
81. Andersson J. Whole grain wheat - effects of peeling and pearling on chemical composition, taste and colour; 2011
82. Kadakal Ç, Ekinci R, Yapar A. The effect of cooking and drying on the water-soluble vitamins content of bulgur. Food Sci Technol Int [Internet]. 2007;13:349–54 [cited 2018 Nov 9]. Available from: http://journals.sagepub.com/doi/10.1177/1082013207085688
83. Özboy Ö, Köksel H. Preliminary communication dietary fiber content of bulgur as affected by wheat variety. Acta Aliment [Internet]. 2001;30:407–14 [cited 2018 Nov 9]. Available from: http://www.akademiai.com/doi/abs/10.1556/AAlim.30.2001.4.9
84. Bird A, Stevens MJ. Toward an emergent global culture and the effects of globalization on obsolescing national cultures. J Int Manag. 2003;9:395–407.
85. Özboy Ö, Özkaya B, Özkaya H, Köksel H. Effects of wheat maturation stage and cooking method on dietary fiber and phytic acid contents of firik, a wheat-based local food. Nahrung/Food [Internet]. John Wiley & Sons, Ltd; 2001;45:347 [cited 2019 Jul 11]. Available from: http://doi.wiley.com/10.1002/1521-3803%2820011001%2945%3A5%3C347%3A%3AAID-FOOD347%3E3.0.CO%3B2-T
86. Özkaya B, Turksoy S, Özkaya H, Baumgartner B, Özkeser İ, Köksel H. Changes in the functional constituents and phytic acid contents of firiks produced from wheats at different maturation stages. Food Chem [Internet]. 2018;246:150–5 [cited 2019 Jul 11]. Available from: http://www.ncbi.nlm.nih.gov/pubmed/29291833.
87. Köksel H, Edney MJ, Özkaya B. Barley bulgur: effect of processing and cooking on chemical composition. J Cereal Sci [Internet]. 1999;29:185–90. Academic Press; [cited 2018 Nov 9]. Available from: https://www.sciencedirect.com/science/article/pii/S0733521098902302
88. Maskan M, İbanoğlu Ş. Hot air drying of cooked and uncooked tarhana dough, a wheat flour-yoghurt mixture. Eur Food Res Technol [Internet]. 2002;215:413–8 [cited 2018 Nov 9]. Available from: http://link.springer.com/10.1007/s00217-002-0572-4
89. Alldrick AJ. Food safety aspects of grain and cereal product quality. Cereal Grains [Internet]. 2017;393–424. Woodhead Publishing; [cited 2019 Jul 11]. Available from: https://www.sciencedirect.com/science/article/pii/B9780081007198000152
90. Wynants E, Frooninckx L, Van Miert S, Geeraerd A, Claes J, Van Campenhout L. Risks related to the presence of Salmonella sp. during rearing of mealworms (Tenebrio molitor)

for food or feed: survival in the substrate and transmission to the larvae. Food Control. 2019;100:227–34.
91. Lähteenmäki L. Claiming health in food products. Food Qual Prefer. 2013;27:196–201.
92. Losio MN, Dalzini E, Pavoni E, Merigo D, Finazzi G, Daminelli P. A survey study on safety and microbial quality of "gluten-free" products made in Italian pasta factories. Food Control. 2017;73:316–22.
93. Laca A, Mousia Z, Díaz M, Webb C, Pandiella SS. Distribution of microbial contamination within cereal grains. J Food Eng [Internet]. 2006;72:332–8 [cited 2018 Nov 9]. Available from: http://linkinghub.elsevier.com/retrieve/pii/S0260877405000154
94. Corsetti A, Settanni L, Braga TM, de Fatima Silva Lopes M, Suzzi G. An investigation of the bacteriocinogenic potential of lactic acid bacteria associated with wheat (Triticum durum) kernels and non-conventional flours. LWT - Food Sci Technol. 2008;41(11):1173–82.
95. Vermeulen P, Fernandez Pierna JA, Abbas O, Rogez H, Davrieux F, Baeten V. Authentication and traceability of agricultural and food products using vibrational spectroscopy. Food Traceabil Auth [Internet]. 2017. Boca Raton, FL: CRC Press. | Series: Food biology series | "A science publishers book.": CRC Press; [cited 2019 Aug 4]. p. 298–331. Available from: https://www.taylorfrancis.com/books/9781498788434/chapters/10.1201/9781351228435-16
96. Khande R, Sharma SK, Ramesh A, Sharma MP. Zinc solubilizing Bacillus strains that modulate growth, yield and zinc biofortification of soybean and wheat. Rhizosphere. 2017;4:126–38.
97. Gill A, Carrillo C, Hadley M, Kenwell R, Chui L. Bacteriological analysis of wheat flour associated with an outbreak of Shiga toxin-producing Escherichia coli O121. Food Microbiol. 2019;82:474–81.
98. Guinebretière M-H, Thompson FL, Sorokin A, Normand P, Dawyndt P, Ehling-Schulz M, et al. Ecological diversification in the Bacillus cereus Group. Environ Microbiol [Internet]. 2008;10:851–65 [cited 2018 Nov 9]. Available from: http://doi.wiley.com/10.1111/j.1462-2920.2007.01495.x
99. Juan C, Ritieni A, Mañes J. Occurrence of Fusarium mycotoxins in Italian cereal and cereal products from organic farming. Food Chem. 2013;141:1747–55.
100. Arroyo-Manzanares N, Huertas-Pérez JF, Gámiz-Gracia L, García-Campaña AM. Simple and efficient methodology to determine mycotoxins in cereal syrups. Food Chem. 2015;177:274–9.
101. Pairaud S, Grout J, Vignaud M-L, Cadel-Six S, Herbin S, Glasset B, et al. Bacillus cereus-induced food-borne outbreaks in France, 2007 to 2014: epidemiology and genetic characterisation. Eur Secur. 2016;
102. Sadiq FA, Li Y, Liu T, Flint S, Zhang G, Yuan L, et al. The heat resistance and spoilage potential of aerobic mesophilic and thermophilic spore forming bacteria isolated from Chinese milk powders. Int J Food Microbiol. 2016;238:193–201.
103. Senthilkumar T, Jayas DS, White NDG, Fields PG, Gräfenhan T. Detection of ochratoxin A contamination in stored wheat using near-infrared hyperspectral imaging. Infrared Phys Technol. 2016;65:30–9.
104. da Luz SR, Pazdiora PC, Dallagnol LJ, Dors GC, Chaves FC. Mycotoxin and fungicide residues in wheat grains from fungicide-treated plants measured by a validated LC-MS method. Food Chem Elsevier Ltd. 2017;220:510–6.
105. Savi GD, Piacentini KC, Rocha LO, Carnielli-Queiroz L, Furtado BG, Scussel R, et al. Incidence of toxigenic fungi and zearalenone in rice grains from Brazil. Int J Food Microbiol. Elsevier B.V. 2018;270:5–13.
106. Piacentini KC, Rocha LO, Savi GD, Carnielli-Queiroz L, Almeida FG, Minella E, et al. Occurrence of deoxynivalenol and zearalenone in brewing barley grains from Brazil. Mycotoxin Res Springer Verlag. 2018;34:173–8.
107. Pestka JJ. Deoxynivalenol: mechanisms of action, human exposure, and toxicological relevance. Arch Toxicol. 2010;64:663–79.

108. Vidal A, Mengelers M, Yang S, De Saeger S, De Boevre M. Mycotoxin biomarkers of exposure: a comprehensive review. Compr Rev Food Sci Food Saf Blackwell Publishing Inc. 2018;17:1127–55.
109. Dellafiora L, Dall'Asta C. Forthcoming challenges in mycotoxins toxicology research for safer food-a need for multi-omics approach. Toxins (Basel). MDPI AG. 2017;9:18.
110. Tittlemier SA, Roscoe M, Blagden R, Kobialka C. Occurrence of ochratoxin A in Canadian wheat shipments, 2010-12. Food Addit Contam A Chem Anal Control Expo Risk Assess. 2014;31:910–6.
111. Pusztahelyi T, Holb IJ, Pócsi I. Secondary metabolites in fungus-plant interactions. Front Plant Sci Front Res Found. 2015;6:573.
112. El Darra N, Gambacorta L, Solfrizzo M. Multimycotoxins occurrence in spices and herbs commercialized in Lebanon. Food Control. 2019;95:63–70.
113. Stanciu O, Juan C, Miere D, Berrada H, Loghin F, Mañes J. First study on trichothecene and zearalenone exposure of the Romanian population through wheat-based products consumption. Food Chem Toxicol. 2018;121:336–42.
114. Bensassi F, Zaied C, Abid S, Hajlaoui MR, Bacha H. Occurrence of deoxynivalenol in durum wheat in Tunisia. Food Control. 2010;21:281–5.
115. Lee HJ, Kim S, Suh HJ, Ryu D. Effects of explosive puffing process on the reduction of ochratoxin A in rice and oats. Food Control [Internet]. 2019;95:334–8. Elsevier; [cited 2019 Jan 4]. Available from: https://www.sciencedirect.com/science/article/pii/S0956713518304006
116. Meneely JP, Ricci F, van Egmond HP, Elliott CT. Current methods of analysis for the determination of trichothecene mycotoxins in food. TrAC Trends Anal Chem. 2011;30:192–203.
117. Blesa J, Moltó J-C, El Akhdari S, Mañes J, Zinedine A. Simultaneous determination of Fusarium mycotoxins in wheat grain from Morocco by liquid chromatography coupled to triple quadrupole mass spectrometry. Food Control. 2014;46:1–5.
118. Ennouari A, Sanchis V, Marín S, Rahouti M, Zinedine A. Occurrence of deoxynivalenol in durum wheat from Morocco. Food Control. 2013;32:115–8.
119. Juan C, Covarelli L, Beccari G, Colasante V, Mañes J. Simultaneous analysis of twenty-six mycotoxins in durum wheat grain from Italy. Food Control. 2016;62:322–9.
120. Olagunju O, Mchunu N, Durand N, Alter P, Montet D, Ijabadeniyi O. Effect of milling, fermentation or roasting on water activity, fungal growth, and aflatoxin contamination of Bambara groundnut (Vigna subterranea (L.) Verdc). LWT. 2018;98:533–9.
121. Oteiza JM, Khaneghah AM, Campagnollo FB, Granato D, Mahmoudi MR, Sant'Ana AS, et al. Influence of production on the presence of patulin and ochratoxin A in fruit juices and wines of Argentina. LWT. 2017;80:200–7.
122. Mousavi Khaneghah A, Fakhri Y, Gahruie HH, Niakousari M, Sant'Ana AS. Mycotoxins in cereal-based products during 24 years (1983–2017): a global systematic review. Trends Food Sci Technol [Internet]. 2019;91:95–105. Elsevier [cited 2019 Jul 11]. Available from: https://www.sciencedirect.com/science/article/pii/S0924224418305715
123. Giménez I, Blesa J, Herrera M, Ariño A. Effects of bread making and wheat germ addition on the natural deoxynivalenol content in bread. Toxins (Basel) [Internet]. 2014;6:394–401. Multidisciplinary Digital Publishing Institute (MDPI); [cited 2018 Mar 4]. Available from: http://www.ncbi.nlm.nih.gov/pubmed/24451845.
124. Whitaker TB. Sampling foods for mycotoxins. Food Addit Contam [Internet]. 2006;23:50–61 [cited 2019 Jul 11]. Available from: http://www.tandfonline.com/doi/abs/10.1080/02652030500241587
125. Sassi AA, Sowan AR, Barka MA, Zgheel FS. Presence of ochratoxin A in human urine from Al-Jafara region, Libya: A preliminary study. J Basic Appl Mycol. 2010;1:35–8.
126. Oueslati S, Romero-González R, Lasram S, Frenich AG, Vidal JLM. Multi-mycotoxin determination in cereals and derived products marketed in Tunisia using ultra-high performance liquid chromatography coupled to triple quadrupole mass spectrometry. Food Chem Toxicol. 2012;50:2376–81.

127. Mahohi A, Raiesi F. Functionally dissimilar soil organisms improve growth and Pb/Zn uptake by Stachys inflata grown in a calcareous soil highly polluted with mining activities. J Environ Manag. 2019;247:780–9.
128. Lim J-H, Cui M-C, Moon D-H, Khim J-H. Stabilization of heavy metal contaminated soil amended with waste cow bone. J Environ Sci [Internet]. 2010;19:255–60 [cited 2018 Mar 2]. Available from: http://koreascience.or.kr/journal/view.jsp?kj=KHGGB3&py=2010&vnc=v19n2&sp=255
129. Zhang X, Wang H, He L, Lu K, Sarmah A, Li J, et al. Using biochar for remediation of soils contaminated with heavy metals and organic pollutants. Environ Sci Pollut Res. 2013;20:8472–83.
130. Clemens S, Ma JF. Toxic heavy metal and metalloid accumulation in crop plants and Foods. Annu Rev Plant Biol Ann Rev. 2016;67:489–512.
131. Song D, Zhuang D, Jiang D, Fu J, Wang Q. Integrated health risk assessment of heavy metals in Suxian county, South China. Int J Environ Res Public Health. MDPI AG. 2015;12:7100–17.
132. Chang CY, Yu HY, Chen JJ, Li FB, Zhang HH, Liu CP. Accumulation of heavy metals in leaf vegetables from agricultural soils and associated potential health risks in the Pearl River Delta, South China. Environ Monit Assess Kluwer Academic Publishers. 2014;186:1547–60.
133. Yao A, Wang Y, Ling X, Chen Z, Tang Y, Qiu H, et al. Effects of an iron-silicon material, a synthetic zeolite and an alkaline clay on vegetable uptake of As and Cd from a polluted agricultural soil and proposed remediation mechanisms. Environ Geochem Health Springer Netherlands. 2017;39:353–67.
134. Bermudez GMA, Jasan R, Plá R, Pignata ML. Heavy metal and trace element concentrations in wheat grains: assessment of potential non-carcinogenic health hazard through their consumption. J Hazard Mater [Internet]. 2011;193:264–71 [cited 2019 Sep 1]. Available from: http://www.ncbi.nlm.nih.gov/pubmed/21835546.
135. Chen D, Zhang H, Li Y, Pang Y, Yin Z, Sun H, et al. Spontaneous formation of noble- and heavy-metal-free alloyed semiconductor quantum rods for efficient photocatalysis. Adv Mater Wiley-VCH Verlag. 2018;30:1803351.
136. Bi C, Zhou Y, Chen Z, Jia J, Bao X. Heavy metals and lead isotopes in soils, road dust and leafy vegetables and health risks via vegetable consumption in the industrial areas of Shanghai, China. Sci Total Environ [Internet]. 2018;619–620:1349–57 [cited 2019 Sep 1]. Available from: http://www.ncbi.nlm.nih.gov/pubmed/29734612.
137. Bermudez GMA, Moreno M, Invernizzi R, Plá R, Pignata ML. Heavy metal pollution in topsoils near a cement plant: the role of organic matter and distance to the source to predict total and hcl-extracted heavy metal concentrations. Chemosphere [Internet]. 2010;78:375–81 [cited 2019 Sep 1]. Available from: http://www.ncbi.nlm.nih.gov/pubmed/19962174.
138. Brunetti G, Farrag K, Soler-Rovira P, Ferrara M, Nigro F, Senesi N. Heavy metals accumulation and distribution in durum wheat and barley grown in contaminated soils under Mediterranean field conditions. J Plant Interact. 2012;7:160–74.
139. Mihucz VG, Silversmit G, Szalóki I, de Samber B, Schoonjans T, Tatár E, et al. Removal of some elements from washed and cooked rice studied by inductively coupled plasma mass spectrometry and synchrotron based confocal micro-X-ray fluorescence. Food Chem. 2010;121:290–7.
140. Sharma S, Nagpal AK, Kaur I. Heavy metal contamination in soil, food crops and associated health risks for residents of Ropar wetland, Punjab, India and its environs. Food Chem [Internet]. 2018;255:15–22 [cited 2019 Sep 1]. Available from: http://www.ncbi.nlm.nih.gov/pubmed/29571461.
141. Mathebula MW, Mandiwana K, Panichev N. Speciation of chromium in bread and breakfast cereals. Food Chem Elsevier Ltd. 2017;217:655–9.
142. Esmaili A, Noroozi Karbasdehi V, Saeedi R, Javad Mohammadi M, Sobhani T, Dobaradaran S. Data on heavy metal levels (Cd, Co, and Cu) in wheat grains cultured in Dashtestan County, Iran. Data Br. Elsevier Inc. 2017;14:543–7.

143. Fiłon J, Ustymowicz-Farbiszewska J, Górski J, Karczewski J. Contamination of cereal products with lead and cadmium as a risk factor to health of the population in the province of podlasie (województwo podlaskie). J Elemntol Polish Society for Magnesium Research; 2012;
144. Bayram M, Öner MD, Eren S. Effect of cooking time and temperature on the dimensions and crease of the wheat kernel during bulgur production. J Food Eng. 2004;64:43–51.
145. Faria CB, dos Santos FÇ, de Castro FF, Sutil AR, Sergio LM, Silva MV, et al. Occurrence of toxigenic Aspergillus flavus in commercial bulgur wheat. Food Sci Technol. Sociedade Brasileira de Ciencia e Tecnologia de Alimentos, SBCTA. 2017;37:103–11.
146. Faria CB, Almeida-Ferreira GC, Gagliardi KB, Alves TCA, Tessmann DJ, Machinski Junior M, et al. Use of the polymerase chain reaction for detection of Fusarium graminearum in bulgur wheat. Food Sci Technol. FapUNIFESP (SciELO). 2012;32:201–8.
147. Antonios D, Guitton V, Darrozes S, Pallardy M, Azouri H. Monitoring the levels of deoxynivalenol (DON) in cereals in Lebanon and validation of an HPLC/UV detection for the determination of DON in crushed wheat (bulgur). Food Addit Contam Part B, Surveill [Internet]. 2010;3:45–51 [cited 2019 Sep 1]. Available from: http://www.ncbi.nlm.nih.gov/pubmed/24785315.
148. Oguz PDH, Aydın H. Determination of aflatoxin existence in mixed feed, wheat flour and bulgur samples. Eur J Vet Sci. 2011;27(3):171–5.
149. Turhan İ, Büyükünal SK, Şakar FŞ. Occurrence of aflatoxin B1, total aflatoxin and ochratoxin a in bulgur commercialized in Turkey. Acta Alimentar. 2015;49:118–24.
150. Ye XX, Sun B, Yin YL. Variation of as concentration between soil types and rice genotypes and the selection of cultivars for reducing as in the diet. Chemosphere. 2012;87:384–9.
151. Guo G, Lei M, Chen T, Yang J. Evaluation of different amendments and foliar fertilizer for immobilization of heavy metals in contaminated soils. J Soils Sediments Springer Verlag. 2018;18:239–47.
152. Sofuoglu SC, Güzelkaya H, Akgül Ö, Kavcar P, Kurucaovalı F, Sofuoglu A. Speciated arsenic concentrations, exposure, and associated health risks for rice and bulgur. Food Chem Toxicol [Internet]. 2014;64:184–91. Pergamon [cited 2018 Nov 9]. Available from: https://www.sciencedirect.com/science/article/pii/S0278691513007916
153. Williams PN, Raab A, Feldmann J, Meharg AA. Market Basket Survey Shows Elevated Levels of As in South Central U.S. Processed Rice Compared to California: Consequences for Human Dietary Exposure. Environ Sci Technol [Internet]. 2007;41:2178–83 [cited 2018 Nov 9]. Available from: http://pubs.acs.org/doi/abs/10.1021/es061489k
154. Williams PN, Villada A, Deacon C, Raab A, Figuerola J, Green AJ, et al. Greatly enhanced arsenic shoot assimilation in rice leads to elevated grain levels compared to wheat and barley. Environ Sci Technol [Internet]. 2007;41:6854–9 [cited 2018 Nov 9]. Available from: http://pubs.acs.org/doi/abs/10.1021/es070627i
155. Adomako EE, Williams PN, Deacon C, Meharg AA. Inorganic arsenic and trace elements in Ghanaian grain staples. Environ Pollut. 2011;159:2435–42.
156. Sait E, Isil A, Zahir DM, Firat A, Serdar ÇK, Osman A, Hamamci C. Simultaneous multielement determination of Al, As, Cd, Cr, Cu, Fe, Hg, Mn, Ni, Pb, Sn, and Zn in bulgur wheat by ICP-OES. At Spectrosc. 2015;36:210–5.
157. Sofuoglu SC, Sofuoglu A. An exposure-risk assessment for potentially toxic elements in rice and bulgur. Environ Geochem Health [Internet]. 2018;40:987–98 [cited 2019 Sep 1]. Available from: http://www.ncbi.nlm.nih.gov/pubmed/28397064.
158. Purlis E. Browning development in bakery products – a review. J Food Eng. 2010;99:239–49.
159. Anese M, Suman M, Nicoli MC. Technological strategies to reduce acrylamide levels in heated Foods. Food Eng Rev. Springer Science and Business Media LLC. 2009;1:169–79.
160. Wang Y, Song C, Yu X, Liu L, Han Y, Chen J, et al. Thermo-responsive hydrogels with tunable transition temperature crosslinked by multifunctional graphene oxide nanosheets. Compos Sci Technol. 2017;151:139–46.
161. Mariotti M, Granby K, Fromberg A, Risum J, Agosin E, Pedreschi F. Furan occurrence in starchy food model systems processed at high temperatures: effect of ascorbic acid and

heating conditions. J Agric Food Chem [Internet]. 2012;60:10162–9 [cited 2019 Jan 14]. Available from: http://pubs.acs.org/doi/10.1021/jf3022699
162. Halder A, Dhall A, Datta AK. An improved, easily implementable, porous media based model for deep-fat frying: part i: model development and input parameters. Food Bioprod Process [Internet]. 2007;85:209–19 [cited 2019 Jan 4]. Available from: https://linkinghub.elsevier.com/retrieve/pii/S0960308507705989
163. Zamani E, Shaki F, AbedianKenari S, Shokrzadeh M. Acrylamide induces immunotoxicity through reactive oxygen species production and caspase-dependent apoptosis in mice splenocytes via the mitochondria-dependent signaling pathways. Biomed Pharmacother. 2017;94:523–30.

Chapter 6
Cereals and Pulses: A Duet of the Mediterranean Diet for a Healthier Future

Ozge Kurt Gokhisar and Mahir Turhan

Abstract The Mediterranean diet (MD) is known to be one of the healthiest dietary patterns due to its plant-based consumption, where vegetables, fruits, cereals (preferably as whole grain) and pulses consumed in high amount and frequency. Besides, MD has been a permanent laboratory for humankind to show the ability to combine pulses and cereals in foods preparation. Mediterranean people consume cereal-pulse based composites for thousands of years in their diet. They have derived many ways to consume them together such as soups, salads, pastas, pilafs, main dishes etc. In this light, this chapter aims to provide the reader an overview on the importance of cereals, pulses and their blends in MD through (1) defining and characterizing the elements of MD; (2) focusing on the importance, composition and health benefits of pulses and cereals in the MD; (3) addressing the applications of cereal-legume based composites in culinary applications; and (4) discussing the challenges and opportunities of cereal/pulse blends for designing new products.

Keywords Cereals · Pulses · Health benefits · Food design · Mediterranean diet

1 Introduction

The principal challenge for the food and agricultural sector is to provide simultaneously enough food, in quantity and quality, for present and future generations to meet nutritional needs and to conserve natural resources. Food and Agriculture Organization (FAO) estimates that to satisfy the needs of a growing and richer population with an increased demand for animal products, food production will have to increase by 60% towards 2050 [1]. This figure can be reduced by improving production efficiency, changing diets and decreasing food losses and waste. In this perspective, the Mediterranean Diet (MD) might be a valuable alternative, which has been well characterized scientifically and accepted as a healthier and more

O. K. Gokhisar (✉) · M. Turhan
Department of Food Engineering, University of Mersin, Mersin, Turkey

F. Boukid (ed.), *Cereal-Based Foodstuffs: The Backbone of Mediterranean Cuisine*, https://doi.org/10.1007/978-3-030-69228-5_6

sustainable dietary pattern than several occidental diets [2–4]. For these reasons, the Mediterranean diet was recognized by United Nations Educational, Scientific and Cultural Organization (UNESCO) as an intangible cultural heritage [5].

MD is a cultural model that involves the ways of how foods are selected, produced, processed and consumed, such as the prioritization of fresh, local, and seasonal food, culinary activities and socialization at meals, regular physical activity, rest in the form of afternoon naps, and a whole way of life that is part of the cultural heritage of the Mediterranean countries [4]. Thus, Mediterranean countries share a number of geographical and climatic factors that have favored this common cultural and agri-food framework. Moreover, the MD represents a sustainable dietary pattern in which nutrition, food, environment, culture, and sustainability all interact, and giving rise to a new model of sustainable diet that takes not only human health into account but also the environment [2, 6].

The traditional MD dates back to ancient times (5000–2000 BC) when the key defining ingredients were the trio of whole grains, wine and olive oil, as well as wild edible leafy greens and pulses. The traditional MD (Archetypal Cretan MD) was first described in the scientific literature by Ancel Keys and co-workers [3], who investigated the link between diet, lifestyle behaviors and the causes of death from heart disease in the seven different countries. The researchers found that the group from Crete had almost no deaths from heart disease after 15 years of follow-up, which was linked to their plant-based olive oil rich diet. The traditional MD has become the most researched dietary pattern worldwide today. The results of 13 meta-analyses of observational studies and 16 meta-analyses of randomized controlled trials investigating the link between Mediterranean diet and 37 different health outcomes (covering more than 12.8 million people) has also showed that the adherence to a MD is protective against deaths from non-communicable diseases such as heart disease, stroke, diabetes, some cancers, neurodegenerative diseases and Alzheimer's and other dementias [7]. The Mediterranean dietary pattern was awarded the best diet of 2019 in terms of healthy eating, as an emerging plant-based diet, and as the best diet for diabetes and the easiest diet to follow by the US News and World Report recently [8]. MD is based on the abundant use of olive oil as the major culinary fat, a high consumption of plant foods (nuts, fruits, vegetables, pulses and whole grains), moderate to high fish consumption and a moderate consumption of wine. In contrast, consumption of red meats, processed meats, meat products and butter is very low [4]. This pattern fits well with the current paradigm of assessing overall food patterns instead of isolated foods or nutrients. The profile of nutrient intake of the Mediterranean diet includes a high monounsaturated: saturated fat ratio, a high intake of alpha-linolenic acid, moderate alcohol intake, high intakes of fiber, vitamins, folate and natural antioxidants, and a low intake of animal protein [9]. Especially pulses, cereals and their combined use, which are considered key features maintaining healthy and sustainable diet.

Therefore, this chapter aims to provide an overview on the importance of pulses, cereals and their blends in MD through (1) defining and characterizing the elements of MD; (2) focusing on the importance, composition and health benefits of cereals and pulses in the MD; (3) addressing the applications of cereal-pulse based composites in culinary applications; and (4) discussing the challenges and opportunities to develop new food products to feed the future.

2 Pulses and Cereals in MD: Importance, Health Promoting Effects and their Complementary Applications

Mediterranean dietary pattern is plant-centric and comes into prominence thanks to its abundant and combined usage of minimally processed whole grains (such as whole wheat, brown rice, oats and bulgur) and pulses (dried beans, lentils and peas) [10]. Plant-based foods including cereals and pulses, positioned at the base of the Mediterranean Food Pyramid, provide key nutrients and protective substances that contribute into the general well-being through maintaining a balanced diet [10, 11]. Papanikolaou and Fulgoni [11] recommended the consumption of pulses and cereals to get advantage of their nutritious composition rich in macronutrients, micronutrients, and phytonutrients [12, 13]. Noteworthy, cereals and pulses differ in their structural and physicochemical properties and have varying amounts of fiber, resistant starch, vitamins, minerals, and other bioactive components. However, they can complement each other in terms of amino acid composition [14, 15]. Thus, traditional foods in MD combining pulses and cereals (red beans and rice, burritos with refried beans, and hummus with pita bread, bulgur and lentil pilaf etc.) provide an improved protein quality compared to the individual foods because of their complementary amino acid profiles [10].

2.1 *Pulses*

Pulses are ancient crops for modern times those can be traced back thousands of years. Ancient civilizations in Mesopotamia grew peas, beans and lentils as far back as 8000 BC, and researchers recently discovered evidence of faba beans cultivated in northern Israel over 10,000 years ago [16, 17]. These staple crops have been an integral part of not only MD but also other diets for millennia, and today are an important crop both food security and also for combatting malnutrition, alleviating poverty improving human health and enhancing agricultural sustainability [17]. While Egypt and India consume the largest quantity of pulses, in Europe, 60% of pulses are consumed mainly by Mediterranean countries such as Spain and France [18]. In the light of recent studies, pulses are started to be announced as –perfect foods- that can help solving global food security challenges and meet the nutritional requirements of global population as an inexpensive, sustainable source of proteins and other key nutrients [16, 17]. In this context, The United Nations General Assembly declared 2016 as the International Year of Pulses (IYP) to raise awareness about the importance of these plants and to highlight their role in healthy diets and family farmers' livelihoods as well as their contribution into sustainability [17].

2.1.1 Nutritional Properties

Pulses are dry edible seeds of plants belonging to the *Fabaceae* (*Leguminoseae*) family, which include field peas, dry beans, lentils, chickpeas and faba beans. The contemporary definition of pulses excludes oilseed grains and legumes consumed in immature form [16, 17]. Pulses are the backbone of Mediterranean diet, appearing as main ingredient in numerous traditional Mediterranean dishes such as hummus, soups, salads etc. In MD Food pyramid, at least two servings of 100 g of cooked legumes a week [18] is recommended.

Pulses are nutritionally valuable plant foods, providing proteins (20–45%) with essential amino acids, complex carbohydrates (60%), dietary fiber (5–37%), unsaturated fats, vitamins, essential minerals and bioactive compounds (isoflavones, lignans, protease inhibitors, trypsin and chymotrypsin inhibitors, saponins, alkaloids, phytoestrogens and phytates) for the human diet [10, 19, 20]. Along with being a highly nutritious food, it was reported that pulses could play an important role in the prevention of several health conditions such as type 2 diabetes, hypertension, hyperlipidemia, cardiovascular diseases, cholesterol, cancers and weight management etc. [11, 21]. Beside their nutritional benefits, pulses have also been ascribed economic, cultural, physiological and medicinal roles owing to their possession of beneficial bioactive compounds [19]. Because of their composition, they are attractive to health conscious consumers, celiac and diabetic individuals as well as consumers concerned with weight management. Recent studies demonstrate that the incorporation of legumes in diets, especially in developing countries, could play a major role in eradicating protein energy malnutrition and maintain wellbeing [19, 22].

2.1.2 Limitations

The global consumption of pulses is still low compared to that of cereals due to several limiting factors such as the presence of anti-nutrients, myths about legume consumption, allergy towards pulses resulting in bloating and flatulence, hard-to-cook phenomenon, and off flavors [19]. To fully get advantage of the benefits of pulses, Table 6.1 summarizes some strategies to overcome these constraints regarding their production and consumption. Future studies are required to enhance the flavor and the palatability of pulses and to develop innovative and inexpensive products. To achieve this, there is a need to direct breeding toward the selection of pulses with better attributes (higher nutritional quality and lower anti-nutrients and off-flavors) to develop cheap healthy, tasty and innovative value-added products from pulses that meet consumers requirements.

Table 6.1 Limitation factors of pulse use [19]

Limting factor	Negative effect	Possible solution
Anti-nutrients Such as trypsin inhibitors and amylase inhibitors Phytate, lectins, saponins and oligosaccharides	Decreases protein digestibility and starch digestibility Chelates with minerals resulting in poor mineral bioavailability Reduced bioavailability of nutrients Flatulence and bloating	Dehulling, soaking, boiling, steaming, sprouting, roasting, fermentation, autoclaving, gamma irradiation
Hard-to-cook phenomenon	Energy and time consumption	Soaking prior to cooking
Lack of convenient food applications	Boredom of eating the same food repeatedly	New product development of innovative legume products
Low levels of Sulphur-containing amino acids	Incomplete protein source	Consumed in combination with cereals
Lack of awareness of nutritional value of legumes	Low intake of legumes	Increasing consumer awareness of the nutritional profile of legumes
Beliefs and taboos–for example, eating groundnuts can cause stomach upset	Low intake of legumes	Increasing consumer awareness of the nutritional profile of legumes and of methods to get rid of anti-nutrients and oligosaccharides
Reluctance to try a new kind of food or to change eating habits	Low intake of legumes	Development of innovative, attractive legume-based products to entice consumers
Low iron bioavailability	Poor source of iron	Consumed in combination with vitamin C rich foods, the absorption of iron would be increased

2.2 Cereals

Cereals belong to the *Gramineae* family of grasses and include wheat, rice, barley, corn, rye, oats, millets, sorghum and etc. Whole-grain cereals are the major foods in MD pyramid and appear at the base [10]. Most consumed staple foods in the MD are based on wholegrain cereals are bread, pasta (erişte, couscous, orzo, etc.), bulgur etc. In MD food pyramid, 85 g of whole grain cereals are recommended each day [18].

Whole grain cereals mainly consist of starch [80–85%], protein [2–15%], fat [<5%] and other nutrients [23]. Cereals are rich in nutrients with recognized health benefits, such as dietary fiber, antioxidants, including phenolic compounds, phytoestrogens including lignans, vitamins (vitamin-B complex, vitamin E) and minerals (Fe, K, Mg, Zn, Se) and other bioactive components, which are distributed heterogeneously within the grain components (bran, germ and endosperm) [23, 24]. Cereals also contain various non-amylaceous polysaccharides, namely cellulose, pentosans and beta glucans. These compounds are hydrolyzed by endogenous digestive enzymes and, being cell-wall constituents, abound in the external parts of the grain. Most of health beneficial compounds are located in the germ and bran of wholegrain cereals [23]. Thus, research consistently indicates that the regular consumption of wholegrain cereals reduces the risk of cardiovascular diseases, type 2 diabetes mellitus and certain types of cancer, as well as several gastrointestinal pathologies [24] compared to refined cereals. Such recommendation aligns with the Mediterranean's traditional diet favoring the consumption of whole grains.

2.3 *Chemistry of the Synergetic Effect of Pulses and Cereals in MD*

In the MD, the combined use of pulses and cereals are one of the most important healthy food consumption patterns that makes it superior compared to other diets. As stated above, pulses are an excellent source of good quality protein with 20–45% protein (nearly two fold of cereal protein content) that is generally rich in essential amino acids particularly lysine generally lacking in cereals [19]. Pulse proteins (except soy protein) are, however, low in the essential sulphur-containing amino acids (SCAA) such as methionine, cystine and cysteine as well as in tryptophan thereby considered as an incomplete source of proteins [10]. The main fractions of pulse protein are albumins and globulins, which can be divided into two groups, namely vialin and legumin. Vialin is the major protein group in most pulses, which is characterized by a low content of SCAA, thus explaining the low levels of SCAA in pulses [19, 25]. Pulses and cereals complement each other in terms of protein quality as cereals are high in SCAA (low in pulses) and have low in lysine (high in pulses) [19]. As such, protein quality can be significantly improved when pulses are blended with cereals. For a better nutritional balance, pulses and cereals are recommended to be consumed in the ratio 35:65 [25]. According to a recent study, the addition of pulses to either wheat or rice increases the overall Protein Digestibility Corrected Amino Acid Score (PDCAAS) values from 0.43 to 0.67 (rice), 0.71 (wheat + pulses) and 0.75 (rice + pulses) in the blends, respectively [26].

Considering the above-mentioned facts, traditional culinary combined use of pulses and cereals in MD stand as a good example for shaping further dietary recommendations.

2.4 *Culinary Applications of Blending Pulses and Cereals in MD as a Model for Designing New Foods*

MD has been a permanent laboratory for humankind to show the ability to combine pulses and cereals in preparation of foods such as soups, salads, pastas, pilafs, main dishes, traditional wraps etc. [27]. Some of the traditional pulse-cereal blend foods

Table 6.2 Traditional pulse-cereal blends-based foods consumed in Mediterranean cuisines

Food type	Main ingredients	Cuisine origin
Salads	Rice and bean salad	Greek
	Bulgur and lentil salad	Turkish and Syrian
	Portuguese breadcrumb black bean kale salad	Portuguese
Soups	Yüksük: Meat stuffed pasta (Manti) and chickpea	Turkish
	Lebeniye: Yoghurt soup with stuffed meatball, chickpea and wheat and rice or bulghur	Turkish
	Ayran Aşı: Cold yogurt soup with chickpea and wheat	Turkish
	Pasta e fagoli: Pasta with beans	Italian
	Fažol i testo: Beans and homemade noodles	Croatian
	Harira: Tomato-based soup with lentils, chickpeas and rice or fine broken noodles (chaariya)	Moroccan
Pilafs	Mercimekli bulgur pilavı: Bulghur with lentils.	Turkish
	Mujadara: Lentils and rice with caramelized onions	Lebanon
	Fakorizo: Lentils with rice	Greek
	Koshari: Rice, lentils, pasta with chickpea	Egyptian
Main dishes	Alubias con arroz: Dry beans (pinto, black or red) with rice	Spanish
	Rfissa Medhoussa: Moroccon day-old bread (msemen) with chicken and chickpeas, lamb with vegetables	Moroccan, Tunisia, Algeria
	Kesksoo: Couscous with meat, vegetables, chickpeas	Libyan, Moroccan, Tunisia, Algeria
Traditional wraps or sandwiches	Hummus with pita bread: Hummus in the corn/wheat pita breads	Greek
	Chickpea wrap: Chickpeas and various vegetables in wheat wrap	Turkish Greek
	Spanish bruschetta: Garlic and white bean puree on the bread slice	Spanish

used in the different Mediterranean cuisines are summarized in Table 6.2. Eating pattern of cereal/pulse blends in the Mediterranean countries are quite similar with minor modifications among the cultures. Soups and pilafs are the main culinary applications of pulse/cereal blends in the MD since they are cheap and balanced (in term of energy and proteins) basic food of many families in the past and present [10]. In the past, pulses were considered as the "poor man's meat" but currently considered as healthy and sustainable protein sources for everybody [4]. The cereal/pulse blend Mediterranean dishes can be a dissemination point of millenary routes to recreate today's dishes and therefore diets to provide a healthier future [28, 29].

3 Opportunities of Pulses and Cereals for Designing New Foods

Although pulses and whole grain cereals are the primary sources of human nutrition, many developed countries still have a low intake of them in their routine diet and a greater reliance on processed food material [30–32]. This reliance on high processed foods may reduce the nutritional quality of food such as the use of refined wheat and/or highly polished rice in bread production results in lower fiber intake in the regular diet [33, 34]. The consumption of highly processed foods was associated with the escalating incidences of health-related issues such as diabetes and obesity [33]. Correspondingly, consumers' perception towards healthy food habits is increasing in recent years with growing understanding regarding the relation between food composition, nutrition and health [35]. Beside healthiness, the global transition from animal-based foods to plant-based foods (especially pulses and/or whole grain cereals) might be also attributed to economic motives, environmental concerns, and ethical aspects related to animal welfare [33, 34]. Many national/international health organizations particularly urged the utilization of pulses and/or whole grain cereals plants owing to their nutritional composition, availability and cost efficiency as compared to animal nutrients [36]. Mediterranean Diet, which is a kind of laboratory for humankind, is an illustration of how to use pulses and/or cereals in the preparation of a wide array of foods.

4 Potential of the Pulses for the Enhancement of Cereal Based Foodstuff

Reformulating cereal-based foods with pulses has the potential to increase the protein quality of the overall food product [20]. Pulse-enriched cereal flours (mainly wheat) represent a potential way to increase the nutritional properties of cereal-based foods due to their well-known complementary amino acid composition and synergistic effect of phytochemicals [31]. Most of the researches are focused on reformulating wheat bread, mainly with lentils [31, 37, 38] chickpeas [38], and peas [39]. Protein concentrates and protein isolates of peas, lentils and chickpeas have been successfully incorporated in baked products [31, 38]. Several studies established the positive impact of the addition of different types of pulses such as chickpeas (up to 10%) [40], lupins (up to 20%) [40], green lentils (from 25 to 100%) [41], and navy beans (up to 30%) on cookies' technological and nutritional properties [42]. These studies showed that, in cookies, the protein content increased proportionally with the addition of pulse flour while reducing dough spread. Malcomson et al. [43] showed that adding 20% yellow pea flour to rice flour and tapioca starch did not modify the characteristics of cookies in terms of acceptability and texture. High levels (beyond 20%) of addition of pulse flour can lead to darker surface color and a proportional increase in the hardness of the cookies.

Cakes were also fortified with several types of pulse flours, such as chickpeas [44] or peas [45]. However, Gomez et al. [45] showed that pulse proteins addition decreased cake volume and resulted in heterogeneous crumb texture. As a result, firmness increased while springiness and cohesiveness decreased. Gomez et al. [44] mentioned that it is possible to use chickpea flour to totally substitute wheat flour in the elaboration of different kinds of cakes. Besides, chickpea flour suspensions are widely used in the preparation of oriental traditional sweet and savory snacks using deep-frying. The droplets of chickpea flour slurry become twice in size after incorporation of air bubbles at high temperature during frying [46]. The chickpea flour has also promising characteristics for identical preparation of products as tortilla chips and it is also used in extruded products with the combination of corn and rice flour by using the fried extruder [36]. Moreover, nutritious snacks were formulated by blending proteins from different sources (peas, lentils) with different starch sources (wheat, barley and rice) to enhance both functional and nutritional features of conventional products [47]. For instance, protein rich biscuits were developed by adding different levels (5, 10, 15, 20, and 25%) of pigeon pea flour to traditional wheat flour and obtained promising results in terms of quality [48]. Puffed snacks made from blends of navy and red bean flours and corn starch at different levels of substitution (15, 30, and 45%) had increased protein and starch contents compared to the control [49]. Snacks, made of 100% pulses, are also investigated in several studies [44, 46]. For instance, a high protein snack food, prepared by soaking mung bean in water overnight and immediate deep-frying after removal of water, resulted in high sensorial acceptability [46]. Researchers mentioned that, the starch and proteins, present in the pulses display good pasting, foaming, and emulsifying properties to make it suitable for deep-frying to produce pulse snacks. In another work, chickpea flour is used for production of imitation milk and product formulations thanks to its suspension's flow behavior. Reducing apparent viscosity to enhance rheological properties by incorporating additives such as salt, oil, or the carboxy methyl cellulose has brought the flour from household products to industrial levels [50].

As for crackers, Malcomson et al. [43] added crackers supplemented with lentil flour (30%) result in a protein-rich cracker with twice the total dietary fiber of wheat crackers. However, enriched crackers with lentil flour were darker in color, but their crisp texture and peppery flavor were considered acceptable and comparable to the control. Considering the non-essential role of gluten, crackers are suited for a complete replacement with pulses including chickpea, green, red lentils pinto bean, navy bean, and yellow pea flours. Indeed, the gluten-free cracker made with pulses have high nutritional value and were appreciated by the consumers since their color, texture and taste were comparable to commercial products [51].

The addition of black grams improved the cooking properties (cooking loss, water absorption and texture) and nutritional value (protein content and potential digestibility) of pasta [52]. Two recent studies also demonstrated that it is possible to produce 100% pulse pasta [53, 54] using lentil and faba bean. The organoleptic attributes all-lentil (100%) pasta showed the highest sensorial scores among the pastas containing pulses in their formulation up to now [53]. Researchers also reported that all-lentil pasta has not reached the semolina pasta scores in terms of the criteria investigated [53].

In the case of bread making, pulse proteins are not able to form gluten networks, and consequently, the weak interactions between pulse and wheat proteins reduce the formation of viscoelastic dough and gas retention during leavening, resulting in bread with poor crumb structure and texture [31, 55]. Thus, the addition of chickpea or peas flour is limited to percentages below 10–15% [31]. For gluten-free bread, Miñarro et al. [55] used chickpea flour, pea isolate, carob germ flour or soya flours and pea isolate. Chickpea bread showed the best baking characteristics.

Pulses have also been used in yoghurt preparations to enhance the prebiotic as well as probiotic content thanks to their oligosaccharide content. They also successfully used for diverse applications such as: protecting the probiotic bacteria using pulse proteins as encapsulating agents, using high-fiber micronized pulse flours for developing low fat bakery products with increased fiber content [56].

In a framework of replacement of raw materials with pulses is appealing to fulfill the nutritious and eco-friendly food expectations of the millennium consumers.

The functional properties of protein, starch and fiber fractions of pulses such as water binding, oil binding, emulsion stabilization and gelling could also be harnessed in the development of various food products. Functional properties of pulse fractions and their applications in foods are summarized in Table 6.3. Protein ingredients originating from pulses seem to be a promising and valid option for developing gluten-free (GF) pasta and bakery products because of good nutritional profiles that can be further enhanced by fermentation, functional properties such as water-binding, foaming, and emulsifying capacities, suitability for gluten-free and vegan/vegetarian diet, the lack of association with diseases such as bird flu and mad cow, the low-cost advantage over animal proteins, and low carbon footprint [57]. The functional properties of pulse proteins have been exploited in the preparation and development of products such as soups, extruded products and ready to eat snacks [58]. Dietary fiber fractions from pulses have found use in the bakery, meat, extruded products and beverage industries as stabilizers, texturing agents, fortifiers, bulking agents, fat replacers and emulsion stabilizers [59]. Pulse starches can be used in processed meat products, particularly where extensive heating and mechanical stirring is required. Canned foods, cooked sausages, soups, sauces, noodles and vermicelli prepared using pulse starches have better sensory attributes than those made from starches from traditional sources. The thickening effect of pulse starches is used to increase viscosity of different food products [58, 59].

5 Conclusion

Growing number of consumers are increasingly conscious of healthy eating habits, dietary needs and eco-friendly food consumption. As such, the request for ingredients with nutritional benefits keeps increasing due to the considerable influence of food journalists, online recipes, celebrities and social media. This influence turns

Table 6.3 Functional properties of pulse fractions and their applications in foods

Pulse ingredient	Functional property	Mechanism	Effects	Foods
Pulse protein	Solubility	Solubility	High solubility positively influences emulsification, foaming, and gelling properties	Bread, cake, muffin, pasta
	Water holding	Hydrogen bonding, entrapment of water	Providing resistance to dough expansion	Bread, cake, muffin
	Fat absorption	Binding of free fat	Preventing coalescence	Cake, muffin
	Emulsification	Formation and stabilization of fat emulsions	Well-developed crumb	Bread, cake, muffin
	Foaming	Form stable film to entrap gas	Providing good crumb structure and loaf volume	Bread, cake, muffin
	Gelation	Gel matrix formation	Texturizing agent	Bread, cake, muffin, pasta
Pulse starch	Stability	Stabil against heat and mechanical shear	Texturizing agent	Processed meat products
	Thickening	Gel matrix formation	Increasing viscosity	Soups, sauces
Pulse fiber	Emulsification	Bind high amounts of water	Low water activity	Bakery products, soups, dairy products
	Thickening	Gel matrix formation	Texturizing agent	Soups, bakery products

ingredients into popular "superfoods". Consumers are also better informed about food ingredients nowadays than in the past [60]. Consumer awareness gave attraction to grains and pulses to a large extent since they can have natural organic, residue-free, raw food, unprocessed or minimally processed, plant-based (vegan products, plant-based protein, vegetarianism, and flexitarians), functional (functional foods, superfoods, health benefits, nutrients, dietary fibers, vegetable and pulse protein), free-from (free from allergens, less sugar, gluten-free, nut-free, clean label, no additives) labels. These labels provide commercial potential for a large range of grains and pulses. In this manner, the MD, sustainable plant centric (mainly pulses and whole grain cereals oriented) dietary pattern in which nutrition, food, environment, culture, and sustainability all interact, appreciated by the consumers as the best diet in the world [8] and an intangible cultural heritage by UNESCO [5]. In this frame, one of the key features of MD is the fulfillment of the consumers' above mentioned expectations by combining pulses and whole grain cereals in their dietary pattern. This combination synergizes to foster favorable changes in intermediate pathways of non-communicable diseases' risk, as supported by randomized trials [10]. Therefore, more Research and Development towards the blending pulses and wholegrain cereals, usage of pulses by oneself, understanding the effect of processing on the nutritional and functional properties of the blend and pulses are needed. These approaches can boost the demand for pulses through establishment of a continuous "innovation funnel" delivering new scientifically validated product concepts [59]. All the relevant stakeholders must promote and implement appropriate policies, and increase investment in research and development, and extension services, focusing on development of pulse-based products, which can play a key role for future global food security and environmental challenges, as well as alleviating the global burden of non-communicable diseases which is a major global public health goal.

Finally, the MD is not only a way of eating, but a lifestyle in itself. It is conviviality, sharing a good meal and the pleasure of eating with friends and family, taking your time, slowing down, but it is also about an active life, involving exercise and vitality. It is happiness, enjoying life. The Mediterranean also evokes the hospitality and generosity, which is expressed by sharing a good meal even with a stranger [61]. From a holistic point of view, all the physiological and social aspects of MD make it an achievable and sustainable lifestyle that should be promoted to feed the future.

References

1. Food and Agriculture Organization of United Nations [FAO]. International year of legumes: nutritious seeds for a sustainable future. 2016. http://www.fao.org/pulses-2016/resources/fao-publications/en/. Accessed 2 Mar 2020.
2. Corella D, Coltell O, Macian F, Ordovas MJ. Advances in understanding the molecular basis of the Mediterranean diet effect. Food Sci Technol. 2018;9:227–49.
3. Keys A. Seven countries: a multivariate analysis of death and coronary heart disease. Cambridge: Harvard University Press; 1980.
4. Renna M, Rinaldi VA, Gonnella M. The Mediterranean diet between traditional foods and human health: the culinary example of Puglia [southern Italy]. Int J Gastron Food Sci. 2015;2:63–71.
5. International Centre for Advanced Mediterranean Agronomic Studies [CIHEAM]. Final declaration- 9th meeting of the Ministers of Food. Agriculture and Fisheries of the Member Countries of CIHEAM. 2012. https://www.ciheam.org/uploads/attachments/111/Final_declaration_RMC_2012.pdf. Accessed 22 May 2020.
6. Dernini S, Berry EM. Mediterranean diet: from a healthy diet to a sustainable dietary pattern. Front Nutr. 2015;2:15.
7. Mayr HL, Tierney AC, Kucianski T, Thomas CJ, Itsiopoulos C. Australian patients with coronary heart disease achieve high adherence to 6-month Mediterranean diet intervention: preliminary results of the AUDMED heart trial. Nutrition. 2019;61:21–31.
8. US News & World Report. U.S. news reveals best diets rankings for 2018. 2018. https://www.usnews.com/info/blogs/press-room/articles/2018-01-03/us-news-reveals-best-diets-rankings-for-2018. Accessed 1 Jun 2020.
9. Hu FB. Globalization of food patterns and cardiovascular disease risk. Circulation. 2008;118:1913–4.

10. Rebello CJ, Greenway FL, Finley JW. A review of the nutritional value of legumes and their effects on obesity and its related co-morbidities. Obes Rev. 2014;15(5):392–407.
11. Papanikolaou Y, Fulgoni VL. Bean consumption is associated with greater nutrient intake, reduced systolic blood pressure, lower body weight, and a smaller waist circumference in adults: results from the National Health and Nutrition Examination survey 1999–2002. J Am Coll Nutr. 2008;27:569–76.
12. Koh-Banerjee P, Rimm EB. Whole grain consumption and weight gain: a review of the epidemiological evidence, potential mechanisms and opportunities for future research. Proc Nutr Soc. 2003;62:25–9.
13. He M, van Dam RM, Rimm E, Hu FB, Qi L. Whole-grain, cereal fiber, bran, and germ intake and the risks of all-cause and cardiovascular disease-specific mortality among women with type 2 diabetes mellitus. Circulation. 2010;121:2162–8.
14. Tharanathan RN, Mahadevamma S. Grain legumes – a boon to human nutrition. Trends Food Sci Technol. 2003;14:507–18.
15. Fardet A. New hypotheses for the health-protective mechanisms of whole-grain cereals: what is beyond fibre? Nutr Res Rev. 2010;23:65–134.
16. Tyler R, Wang N, Han J. Composition, nutritional value, functionality, processing, and novel food uses of pulses and pulse ingredients. Cereal Chem J. 2017;94:1.
17. Bach-Faig A, Berry EM, Lairon D, Reguant J, Trichopoulou A, Dernini S, Medina FX, Battino M, et al. Mediterranean diet pyramid today: science and cultural updates. Public Health Nutr. 2011;14(12A):2274–84.
18. Barilla, The Mediterranean Nutrition Model. https://www.barillagroup.com/sites/default/files/si_mediterraneo_first_booklet_ENG.pdf. Accessed 19 May 2020.
19. Maphosa Y, Jideani VA. The role of legumes in human nutrition. In: Hueda MC, editor. Improve health through adequate food. London: Intech Open; 2017. p. 103–9.
20. Di Daniele N, Noce A, Vidiri MF, Moriconi E, Marrone G, Annicchiarico-Petruzzelli M, et al. Impact of Mediterranean diet on metabolic syndrome, cancer and longevity. Oncotarget. 2017;8(5):8947–79.
21. Messina MJ. Legumes and soybeans: overview of their nutritional profiles and health effects. Asia Pac J Clin Nutr. 2016;25(1):1–17.
22. Mlyneková Z, Chrenková M, Formelová Z. Cereals and legumes in nutrition of people with celiac disease. Int J Celiac Dis. 2014;2(3):105–9.
23. Gill A, Ortega RM, Maldonado J. Wholegrain cereals and bread: a duet of the Mediterranean diet for the prevention of chronic diseases. Public Health Nutr. 2011;14(12A):2316–22.
24. Slavin J. Whole grains and human health. Nutr Res Rev. 2004;17:99–110.
25. Leonard E. Cultivating good health. In: Grains and legumes nutrition council. Adelaide: Cadillac Printing; 2012. p. 3–18.
26. Pulse Canada. Protein quality of cooked pulses. 2017. http://www.pulsecanada.com/wp-content/uploads/2017/09/Pulses-and-Protein-Quality.pdf. Accessed 23 Feb 2020.
27. Tumo IG. The mediterranean diet: consumption, cuisine and food habits. MediTERRA Presses de Sciences Po Annuels; 2012. p. 115–132.
28. Miguel A, Martinez-Gonzalez AG. Mediterranean diet: the whole is more than the sum of its parts. Br J Nutr. 2012;108:577–8. https://doi.org/10.1017/S0007114512001833.
29. Food and Agriculture Organization of United Nations [FAO]. The International year of Pulses-final report. 2019. http://www.fao.org/3/CA2853EN/ca2853en.pdf. Accessed 10 Sep 2019.
30. Mudryj AN, Yu N, Aukema HM. Nutritional and health benefits of pulses. Appl Physiol Nutr Metab. 2014;39(11):1197–204. https://doi.org/10.1139/apnm-2013-0557.
31. Bresciani A, Marti A. Using pulses in baked products: lights, shadows, and potential solution. Foods. 2019;8:451. https://doi.org/10.3390/foods8100451.
32. Garden-Robinson J. Pulses: the perfect food. 2017. https://www.ag.ndsu.edu/publications/food-nutrition/pulses-the-perfect-food-healthy-to-eat-healthy-to-grow-peas-lentils-chickpeas. Accessed 19 May 2020.

33. Jemal A. Trends in the leading causes of death in the United States, 1970-2002. JAMA. 2005;294:1255–9.
34. Kendall CWC, Esfahani A, Jenkins DJA. The link between dietary fibre and human health. Food Hydrocoll. 2010;24(1):42–8. https://doi.org/10.1016/j.foodhyd.2009.08.002.
35. Brennan CS, Tudorica CM. Evaluation of potential mechanisms by which dietary fibre additions reduce the predicted glycaemic index of fresh pastas. Int J Food Sci Technol. 2008;43(12):2151–62.
36. Patil SS, Brennan MA, Mason SL, Brennan CS. The potential of combining cereals and legumes in the manufacture of extruded products for a healthy lifestyle. EC Nutr. 2016;5(2):1120–7.
37. Kohajdová Z, Karovicová J, Magala M. Effectect of lentil and bean flours on rheological and baking properties of wheat dough. Chem Pap. 2013;67:398–407.
38. Aïder M, Sirois-Gosselin M, Boye JI. Pea, lentil and chickpea protein application in bread making. J Food Res. 2012;1:160–73.
39. Dabija A, Codina GG, Fradinho P. Effect of yellow pea flour addition on wheat flour dough and bread quality. Rom Biotech Lett. 2017;22:12888–97.
40. Hegazy NA, Faheid S. Rheological and sensory characteristics of doughs and cookies based on wheat, soybean, chickpea and lupine flour. Food Nahrung. 1990;34:835–41.
41. Zucco F, Borsuk Y, Arntfield SD. Physical and nutritional evaluation of wheat cookies supplemented with pulse flours of different particle sizes. LWT Food Sci Technol. 2011;44:2070–6.
42. Hoojjat P, Zabik ME. Sugar-snap cookies prepared with wheat-navy bean-sesame seed flour blends. Cereal Chem. 1984;61:41–4.
43. Malcolmson L, Boux G, Bellido AS, Fröhlich P. Use of pulse ingredients to develop healthier baked products. Cereal Foods World. 2013;58:27–32.
44. Gómez M, Oliete B, Rosell CM, Pando V, Fernández E. Studies on cake quality made of wheat–chickpea flour blends. LWT Food Sci Technol. 2008;41:1701–9.
45. Gómez M, Doyagüe MJ, De La Hera E. Addition of pin-milled pea flour and air-classified fractions in layer and sponge cakes. LWT Food Sci Technol. 2012;46:142–7.
46. Asif M, Rooney LW, Ali R, Riaz MN. Application and opportunities of pulses in food system: a review. Crit Rev Food Sci Nutr. 2013;53(11):1168–79.
47. Boye JI, Aksay S, Roufik S, Ribereau M, Farnworth ME, et al. Comparison of the functional properties of pea, chickpea and lentil protein concentrates processed using ultrafiltration and isoelectric precipitation technique. Food Res Int. 2010;43(2):537–46.
48. Kumar TB, Brennan CS, Jaganmohan R, Surabi AK. Alagusundaram utilisation of pigeon pea [Cajanus cajan L] byproducts in biscuit manufacture. LWT Food Sci Technol. 2011;44(6):1533–7.
49. Anton AA, Fulcher GR, Arntfield SD. Physical and nutritional impact of fortification of corn starch-based extruded snacks with common bean [Phaseolus vulgaris L.] flour: effects of bean addition and extrusion cooking. Food Chem. 2009;113(4):989–96.
50. Keshava KB, Bhattacharya S. Deep fat frying characteristics of chickpea flour suspensions. Food Sci Technol. 2001;36:499–507.
51. Han J, Janz JA, Gerlat M. Development of gluten-free cracker snacks using pulse flours and fractions. Food Res Int. 2010;43:627–33.
52. Madhumitha S, Prabhasankar P. Influence of additives on functional and nutritional quality characteristics of black gram flour incorporated pasta. J Texture Stud. 2011;42(6):441–50.
53. Kurt Gokhisar O. Investigation of red lentil [lens culinaris] pasta production. University of Mersin; Institute of Life Sciences. Mersin, Turkey. PhD. Thesis. 2019. pp. 72–88.
54. Rosa-Sibakov N, Heiniö RL, Cassan D, Holopainen-Mantila U, Micard V, Lantto R, et al. Effect of bioprocessing and fractionation on the structural, textural and sensory properties of gluten-free faba bean pasta. LWT Food Sci Technol. 2016;67:27–36.
55. Miñarro B, Albanell E, Aguilar N, Guamis B, Capellas M. Effect of legume flours on baking characteristics of gluten-free bread. J Cereal Sci. 2012;56(2):476–81.
56. Zare F, Boye J, Orsat V, Champagne C, Simpson B. Microbial, physical and sensory properties of yogurt supplemented with lentil flour. Food Res Int. 2011;44(8):2482–8.

57. Foschia M, Horstmann SW, Arendt EK, Zannini E. Ingredients in gluten-free bakery and pasta products. Annu Rev Food Sci Technol. 2017;8:75–96.
58. Boye J, Zare F, Pletch A. Pulse proteins: processing, characterization, functional properties and applications in food and feed. Food Res Int. 2010;43(2):414–31. https://doi.org/10.1016/j.foodres.2009.09.003.
59. Mazumdar SD, Durgalla P., Gaur PM. Utilization of pulses – value addition and product development. In Gurung TR, Bokthiar SM, editors. SAARC Agriculture Centre, Bangladesh; 2017. pp. 65–97. http://www.sac.org.bd/archives/publications/Pulses%20for%20Nutrition%20Security.pdf Accessed 15 Mar 2020.
60. Dennet C. Key ingredients of the Mediterranean diet — the nutritious sum of delicious parts. 2016. https://www.todaysdietitian.com/newarchives/0516p28.shtml. Accessed 12 May 2020.
61. Conner Middelmann-Whitney. The Mediterrenean anti-cancer diet. 2011. https://modernmediterranean.com/wpcontent/uploads/2013/09/Zest_for_Life_Mediterranean_Anti-Cancer_Diet.pdf. Accessed 19 Nov 2019.

Chapter 7
Snacking: Ingredients, Processing and Safety

Nicola Gasparre and Cristina M. Rosell

Abstract Urbanization has changed worldwide consumers' lifestyle and consequently their dietary habits. Consuming foods or snacking between main meals is becoming a growing practice in people's lives, initially for reducing the starving sensation but later on as mainstream meal. This upward trend is confirmed by the exponential increase in sales of the snacks industries. The term snack includes a great variety of small pieces foods. Nevertheless, this chapter will be focused on the cereals-based snacks originally from the Mediterranean area, as well as a reference to other initially autochthonous snacks that are becoming globally popular is included. In this chapter, main cereal-based snacks are considered, explaining their production process, ingredients interactions, quality characteristics and safety issues, with especial emphasis on the extrusion process because of its importance in the snacks market. Likewise, an overview of the trends driving snacks innovation is presented, particularly related to the search of alternative ingredients and new technologies applied to this food category that is very prone to adapt towards an ever-changing future.

Keywords Snacking · Extruded snacks · Expanded snacks · Fried snacks · Snack bar · Food innovation

1 Introduction

Snack and snacking are now very common terms adapted to consumers lifestyle. Nevertheless, meaning of "snack" and "snacking" is still very confusing. During the last years, several studies have been carried out trying to clearly describe snack food and snacking behavior, but until now there is a lack of an unanimously recognized definition. According to some publications in scientific literature, "snack" meanings are principally focused on the daily moment of an eating event [1, 2], kind and

N. Gasparre · C. M. Rosell (✉)
Institute of Agrochemistry and Food Technology (IATA CSIC), Valencia, Spain
e-mail: crosell@iata.csic.es

F. Boukid (ed.), *Cereal-Based Foodstuffs: The Backbone of Mediterranean Cuisine*, https://doi.org/10.1007/978-3-030-69228-5_7

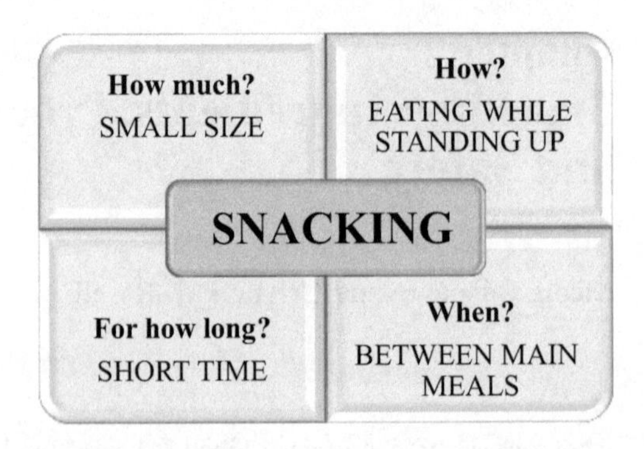

Fig. 7.1 Snacking features. Adapted from Drapeau et al. [4]

quantity of eaten food, place of the eating occasion and on a conjunction of these factors [3]. Therefore, the term snack would be related to the rapid consumption of a small food between meals that does not require sitting (Fig. 7.1). Frequently, the global snack food is segmented into salted snacks, bakery snacks, confectionary, specialty and frozen snacks. Nevertheless, the highest market segment is covered by bakery snacks.

Snacks food market is rapidly growing as a consequence of changes in consumer lifestyle, rapid urbanization, and most of it owing to the increasing demand for on-the-go convenience foods. Even distribution channels have fueled this trend extending accessibility to these products through specialty stores, independent stores, on-line sales, convenience stores, supermarkets, malls, and also vending machines. This is a global trend with USD 605 billion in global sales in 2018 and with a projection of growing at a compound annual growth rate (CAGR) of 5.34% during 2020–2025 period [5]. Snacks food are not immune to the healthy and environmentally friendly drivers, and functional ingredients, organic foods or green packaging technology are fueling the innovation in this sector. Considering that urbanization and consumer lifestyle are the main reasons that have motivated the expansion of snacks, Europe is leading the market closely followed by North America. According to the data presented by Forecast [6] in August 2019, European snacks market reached the value of USD 107.07 billion in 2016 and it is expected to increase at a CAGR of 5.2%, to achieve USD 220 billion by 2021. In Europe, the average per capita consumption amounted to nearly 4.6 kg in 2020 [7].

The Mediterranean basin is recognized for its rich gastronomic history and the beneficial effects associated with it [8–10]. Across the Mediterranean area, “snack” principally refers to a small food ration consumed in a short period time (eating while standing up) between main meals (breakfast, lunch and dinner), with a purpose of reducing the sense of hunger until the next regular meal. The rich gastronomy that characterizes the Mediterranean area is also extended to the snacks concept. Nevertheless, despite the variety of foods that can be consumed as snacks, this chapter will be focused on those snacks that are based on cereals. Specifically, in the Mediterranean area, a variety of grains including wheat, maize, oats and rice have typically been used as snacks ingredients, and in rather minor proportion others like rye, sorghum, millet and triticale. Moreover, in the last 15 years, food

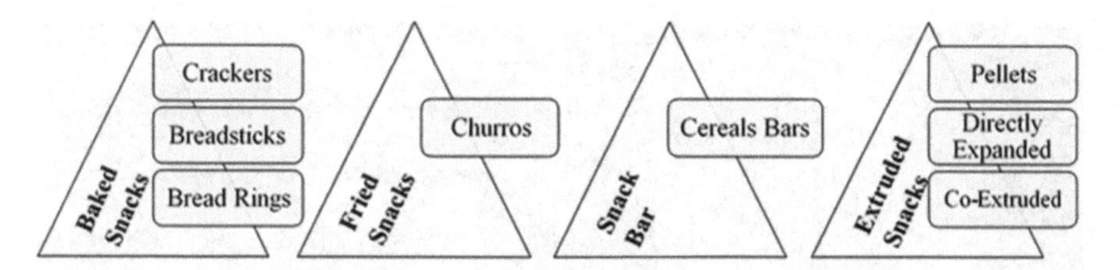

Fig. 7.2 Technological classification of cereals-based snacks

industry has started adding pseudocereals and pulses to maize and rice to produce gluten free snacks, one of the growing food niches. The present chapter discusses the most popular cereal-based snacks from the Mediterranean area with an emphasis on ingredients, processing and safety. It also examines the nutritional quality, the improvements occurred during the last years and the new trends in terms of snacking. Given the wide variety of cereals-based snacks, a technological classification (Fig. 7.2) based on their production process has been used to better categorize them. In the following sections, the production process and the characteristics of the snacks included within each category are described.

2 Baked Snacks

Many bakery cereal-based products fit in the category of snacks, like crackers, breadsticks and bread rings. Generally, they are produced from wheat following the main breadmaking stages: mixing, fermentation (yeast or chemical) and baking. This last step, baking, is the one that mostly characterizes this type of products. Through baking process, heating converts a raw dough into a crunchy crumb. Volume expansion, non-enzymatic browning reactions, yeast and enzymatic activities inactivation, protein denaturation, starch gelatinization and moisture loss represent the key transformations that occur during baking [11]. In the following sections, specific mention to the different baked snacks is included.

2.1 Crackers

Generally, crackers (Fig. 7.3a) are prepared from wheat flour but can be divided into three types (saltine, savory and chemically leavened) depending on the production process. Saltine requires a prolonged yeast fermentation (about 19 h) to get the typical dough sponge. After yeast fermentation, additional ingredients might be added, like sodium bicarbonate to lower pH, and then dough is left to rest for additional 4–6 h. Process continues with dough sheeting that is carried out using sheeting rolls with docking pins leading to a sheeted dough with 6–8 layers. The use of docking

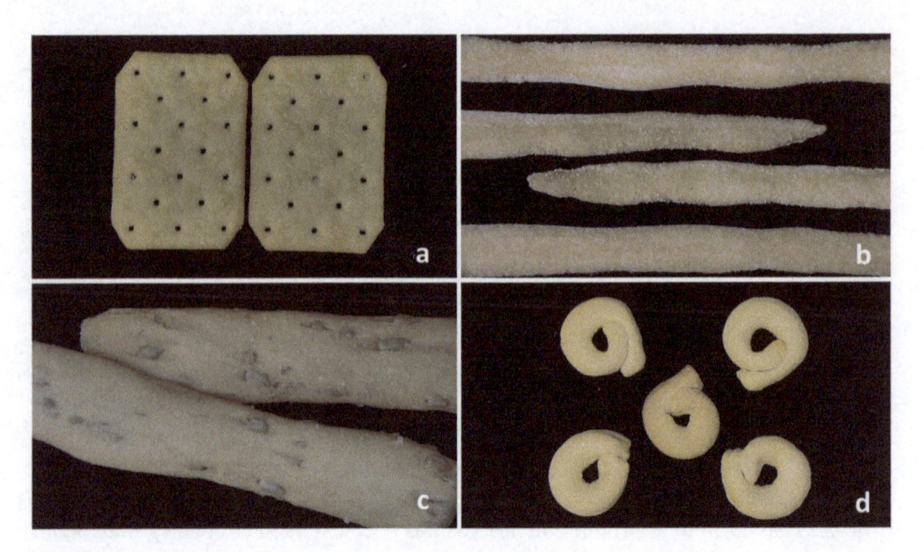

Fig. 7.3 Baked snacks. (**a**) crackers; (**b**) bread sticks "grissini"; (**c**) bread sticks "rosquilletas"; (**d**) bread rings "tarallini"

pins is needed to produce the dough cavities that facilitate the migration and evaporation of inner water during baking. Laminating process is essential for this type of crackers because it allows: improving the gluten network development, obtaining a characteristic layered structure and introducing other ingredients, like fat to generate the layered dough [12]. During baking, the thin dough sheets are subjected to high oven temperature (around 250 °C) for short-time (5–6 min) causing a fast-vertical expansion that results in the typical crumbly structure after baking.

Conversely, chemically leavened and savory crackers do not require fermentation time because chemical leavening agents (usually sodium bicarbonate, monocalcium phosphate) are used to get the desired expansion. In this case, the volume increase is primarily due to the chemical reaction of sodium bicarbonate with acid, followed by the thermal decomposition of soda alone and the water loss in form of steam during baking [13].

To imprint some specific sensory characteristics, some oils and flavor (savory snacks) can be sprayed after baking. Final moisture content of most crackers is around 2%. From nutritional point of view, crackers are characterized for a very low amount of sugar, a moderate fat level (10–20%) and low dietary fiber content [13].

2.2 *Breadsticks*

Among the most appreciated baked snacks in the Mediterranean area, breadsticks take up an important position. These products are known as "grissini" (Fig. 7.3b) in Italy, "rosquilleta" (Fig. 7.3c) in Spain and "kaki" in Tunisia, but the production technology and the raw materials used are rather similar. The basic breadstick is a crispy elongated bread (40–80 cm) slightly salted. The most common varieties are made mixing wheat flour, water, salt, either oil or fat, and yeast or a chemical raising agent [14]. In the case of the Spanish snack, it may also be combined with peanuts, sunflower seeds, cheese or chocolate. Regarding the making process, after mixing the dough is fermented for 90–120 min and then cut into strips ready to be baked at 250 °C for 10 min in the case of grissini [15] and at around 160 °C for 30 min in the case of "Rrosquilleta".

2.3 *Bread Rings*

Bread rings also known as "taralli" or "tarallini" (Fig. 7.3d) are frequently consumed as an appetizer during many occasions in Italy. These savory snacks are originally from the South of Italy (Apulia region) and usually they are flavored with fennel seeds, onion, pepper or simply salt, but also it is easy to find a sweet adaptation of them. These snacks have a rounded shape with a variable diameter from 2–3 cm to about 9–10 cm, and a thickness from 7 to 8 mm to 1.5 cm 1–1.5 cm [16]. As breadsticks texture, "taralli" is appreciated for its crunchy and friable sensation in the mouth. Core ingredients of "taralli" are wheat flour (sometimes re-milled semolina), olive oil, yeast, wine and salt. Ingredients mixing and kneading (20 min) are performed using an arms kneader followed by sheeting and shaping process, although in the past process were more artisanal. The raw "taralli" obtained from the previous process is immediately submerged in salty boiling water for 2 min and then rested for 30 min before baking at 230 °C for approximately 30 min. In past times, boiling process allowed stopping spontaneous fermentation before baking, because "taralli" were produced in public municipality bakeries [17]. Nowadays, the process has been automatized and dough hand-shaping is frequently substituted by a "tarallatrice", which is a specific machinery to shape rolled strips of dough enhancing the productivity [18]. To guarantee their characteristic crispy texture, "taralli" snacks contained around 20% fatty materials, and thus, prone to oxidations. In fact, Caponio et al. [16] studied the impact of processing and storage on the fatty fractions of "taralli" confirming that oxidation led to the formation of volatile compounds like aldehydes and polymerization complexes, both of them impairing negative characteristics to this snack. Specifically, aldehydes because they are related with off-flavors that reduce the product shelf life and consumers acceptance, while the polymerization compounds may have adverse effects on human's health. Because of the impact of fats on the quality on this snack, the effectiveness of different oils have been compared [19]. The effect of four different types of oil (extra virgin olive oil, olive oil, olive-pomace oil and refined palm oil) was evaluated on the final quality of this bread rings. Regarding visual appearance and odor, "taralli" containing extra virgin olive oil reached the highest score in the sensory test. Furthermore, samples with extra virgin olive oil had the lowest content of triacylglycerol oligopolymers and no presence of trans fatty acid isomers was found with respect to the other oils used in this study [19].

3 Fried Snacks

Other type of snacks is obtained applying frying, in which fat or oil are the heat transfers to food. The direct contact between food and hot oil (160–200 °C) and fast heat transfer are responsible for the characteristic dual structure of these snacks, with a dry, crunchy and golden crust and a soft inner core. Principal physio-chemical modifications in fried snacks include starch gelatinization, protein coagulation, Maillard reaction, moisture loss with a concomitant oil uptake and texture changes [20]. Plenty of different snacks could be identified worldwide within this category, but following section is restricted to the typical fried snacks located in the Mediterranean area.

3.1 Churros

Sweet fritters are widespread everywhere in Spain. The Spanish word "churro" (Fig. 7.4) refers to a deep-fried sweet dough, especially consumed during breakfast or as a snack. Basic ingredients are wheat flour, water and salt, which are mixed until obtaining a pretty sticky and fluid dough or batter. Then, a press called "churrera" allows extruding the dough that is cut it into small pieces directly submerged into boiling olive or sunflower oil. The raw strips are deep-fried, usually for 3–4 min at 185–200 °C, although those are quite variable, in fact, the golden color and crispy crust appearance are used as process end point marker [21]. Owing to their carbohydrate content and the high temperatures reached during frying, some undesirable and possibly dangerous compounds such as acrylamide or 5-(hydroxymethyl)-2-furfural may be found in the final products [22]. Therefore, some studies have been carried out trying to identify and reduce these complexes molecules. Morales, Arribas-Lorenzo [21] studied the impact of frying temperatures and times on the

Fig. 7.4 Spanish sweet fritters "churros"

levels of harmful compounds. For that, churros were deep fried in sunflower oil at 180, 190, 200 °C for 2, 3, 5 and 7 min and compared with commercial samples. Average values of acrylamide in commercial products were half of that found in the experimental products, while commercial samples contained higher amount of 5-(hydroxymethyl)- 2-furfural. The major increase in the acrylamide level was observed in the range 190–200 °C, suggesting the need for a more accurate control of the frying temperatures to limit the formation of this unfavorable compounds [21]. Mild vacuum frying (about 21 kPa) may be an attractive alternative to cut down the formation of these harmful molecules [23]. In fact, data coming from vacuum frying (21 kPa) at 100, 120 and 140 °C were compared with those acquired from the traditional frying (atmospheric pressure) at 140 and 180 °C, observing a significant reduction in the 5-(hydroxymethyl)- 2-furfural content [23]. Simultaneously, vacuum frying reduced the non-enzymatic browning with hardly impact on the snack color, compared with samples fried following the manufacturer recommended conditions (180 °C, 3 min). Besides, "churros" treated under vacuum conditions were crunchier than those fried at 140 °C under atmospheric pressure [23].

4 Cereal Snack Bar Type

At the beginning of the '90s, a new product, which can be categorized within snack foods, was introduced into the world market. It was presented as a healthy choice for the consumers interested in a wholesome diet [24]. Nowadays, cereal bars (Fig. 7.5) have reaped more relevance and acceptance worldwide. In fact, the global snack bars market has reached the value of USD 20.15 billion in 2018 and is supposed to increase at CAGR of 6.64% from 2019 to 2025 [25]. Generally, cereals bars refer to a rice- or oats-based snack combined with several ingredients such as chocolate, fruits, raisins, nuts or seeds. They have been consumed as a snack but for many occasions even as breakfast, energy food, meal replacement or sometimes for weight control [26]. The basic production process of cereal bars consists in mixing

Fig. 7.5 Cereal snack bar type

the dry and wet raw materials following the dough portioning and baking, varying time and temperature depending on the characteristics of the final snacks. In some type of cereal bar snacks, the baking phase is not needed, and the ingredients blend is just divided in the chosen shape. For both kinds of cereals bars, extra processes can be included, like filling, coating with assorted glazes, drying, etc. [27]. Usually, they are marketed as an individual portion of 25–30 g, thus the last step is packaging.

5 Extruded Snacks

Extrusion process was introduced for the first time in 1797 by Joseph Bramah, for making lead pipe and during the last 250 years this process has been employed to produce plastics and synthetic stuffs. At the beginning of the seventies, extrusion cooking was introduced in the food industry [28]. New opportunities of mixing different ingredients and the design of innovative layouts for extrusion equipment have allowed producing novel foods with new structures, increasing the range of their application in the food industry. Because of that, extrusion process has been extended to the production of pasta, breakfast cereals, puffed rice, texturized protein, instant drinks, meat analogs, pet foods and snacks [29]. In 2019, the global market for extruded snacks was around USD 48.3 billion and according to forecasts in 2026 will reach USD 65.2 with a CAGR of 4.4%. Mostly, the increase is associated with the growing available income and modifying lifestyle among the new generations [30].

In the extrusion-cooking ingredients are mixed and then subjected to heating, high pressure, shear and afterwards forced to flow through a die with a precise shape [31]. Final quality of the extrudate foods is influenced by: raw material type, feed moisture, barrels temperature, screw configuration and speed [32]. Considering the huge number of marketed extruded snacks, their classification is rather difficult, but it has been divided in three major groups: pellets, directly expanded snacks and co-extruded snacks.

5.1 Biochemical Changes Occurring during Extrusion

Food products, which only undergo simple physical treatment have proven to be healthier [33, 34]. Despite extrusion is a short time cooking technology, temperature and shear force reached in the barrel are able to cause important changes in macro and micronutrients of the raw materials.

In cereal based extruded snacks, starch is the main constituent and gelatinization during extrusion of cereal based blends play a crucial role for defining snack features. Starch is a polymer constituted by a mostly linear polymer named amylose (20–30%) and a branched polymer termed amylopectin; and depending on the ratio between these two polysaccharides, starch can assume different properties in terms of viscosity and gel formation [35]. During extrusion amylose and amylopectin are subjected to shear force and thermal energy that may reduce their molecular weight also changing the crystallinity of the structure [36]. In fact, this phenomenon has been observed in maize starch [37], rice starch [38], banana starch [39] and sweet potato starch [40]. The consequence of the rupture of covalent hydrogen bonds within starch molecules is an increase of the starch digestibility [41]. Nevertheless, simultaneous extrusion-cooking may contribute into the formation of amylose-lipid complexes [42] and resistant starch [43], and both of those phenomena resulted in a decrease of the starch digestibility, which is considered nutritionally advantageous. Those phenomena could be promoted by selecting appropriate extrusion conditions, for instance, an increase in barrel temperature results in an increase of starch-lipid complexes, which in turn are strongly correlated to the moisture content of the blend [44].

Other nutrient that has technological impact in the extrusion process, apart from nutritional implications, is the dietary fiber. In general terms, dietary fiber is described as plant polysaccharides and lignin that are resistant to hydrolysis by digestive enzymes in human, and its intake promote beneficial health effects that have been largely demonstrated and worldwide accepted [45]. Based on its solubility in hot water, it can be classified in soluble and insoluble fiber [46]. Extrusion process can affect the dietary fiber functionality, particularly temperature and screw speed. Lately, several studies demonstrated that soluble fiber content increased with the extrusion of high insoluble fiber raw materials [47–49]. For instance, when subjecting whole-grain wheat flour to extrusion, a significant increase in dietary fiber and resistant starch was observed in maize based extrudates, while total and digestible starch decreased [50]. Again, water content of the blends plays an important role during extrusion processing because insoluble and soluble fibers behave differently. In fact, insoluble fiber is responsible for the volume decrease in extruded products while soluble fiber improves the expansion with less effect on bulk density [51].

Regarding proteins, being thermolabile compounds, they are affected by extrusion-cooking. This technological process may facilitate the formation of non-covalent molecular interactions, covalent cross-linking and protein-lipid-starch interactions [52]. Moreover, shear force and high temperature cause a conformational alteration named denaturation, which makes protein sites more available for the proteases with a consequent increase of the *in vitro* protein digestibility [53]. The deactivation of antinutritional factors (protease inhibitors) might be also involved, favoring the accessibility of the proteases [35]. Actually, some studies have reported the destructive effect of extrusion towards antinutritional compounds like phytic acid, polyphenols, oxalates and trypsin inhibitors, highlighting temperature as a key factor in this reduction [54, 55]. This, has a direct consequence on the bioavailability of minerals in extruded foods, which is enhanced by the disruption of some antinutritional compounds through extrusion-cooking [56].

The role of fat in extrusion have been extensively analyzed due to its great impact on processing performances. In food extrusion, fat mainly comes from raw materials and added ingredients. Cereals like wheat and rice have a low fat content (<2%), but content could be higher when including oats (10%) [35]. Raw materials with fat level higher than 6% are not recommended for the extrusion process, especially when the target product is an expanded snack. In fact, high fat content increases creep in the barrel, with consequent torque and pressure reduction, which lead to less expanded product [57]. However, when fat presence is <6%, it behaves as a plasticizer or emulsifier [58]. High temperature and force shear transform fat into liquid oil that is expelled from the die because of the high pressure [59]. Likewise, when high temperatures are reached a lipase and lipoxygenase activity decrease is observed with a reduction of fatty acid oxidation [60]. Total fat content in the extruded food may result lower than in the raw ingredients due to the complex lipid-amylose formation; this aggregation often occurs when raw materials with high free fatty acids and high amylose content are processed [61].

Other compounds that could be highly affected by the extrusion conditions are the bioactive constituents. Food antioxidant activity is related with phenolics compounds and their presence and functionality will depend on the extrusion process conditions. In general, high temperatures and high moisture content are responsible for the decarboxylation of phenolic compounds, thus their loss after extrusion [62]. Moreover, an antioxidant activity reduction of 60–68% and a decrease of 46–60% of total phenolics were found in barley extrudates [63]. Nevertheless, simultaneously extrusion can improve the level of these bioactive molecules. In fact, a progressive enhancement of the total phenolic compounds content was observed in extruded black rice when extrusion temperature increased and die pressure diminished [64].

In the case of vitamins, owing to their different structure and composition, their stability during extrusion may vary. Athar et al. [65] investigated the impact of extrusion on the vitamin retention in extruded foods. Authors reported that high barrel temperature and low feed moisture provoked ascorbic acid degradation in extruded snack. In whole oat grain and maize with pea grits, vitamin B retention was not related with its initial level, but it depended on the cereal type [65]. Among the fat-soluble vitamins, vitamin D and K are relatively stable compared with vitamin A and E [35]. Indeed, in oat, barley, wheat, rye and buckwheat a loss of 30% in tocopherol and tocotrienol was described [66]. Zieliński et al. [67] reported a reduction of about 63% in vitamin E when buckwheat groats were subjected to extrusion.

5.2 *Safety Issues Related to Extrusion Process*

Pertain to the cereal-based product, the issue of food safety is particularly crucial to preserve consumers health. Mycotoxins represent the main risk coming from plant-based commodities; particularly, those globally recognized as economically and toxicologically important: fumonisins, aflatoxins, deoxynivalenol and derivatives, zearalenone and derivatives and ochratoxins. Industrial food processes can somehow reduce the mycotoxins content in the final products. In the extrusion technology, this extent of the reduction may depend on the extruder and screw type, die configuration, initial mycotoxin concentration, barrel temperature, screw speed and raw material feed moisture. In cereals, fumonisins, aflatoxins and zearalenone levels declined by 100, 95 and 83%, respectively; while deoxynivalenol, ochratoxin A and moniliformin content dropped down by 55, 40 and 30%, respectively [68].

Another food safety problem related to extrusion is the acrylamide formation along the extruder barrels caused by high thermal and mechanical energy developed during the process. Acrylamide is one of a resulting products from the Maillard reaction, and it has been categorized as potential carcinogen to humans by the International Agency for Research on Cancer [69]. The amount of acrylamide formation is greatly dependent on the type of cereals used for the extrusion and particularly, their content of free asparagine. In fact, acrylamide formation requires the presence of this amino acid and reducing sugars (glucose) [70]. Because of the higher asparagine content, rye based extrudates contain higher levels of acrylamide compared to extrudates made by rice, maize and wheat [71]. A study carried out by Mulla et al. [72] investigated the effect of the extrusion parameters and some mitigating agents on the acrylamide formation in potato flakes and semolina blends. Acrylamide content was higher in those blends with more potato flour and it increased with low moisture and high die temperature. Calcium chloride at 50 mmol/g was able to reduce by 65% the level of acrylamide.

5.3 *Types of Extruded Foods*

5.3.1 Pellet Snacks

In general, pellet snacks (Fig. 7.6a) have similar characteristics to the starting dough. They are also known as unexpanded snacks. In fact, after extrusion they are considered like a half-product, which need an additional step to get the final expanded shape. For their preparation two kind of ingredients may be used: raw or precooked starch-based materials. During extrusion moisture level is kept around 25–35%, which does not allow the complete swelling of the starch granules and neither their gelatinization. At this stage, temperature ranges from 100 to 120 °C, then the viscous material is cooled passing through the last extrusion module at around 80 °C till reaching the die. In this type of snack, the absence of any air bubble in the final product is of key importance. For that purpose, after the extrusion the pellet is typically cut as a thin sheet (1 mm thick) and gently dried (50–60 °C for

Fig. 7.6 Extruded snacks. (**a**) pellet snacks; (**b**) directly expanded snacks; (**c**) co-extruded snacks

6–7 h) until reaching a hard-glassy consistency with moisture content of 10–12%. For pellet expansion, frying, baking or microwaving are employed to promote rapid heating that will convert water into vapor leading to puffing food [73]. At the end of the process, expanded pellets have a moisture content of 1–2% and they are generally subjected to a flavoring step to enhance the taste and the flavor [74].

5.4 *Directly Expanded Snacks*

This category represents the first form of industrially produced snacks. In the 1940s, Adams company started to produce this kind of snack using a single screw extruder and maize as ingredient. Usually, this extrusion requires low amount of moisture (16—18%). Nowadays, directly expanded snacks (Fig. 7.6b) are attracting more attention due to their sensorial quality like crunchiness. In fact, among the desirable qualities for a snack are high expansion rates, low densities, high specific volumes, light colors, and high freshness [75]. For the production of these snacks, the temperature in the first module of the extruder ranges from 80 to 150 °C, hence starch granules from maize, wheat or rice start to lose their organized crystalline structure becoming amorphous, then they are squeezed in the screw reverse section and dispersed into the new viscous matrix. Pressure reached during the extrusion process is around 40–80 atmospheres [76]. When the hot-viscous material goes out through the die, pressure drop off causing the water vaporization. The high temperature and pressure reached during the process, contribute into the formation of air cells resulting in the expansion of the product. The expansion ratio depends on both the die shape and viscosity of the material in the barrels [28]. When the viscous starch-based matrix cools down, glass transition starts, which usually occurs at 40–60 °C with 5–8% moisture content, resulting in the final brittle puffed structure, typical of the directly expanded snacks [73]. The shape of the snack depends on the die and the cutting action of the knives that makes possible the new tridimensional structures. Conventionally, flavorings of these snacks were coated at the end of the process, but new trends dictate the use of some flavor precursors (reducing sugar, amino acids and peptides) added directly into cereal blends (maize, wheat and rice), which with the high temperatures give rise to new colors and flavors [77, 78].

5.5 *Co-extruded Snacks*

Coextrusion technology involves the combination of two different materials in the extrusion die, coming from two extruders or from an extruder and a pump [79]. This technology extends the characteristics of the extruded foods varying textures, colors and flavors. Generally, the external part is composed by a dried cereal-based mixture, while the filler may be a material with solid, creamy (Fig. 7.6c) or gel-like consistency (fruit jam, ketchup, lemon cream or cheese cream) [80].

6 New Trends in Snacking

For decades, people have mostly considered as snacks, chips, biscuits and chocolate. They consumed this kind of "junk food" thinking of eating something with a reduced nutritional value. But the adoption of new lifestyles is pushing an increasing number of consumers, mainly new generations, to replace traditional meals with on-the-go snacks. Healthier and dietary patterns are driving the consumer selection for these products, making them very appropriate carriers for nutrients or bioactive compounds. For this purpose, food research has put the attention on nutritional enrichment of cereal-based products to improve consumers health. Although breads have received very much attention in the enriching strategies by adding pseudocereals, legumes and lately fruits and vegetables [81], snacks are following the same trend too [82]. With regards to snacks, the enrichment strategy has been quickly extended and it became even easier to find in the market snack foods made by pseudocereals, pulses and seaweed. Those trends have encouraged producers to look for alternative or new ingredients that imprint the innovative character.

From a technological point of view, there is constant innovation in processing technologies for creating alternative foods within the cereals-based snack segment. The implementation of new technologies is allowing to increase efficiency, to reduce the environmental impact and to develop new functional and healthy foods. Moreover, with the recent increase trend of having and sharing new cooking and sensorial experiences, people are more and more attracted towards this short bites' foods for initial tastings of diverse foods. This trend is global, those products are not really from the Mediterranean area, but it is the general consumers' tendency. In fact, most of this products, present new characteristics in terms of texture, color and flavor that contributes into the attention-grabbing features.

6.1 Ingredients

Many of different ingredients have been tested in snack foods, and some recent examples have been selected to highlight their importance. In the search for alternative ingredients, it has been observed two motivation drivers; first, to explore for ingredients that might confer additional healthy benefits to the snack food [82], and secondly, immersed in the sustainability trends, to investigate the revalorization of by-products by using them as unconventional ingredients in the snack production. For instance, some researchers have evaluated the effect of the incorporation of brewer's spent grain in breadsticks [83]. Dietary fiber content increased in breadsticks containing brewer's spent grain, although they were significantly darker, less crispy and had lower volume [83].

Pseudocereals, pulses, algae and fruits and vegetables are being included as ingredients in the snacks production, either as powders, flours or some specific extracted fractions with nutritional interest. Commonly snack bars are composed by a nutrient-poor ingredient and due to this they cannot be classified as functional foods. In order to overcome this weakness, researchers and manufacturers began focusing on the snack bar enrichment with bioactive compounds so as to ensure more healthy choices [26]. For instance, adding bean flour to oats-based snack bars allowed increasing the dietary fiber and protein content and the antioxidant capacity [84]. Gluten-free has been a driving force in the innovation of the food market, reaching also to snacks foods. Many studies have been carried out with this focus, introducing pseudocereals and pulses. With that purpose, buckwheat flour (10, 20 and 30%) was added to maize for making an extruded snack, improving the diameter, redness, phenolic content and antioxidant capacity, while reducing bulk density and water absorption index [85]. Cueto et al. [86] reported the impact of quinoa and chia flour (20 and 5%) on the physical characteristics of extrudates made by maize. Chia flour increased density and crunchiness but lowered expansion index, whereas quinoa extrudates showed smoother structure than those containing chia. Regarding pulses, lentil based extruded snacks show high content of some prebiotics (raffinose and stachyose) [87]. Beans were also incorporated into extruded snacks made with blends of carob fruit and rice with a subsequent enhancement of the phenolic compounds [88]. Other non-conventional seeds with an attractive nutritional profile have also been used like flaxseed, which is rich in α-linolenic acid, lignans and dietary fiber [89]. Flaxseed flour added at different levels (6, 12, and 18%) to oat-based bars affects their color, decreasing lightness and increasing redness, but provides a significant increase in polyunsaturated fat and dietary fiber with good sensory acceptance up to 12% addition [90].

Several studies have been carried out exploring the incorporation of algae in snacks foods. In the study carried out by Batista et al. [91] the influence of four microalgae (*Arthrospira platensis, Chlorella vulgaris Allma, Tetraselmis suecica* and *Phaeodactylum tricornutum*) addition (2 and 6%) was evaluated on crackers quality. Interesting results were obtained, since no differences in terms of structure were observed, thus gas retention was not significantly altered by microalgae adding. Samples with 6% of *Arthrospira platensis* and *Chlorella vulgaris Allma* presented the highest protein content. *Arthrospira platensis* samples reached the highest antioxidant activity and achieved better sensory analysis scores. Conversely, *Tetraselmis suecica* and *Phaeodactylum tricornutum* received the lowest liking score in the sensory test. The effect of algae was also tested in breadsticks by [92]. Authors evaluated the addition (1.5%) of two types of microalgae (*Chlorella vulgaris* and *Arthrospira platensiswere*) to wheat-based breadsticks. Microalgae addition decreased hardness, resilience, crispiness and brittleness of the breadsticks, but nutritionally snacks had high mineral content, specifically iron and selenium [92].

Among the most consumed cereals bars, those containing fruits have great popularity, especially those with certain berries, likely related with the perception of their health benefits besides their appearance and taste [93]. In a recent *in vivo* study, freeze-dried black raspberries (10 and 20%) and cranberry extract (0.5 and 1%) were incorporated into rice crisp cereal bars, comparing the results with those of a reference bar [94]. Raspberries bars dulled postprandial insulin peak and slightly enhanced the glycemic responses to a high carbohydrate food [94]. Likewise, fibers coming from grapes and orange seeds were used (2.9%) to prepare crackers, with the additional benefit of increasing antioxidant capacity and phenolics compounds [95]. Fibers from cooked pear apple co-product were incorporated in cereal bars [96], showing high antioxidant capacity and retaining a good amount of phenolics compounds [97].

More exotic ingredients added to crackers are the residues of *Hibiscus sabdariffa* [98]. When they were incorporated at different levels (1.25, 2.5, 3.75, and 5%), crackers contained less protein and fat while more ash and total dietary fiber compared to the control (0%). Authors found that adding 5% of *Hibiscus sabdariffa* increased phenols and flavonoid content from 5.99 to 17.57 mg/g and from 49.36 to 104.63 mg/g, respectively; consequently, enhancing twice the 2,2-Diphenyl-1-picrylhydrazyl (DPPH) radical scavenging activity. Sensory test decreed that crackers with 3.75% of plant incorporation was the most acceptable sample [98].

One of the biggest challenges for the quality improvement of savory snacks that is still ongoing is salt reduction. In most of salty snacks, sodium chloride is employed as a flavor enhancer. The exceeding of sodium daily intake (10–20 mmol/day) may increase blood pressure and affect cardiovascular health [99]. In the last years, the presence of salt reduced products on the grocery stores shelves has raised, in which sodium has been replaced with potassium, calcium, magnesium and other flavor enhancers [100].

Fat reduction in the extruded snacks have been approached modifying processing technologies and with new ingredients combinations. For instance, a reduction of 15% of oil level was achieved by spraying flavors contained in an aqueous hydrocolloid solution rather than traditional oil-based solutions [101].

6.2 *Technological Innovations*

6.2.1 Supercritical Carbon Dioxide Extrusion

As mentioned before, extrusion cooking is commonly applied in cereal-based snack production because of its flexibility. Nevertheless, high temperatures may originate some disadvantages like Maillard reaction products [102] and antioxidant activity reduction [103]. To avoid these phenomena and to produce high-quality snacks, in the last years a new extrusion system has been implemented in the food industry. Carbon dioxide supercritical fluid has been incorporated to the traditional extruder. This invention allows obtaining a large variety of puffed snacks at lower temperature (<100 °C) [104]. Among the advantages of this technology, energy savings, preservation of highly nutritional compounds and blocking the formation of harmful molecules released at high temperatures are counting the most. Carbon dioxide is a super critical fluid, behaving in between gases and liquids at temperatures in the range 60–80 °C. Carbon dioxide rapidly solubilizes in the blend, generating the nuclei that will produce the air pockets when the pressure drops at the die exit [105]. Thus, gas expansion and diffusion are responsible for the snack expansion, which starts from the core and progressively migrates to the edge. Resulting snacks have smooth and uniform surface and regular shape with internal porous [106]. Many studies have described the application of this technology for the obtention of new cereal-based snacks understanding the process impact on the nutritional quality of the final products. In lentil and pre-gelatinized potato starch extrudates, total phenolics and DPPH radical scavenging activity increased by 30% and 18%, respectively [107]. Bilgi Boyaci et al. [108] demonstrated that CO_2 cold extrusion increased the retention of some thermolabile compounds such as thiamine and riboflavin in maize extrudates. The role of supercritical CO_2 as plasticizer and blowing instrument was also studied in puffed rice fortified with protein, dietary fiber and micronutrients, retaining all added minerals, 55–58% of vitamin A and 64–76% of vitamin C, besides complete bioavailability of lysine (98.6%) that was not blocked by Maillard reaction [109]. Nevertheless, there is much way ahead for this technology to become fully exploited.

6.2.2 3D Printing

Digitalization revolution has also reached snack production and 3D printing technology opens great opportunities to the food industry. This innovative process is digitally controlled by a robotic system that allows manufacturing three-dimensional objects built on a layer-by-layer deposition [110]. Until now, this futuristic approach has been applied in different field as medicine, pharmaceutical, biotechnology, engineering and more recently has been applied in the food industry [111]. Moreover, 3D printing could be the technological tool for making at home tailored made foods and more personalized nutrition [112]. Due to its low cost, ease to use, less waste, versatility and customizability, food 3D extrusion has become one of the most extensively explored technology [113]. Using this modern approach, edible ingredients with soft consistency are loaded into a cylinder and by a piston force, they are extruded through a nozzle in sequential layers that are then closely adhered [114]. The 3D shape is previously designed by a software that communicates with the 3D printer to produce the desired design. Essentially, 3D extrusion process is articulated in five main steps: powder preparation, binding method selection, binder selection, process specifications definition and post processing operations [115]. First applications of 3D printing were based on blends of starch, sugar, maize syrup and yeast mixture [116]. Lately, system optimization has allowed many researchers to carry out new 3D printing food applications. Different 3D products with various matrixes such as cereals dough, sugar powder, processed cheese, meat and fruit and vegetables have been extruded [117]. Krishnaraj et al. [118] applied this technology to produce snacks made by composite flour such as barnyard millet, green gram, fried gram, and ajwain seeds. Less explored ingredients like edible insect have been blended (at 20% addition level) with wheat flour for obtaining high protein snacks [114], while fruit and vegetables have provided nutritious tailored made fruit-based snacks [119]. Nevertheless, the manufacturing of complex food matrixes with this printing technology is still a challenge and a deep understanding of food rheology is needed to boost this food technology [120].

6.3 Other Snacks Globally Consumed

6.3.1 Tortilla Chips

In recent years, tortilla chips (Fig. 7.7a) have achieved a large acceptability among worldwide consumers. Originally produced in Mexico and Central America, they firstly spread to United States and then Europe [121]. In 2018, their global market represented USD 20.28 billion and from 2019 to 2025 it is supposed to reach a CAGR of 4.41% [122]. Mainly, tortilla chips are maize-based and are baked before frying. First step is the production of the "masa" (dough), for that maize kernels are mixed with three parts of water and 1% of lime (maize weight) and subjected to an alkaline (0.2–2% calcium hydroxide maize weight) cooking process (85–100 °C) for 15–45 min [123]. After cooking, the nixtamal is left resting for 8–16 h and then washed with water to remove the excess of lime and most of the pericarp. Washed nixtamal is grounded in a stone mill until obtaining a consistent dough, named "masa". At this point masa can be kneaded using mixers or extruders, before going to a roll sheeter. A rotating cutter gives a typical triangular shape before baking (280–302 °C for 30–45 s) and frying (165–195 °C for 50–90 s). Immediately after, tortilla chips are regularly salted and flavored with cheese, hot/spicy, barbecue, lemon salt or jalapeño. To boost shelf life and texture, the traditional formulations may include also gums, emulsifiers, acidulants and preservatives (e.g., sorbates and/ or propionates) [124].

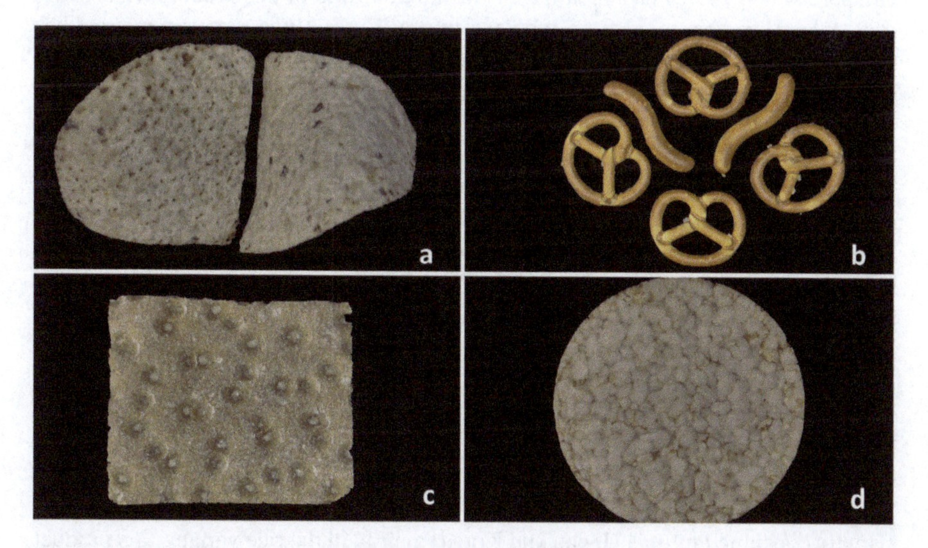

Fig. 7.7 Other snacks globally consumed. (**a**) tortilla chips; (**b**) pretzel; (**c**) crispbread; (**d**) puffed rice cake

6.3.2 Hard Pretzel Snacks

Pretzel (Fig. 7.7b) were introduced in the United States in the late eighteenth century by German-Swiss immigrants. Since then, pretzel consumption started to grow in the global food market and Pennsylvania occupies the first position in hard pretzel snacks production. Traditionally, these snacks are produced by wheat flour and have the distinctive knot shape. Conventional process consists in mixing, forming with a low pressure extruder, cooking in a hot alkali solution and two stages baking [125]. Wheat flour, yeast or baking soda are mixed with hot water (38 °C) to obtain a soft dough. During this step, yeast activity must be controlled to avoid a massive gas production and the subsequent rupture of gluten network. After a short resting time, dough is shaped passing through some extruder dies under low pressure. The dough strands obtained are folded into knot shape and cooked in a boiling alkali solution (1–1.5% sodium hydroxide) for about 10–15 s. During this step, hot alkali solution gelatinizes the superficial starch and causes the protein hydrolysis. Following, coarse salt is sprinkled on the cooked pretzel and baked afterwards. The previously hydrolyzed carbohydrates and proteins participate in the Maillard reaction that takes places during baking, resulting in the exclusive hard texture and glossy brown color. The first phase of baking reduces the moisture up to 8–10% then is further reduced till 4% using a kiln. Pretzels are packed after cooling down [126].

6.3.3 Crispbread

Crispbread (Fig. 7.7c) has been baked for the first time in the Scandinavian countries, where was prepared from wholegrain rye. Currently, a renewed interest has permitted the development of new crispbreads from different cereals such as wheat, rice, maize and pseudocereals. Its worldwide success is principally attributable to its nutritional quality and its long shelf-life (few months). Traditionally, it is prepared by rolling and sheathing the dough before baking, but the advent of the extrusion technology changed the processing. Flours and usually milk powder, vegetables oil, sugar and salt are premixed and conveyed into a co-rotating twin screw extruder. Here, raw ingredients are cooked by high temperature steam under low moisture (10–15%). Products with an alternative fine cell structure are obtained using supercritical fluid extrusion [127]. The extruded product may undergo an additional toasting process, or it may be coated with chocolate. Packing in foil, trays or cardboard boxes represents the last production step [128].

6.3.4 Puffed Grain Cake

Largely consumed in Asia (Japan and Korea) mainly in the rice variant, this product has lately booming up in Europe, where is consumed principally as snack [129]. At the beginning, it was introduced as alternative bread for celiac people, but due to its low calorie profile and being a good source of fiber, many consumers have come close to this snack [130]. Nowadays, there is a wide range of puffed grain cakes. Besides the classic rice (Fig. 7.7d), other cereals such as whole rice, maize and rye are used for their production. Some brief information is following to understand the process. The production of rice puffed cake starts with the soaking of raw rice in water until reaching the target moisture content (16–20%, w/w). After that, moist rice is ready to feed the cast-iron mold of the popping machine. The process core is the rice expansion induced by heat and pressure. Molds are heated (190–250 °C) and rapidly closed by a slide lid, which provides the internal vacuum. Meanwhile the steam pressure builds and after 8–10 s the rice amalgam expands filling all the available space. Once opening the mold, the fast pressure release causes a flash vaporization of the superheated water and steam. Cake is driven through a belt to the cooling zone where salt or flavor are sprayed before packaging [131].

7 Concluding Remarks

In Mediterranean area, cereal-based snacks represent a market segment that is undergoing a fast growth. Technological development and the study of the biochemical interaction among the ingredients have added substantial improvements to the process and product development. The outcomes have been reflected in the launching of snacks with enhanced nutritional quality and food safety. Along with the classic snacks, results of tradition, new ingredients and novel products have appeared in the Mediterranean food market. Concluding, snack products are a constantly evolving foods, and they are strongly related with the costumer's behaviors and on the introduction of new food technologies.

Acknowledgements Authors acknowledge the financial support from Spanish Ministry of Science, Innovation and Universities (RTI2018-095919-B-C21), the European Regional Development Fund (FEDER) and Generalitat Valenciana for financial support (Prometeo 2017/189). N. Gasparre thanks for his predoctoral fellowship Santiago Grisolia (P/2017/104).

References

1. Ovaskainen M-L, Tapanainen H, Pakkala H. Changes in the contribution of snacks to the daily energy intake of Finnish adults. Appetite. 2010;54(3):623–6. https://doi.org/10.1016/j.appet.2010.03.012.
2. Ng SW, Zaghloul S, Ali H, Harrison G, Yeatts K, El Sadig M, Popkin BM. Nutrition transition in the United Arab Emirates. Eur J Clin Nutr. 2011;65(12):1328–37. https://doi.org/10.1038/ejcn.2011.135.
3. Hess JM, Jonnalagadda SS, Slavin JL. What is a snack, why do we snack, and how can we choose better snacks? A review of the definitions of snacking, motivations to snack, contributions to dietary intake, and recommendations for improvement. Adv Nutr. 2016;7(3):466–75. https://doi.org/10.3945/an.115.009571.

4. Drapeau V, Pomerleau S, Pomerleau V. Snacking and energy balance in humans; 2016. https://doi.org/10.1016/B978-0-12-802928-2.00025-4.
5. MordorIntelligence. Snack food market - growth, trends and forecasts (2020–2025). 2020. https://www.mordorintelligence.com/industry-reports/snack-food-market.
6. Forecast MD. Europe Snack Products Market by Type (Salted Snacks, Bakery Snacks, Confectionary, Speciality and Frozen Snacks), by Distribution Channel (Specialist Retailers, Internet Sales, Super Markets, Convenience Stores and Others) and by Region - Industry Analysis, Size, Share, Growth, Trends, and Forecasts (2017–2022). 2019. https://www.marketdataforecast.com/market-reports/europe-snack-products-market. Accessed 18 Feb 2020.
7. Statista. Snack food Europe. 2020. https://www.statista.com/outlook/40110000/102/snack-food/europe. Accessed 6 Apr 2020.
8. Sofi F, Abbate R, Gensini GF, Casini A. Accruing evidence on benefits of adherence to the Mediterranean diet on health an updated systematic review and meta-analysis. Am J Clin Nutr. 2010;92(5):1189–96. https://doi.org/10.3945/ajcn.2010.29673.
9. Knoops KTB, de Groot L, Kromhout D, Perrin AE, Moreiras-Varela O, Menotti A, van Staveren WA. Mediterranean diet, lifestyle factors, and 10-year mortality in elderly European men and women - the HALE project. J Am Med Assoc. 2004;292(12):1433–9. https://doi.org/10.1001/jama.292.12.1433.
10. Serra-Majem L, Roman B, Estruch R. Scientific evidence of interventions using the Mediterranean diet: a systematic review. Nutr Rev. 2006;64(2):S27–47. https://doi.org/10.1111/j.1753-4887.2006.tb00232.x.
11. Rosell CM. Chapter 1 - the science of Doughs and bread quality. In: Preedy VR, Watson RR, Patel VB, editors. Flour and breads and their fortification in health and disease prevention. San Diego: Academic; 2011. p. 3–14. https://doi.org/10.1016/B978-0-12-380886-8.10001-7.
12. Kweon M, Slade L, Levine H, Gannon D. Cookie- versus cracker-baking—what's the difference? Flour functionality requirements explored by SRC and Alveography. Crit Rev Food Sci Nutr. 2014;54(1):115–38. https://doi.org/10.1080/10408398.2011.578469.
13. Hui YH. Handbook of food science, technology, and engineering, vol. 4 set; 2006.
14. Rodriguez-Carrasco Y, Font G, Molto JC, Berrada H. Quantitative determination of trichothecenes in breadsticks by gas chromatography-triple quadrupole tandem mass spectrometry. Food Addit Contam Part A Chem Anal Control Expo Risk Assess. 2014;31(8):1422–30. https://doi.org/10.1080/19440049.2014.926399.
15. Zeppa G, Rolle L, Piazza L. Textural characteristics of typical italian "grissino stirato" and "rubatà" bread-sticks. Ital J Food Sci. 2007;19:449–59.
16. Caponio F, Summo C, Pasqualone A, Paradiso VM, Gomes T. Influence of processing and storage time on the Lipidic fraction of Taralli. J Food Sci. 2009;74(9):C701–6. https://doi.org/10.1111/j.1750-3841.2009.01357.x.
17. Pagani MA, Lucisano M, Mariotti M. Traditional Italian products from wheat and other starchy flours. In: Sinha N, editor. Handbook of food products manufacturing. Hoboken: Wiley-Interscience; 2007. p. 327–81.
18. Barbieri S, Bendini A, Balestra F, Palagano R, Rocculi P, Gallina Toschi T. Sensory and instrumental study of Taralli, a typical Italian bakery product. Eur Food Res Technol. 2018;244(1):73–82. https://doi.org/10.1007/s00217-017-2937-8.
19. Caponio F, Giarnetti M, Summo C, Gomes T. Influence of the different oils used in dough formulation on the lipid fraction of Taralli. J Food Sci. 2011;76(4):C549–54. https://doi.org/10.1111/j.1750-3841.2011.02113.x.
20. Berk Z. Frying, baking, roasting. Food process engineering and technology. 2nd ed. San Diego: Elsevier Academic Press Inc; 2013. https://doi.org/10.1016/b978-0-12-415923-5.00024-1.
21. Morales FJ, Arribas-Lorenzo G. The formation of potentially harmful compounds in churros, a Spanish fried-dough pastry, as influenced by deep frying conditions. Food Chem. 2008;109(2):421–5. https://doi.org/10.1016/j.foodchem.2007.12.042.

22. Delgado-Andrade C, Morales FJ, Seiquer I, Navarro MP. Maillard reaction products profile and intake from Spanish typical dishes. Food Res Int. 2010;43(5):1304–11. https://doi.org/10.1016/j.foodres.2010.03.018.
23. Mir-Bel J, Oria R, Salvador ML. Reduction in hydroxymethylfurfural content in 'churros', a Spanish fried dough, by vacuum frying. Int J Food Sci Technol. 2013;48(10):2042–9. https://doi.org/10.1111/ijfs.12182.
24. Bower JA, Whitten R. Sensory characteristics and consumer liking for cereal bar snack foods. J Sens Stud. 2000;15(3):327–45. https://doi.org/10.1111/j.1745-459X.2000.tb00274.x.
25. GrandViewResearch. Snack bars market size, share & trends analysis report by product (breakfast, energy & nutrition, granola/muesli, fruit), by distribution channel (convenience store, online, super/hypermarket), and segment forecasts, 2019–2025. 2020. https://www.grandviewresearch.com/industry-analysis/snack-bars-market. Accessed 22 Feb 2020.
26. Pinto VRA, Freitas TBD, Dantas MID, Della Lucia SM, Melo LF, Minim VPR, Bressan J. Influence of package and health-related claims on perception and sensory acceptability of snack bars. Food Res Int. 2017;101:103–13. https://doi.org/10.1016/j.foodres.2017.08.062.
27. Constantin O, Istrati D. Functional properties of snack bars. 2018. https://doi.org/10.5772/intechopen.81020.
28. Brennan MA, Derbyshire E, Tiwari BK, Brennan CS. Ready-to-eat snack products: the role of extrusion technology in developing consumer acceptable and nutritious snacks. Int J Food Sci Technol. 2013;48(5):893–902. https://doi.org/10.1111/ijfs.12055.
29. Navarro-Cortez RO, Hernández-Santos B, Gómez-Aldapa CA, Castro-Rosas J, Herman-Lara E, Martínez-Sánchez CE, Juárez-Barrientos JM, Antonio-Cisneros CM, Rodríguez-Miranda J. Development of extruded ready-to-eat snacks using pumpkin seed (Cucurbita pepo) and nixtamalized maize (Zea mays) flour blends. Revista Mexicana de Ingeniera Quimica. 2016;15(2):409–22.
30. MarketsandMarkets. Extruded Snacks Market by Type (Simply extruded, Expanded, Co-extruded), Raw Material (Wheat, Potato, Corn, Oats, Rice, Multigrain), Manufacturing Method (Single-screw, Twin-screw), Distribution Channel, and Region – Global Forecast to 2026. 2016. https://www.marketsandmarkets.com/Market-Reports/extruded-snacks-market-139554331.html.
31. Jing Y, Chi Y-J. Effects of twin-screw extrusion on soluble dietary fibre and physicochemical properties of soybean residue. Food Chem. 2013;138(2):884–9. https://doi.org/10.1016/j.foodchem.2012.12.003.
32. Riaz M. Food extruders; 2013. pp. 427–440. https://doi.org/10.1016/B978-0-12-385881-8.00016-1.
33. Gujral HS, Sharma P, Rachna S. Effect of sand roasting on beta glucan extractability, physicochemical and antioxidant properties of oats. LWT Food Sci Technol. 2011;44(10):2223–30. https://doi.org/10.1016/j.lwt.2011.06.001.
34. Shahidi F. Nutraceuticals and functional foods: whole versus processed foods. Trends Food Sci Technol. 2009;20(9):376–87. https://doi.org/10.1016/j.tifs.2008.08.004.
35. Singh S, Gamlath S, Wakeling L. Nutritional aspects of food extrusion: a review. Int J Food Sci Technol. 2007;42(8):916–29. https://doi.org/10.1111/j.1365-2621.2006.01309.x.
36. Martínez MM, Calviño A, Rosell CM, Gómez M. Effect of different extrusion treatments and particle size distribution on the physicochemical properties of Rice flour. Food Bioprocess Technol. 2014;7(9):2657–65. https://doi.org/10.1007/s11947-014-1252-7.
37. Souza RCR, Andrade CT. Investigation of the gelatinization and extrusion processes of corn starch. Adv Polym Technol. 2002;21(1):17–24. https://doi.org/10.1002/adv.10007.
38. Zhang Y, Liu W, Liu C, Luo S, Li T, Liu Y, Wu D, Zuo Y. Retrogradation behaviour of high-amylose rice starch prepared by improved extrusion cooking technology. Food Chem. 2014;158:255–61. https://doi.org/10.1016/j.foodchem.2014.02.072.
39. Bello-Pérez LA, Ottenhof MA, Agama-Acevedo E, Farhat IA. Effect of storage time on the retrogradation of banana starch extrudate. J Agric Food Chem. 2005;53(4):1081–6. https://doi.org/10.1021/jf048858l.

40. Waramboi JG, Gidley MJ, Sopade PA. Influence of extrusion on expansion, functional and digestibility properties of whole sweetpotato flour. LWT - Food Sci Technol. 2014;59(2, Part 1):1136–45. https://doi.org/10.1016/j.lwt.2014.06.016.
41. Román L, Martínez MM, Rosell CM, Gómez M. Changes in physicochemical properties and in vitro starch digestion of native and extruded maize flours subjected to branching enzyme and maltogenic α-amylase treatment. Int J Biol Macromol. 2017;101:326–33. https://doi.org/10.1016/j.ijbiomac.2017.03.109.
42. Liu Y, Chen J, Luo S, Li C, Ye J, Liu C, Gilbert RG. Physicochemical and structural properties of pregelatinized starch prepared by improved extrusion cooking technology. Carbohydr Polym. 2017;175:265–72. https://doi.org/10.1016/j.carbpol.2017.07.084.
43. Kim JH, Tanhehco EJ, Ng PKW. Effect of extrusion conditions on resistant starch formation from pastry wheat flour. Food Chem. 2006;99(4):718–23. https://doi.org/10.1016/j.foodchem.2005.08.054.
44. De Pilli T, Derossi A, Talja RA, Jouppila K, Severini C. Study of starch-lipid complexes in model system and real food produced using extrusion-cooking technology. Innovative Food Sci Emerg Technol. 2011;12(4):610–6. https://doi.org/10.1016/j.ifset.2011.07.011.
45. Anderson JW, Baird P, Davis RH, Ferreri S, Knudtson M, Koraym A, Waters V, Williams CL. Health benefits of dietary fiber. Nutr Rev. 2009;67(4):188–205. https://doi.org/10.1111/j.1753-4887.2009.00189.x.
46. Rincón-León F. FUNCTIONAL FOODS. In: Caballero B, editor. Encyclopedia of food sciences and nutrition. 2nd ed. Oxford: Academic; 2003. p. 2827–32. https://doi.org/10.1016/B0-12-227055-X/01328-6.
47. Rashid S, Rakha A, Anjum FM, Ahmed W, Sohail M. Effects of extrusion cooking on the dietary fibre content and water solubility index of wheat bran extrudates. Int J Food Sci Technol. 2015;50(7):1533–7. https://doi.org/10.1111/ijfs.12798.
48. Alam MS, Pathania S, Sharma A. Optimization of the extrusion process for development of high fibre soybean-rice ready-to-eat snacks using carrot pomace and cauliflower trimmings. LWT. 2016;74:135–44. https://doi.org/10.1016/j.lwt.2016.07.031.
49. Zhong L, Fang Z, Wahlqvist ML, Hodgson JM, Johnson SK. Extrusion cooking increases soluble dietary fibre of lupin seed coat. LWT. 2019;99:547–54. https://doi.org/10.1016/j.lwt.2018.10.018.
50. Oliveira LC, Rosell CM, Steel CJ. Effect of the addition of whole-grain wheat flour and of extrusion process parameters on dietary fibre content, starch transformation and mechanical properties of a ready-to-eat breakfast cereal. Int J Food Sci Technol. 2015;50(6):1504–14. https://doi.org/10.1111/ijfs.12778.
51. Robin F, Schuchmann HP, Palzer S. Dietary fiber in extruded cereals: limitations and opportunities. Trends Food Sci Technol. 2012;28(1):23–32. https://doi.org/10.1016/j.tifs.2012.06.008.
52. Chanvrier H, Nordström Pillin C, Vandeputte G, Haiduc A, Leloup V, Gumy J-C. Impact of extrusion parameters on the properties of rice products: a physicochemical and X-ray tomography study. Food Struct. 2015;6:29–40. https://doi.org/10.1016/j.foostr.2015.06.004.
53. Arribas C, Cabellos B, Sánchez C, Cuadrado C, Guillamón E, Pedrosa MM. The impact of extrusion on the nutritional composition, dietary fiber and in vitro digestibility of gluten-free snacks based on rice, pea and carob flour blends. Food Funct. 2017;8(10):3654–63. https://doi.org/10.1039/c7fo00910k.
54. Kaur S, Sharma S, Singh B, Dar BN. Effect of extrusion variables (temperature, moisture) on the antinutrient components of cereal brans. J Food Sci Technol. 2015;52(3):1670–6. https://doi.org/10.1007/s13197-013-1118-4.
55. Wani SA, Kumar P. Effect of extrusion on the nutritional, antioxidant and microstructural characteristics of nutritionally enriched snacks. J Food Process Preservat. 2016;40(2):166–73. https://doi.org/10.1111/jfpp.12593.
56. Alonso R, Rubio LA, Muzquiz M, Marzo F. The effect of extrusion cooking on mineral bioavailability in pea and kidney bean seed meals. Anim Feed Sci Technol. 2001;94(1):1–13. https://doi.org/10.1016/S0377-8401(01)00302-9.

57. Moscicki L. Extrusion-cooking techniques. Applications, theory and sustainability. Hoboken: Wiley; 2011.
58. Camire ME. 6 - extrusion and nutritional quality. In: Guy R, editor. Extrusion cooking. Sawston: Woodhead Publishing; 2001. p. 108–29. https://doi.org/10.1533/9781855736313.1.108.
59. Sandrin R, Caon T, Zibetti AW, de Francisco A. Effect of extrusion temperature and screw speed on properties of oat and rice flour extrudates. J Sci Food Agric. 2018;98(9):3427–36. https://doi.org/10.1002/jsfa.8855.
60. Alam MS, Kaur J, Khaira H, Gupta K. Extrusion and extruded products: changes in quality attributes as affected by extrusion process parameters: a review. Crit Rev Food Sci Nutr. 2016;56(3):445–73. https://doi.org/10.1080/10408398.2013.779568.
61. Tumuluru JS, Sokhansanj S, Bandyopadhyay S, Bawa AS. Changes in moisture, protein, and fat content of fish and Rice flour Coextrudates during single-screw extrusion cooking. Food Bioprocess Technol. 2013;6(2):403–15. https://doi.org/10.1007/s11947-011-0764-7.
62. Brennan C, Brennan M, Derbyshire E, Tiwari BK. Effects of extrusion on the polyphenols, vitamins and antioxidant activity of foods. Trends Food Sci Technol. 2011;22(10):570–5. https://doi.org/10.1016/j.tifs.2011.05.007.
63. Altan A, McCarthy KL, Maskan M. Effect of extrusion process on antioxidant activity, total phenolics and beta-glucan content of extrudates developed from barley-fruit and vegetable by-products. Int J Food Sci Technol. 2009;44(6):1263–71. https://doi.org/10.1111/j.1365-2621.2009.01956.x.
64. Hu Z, Tang X, Zhang M, Hu X, Yu C, Zhu Z, Shao Y. Effects of different extrusion temperatures on extrusion behavior, phenolic acids, antioxidant activity, anthocyanins and phytosterols of black rice. RSC Adv. 2018;8(13):7123–32. https://doi.org/10.1039/c7ra13329d.
65. Athar N, Hardacre A, Taylor G, Clark S, Harding R, McLaughlin J. Vitamin retention in extruded food products. J Food Compos Anal. 2006;19(4):379–83. https://doi.org/10.1016/j.jfca.2005.03.004.
66. Tiwari U, Cummins E. Nutritional importance and effect of processing on tocols in cereals. Trends Food Sci Technol. 2009;20(11):511–20. https://doi.org/10.1016/j.tifs.2009.06.001.
67. Zieliński H, Michalska A, Piskuła MK, Kozłowska H. Antioxidants in thermally treated buckwheat groats. Mol Nutr Food Res. 2006;50(9):824–32. https://doi.org/10.1002/mnfr.200500258.
68. Castells M, Marín S, Sanchis V, Ramos AJ. Fate of mycotoxins in cereals during extrusion cooking: a review. Food Addit Contamin. 2005;22(2):150–7. https://doi.org/10.1080/02652030500037969.
69. Tamanna N, Mahmood N. Food processing and Maillard reaction products: effect on human health and nutrition. Int J Food Sci. 2015;2015:526762. https://doi.org/10.1155/2015/526762.
70. Kim CT, Hwang E-S, Lee HJ. Reducing acrylamide in fried snack products by adding amino acids. J Food Sci. 2005;70(5):C354–8. https://doi.org/10.1111/j.1365-2621.2005.tb09966.x.
71. Konings EJM, Ashby P, Hamlet CG, Thompson GAK. Acrylamide in cereal and cereal products: a review on progress in level reduction. Food Addit Contamin. 2007;24(Suppl 1):47–59. https://doi.org/10.1080/02652030701242566.
72. Mulla MZ, Bharadwaj VR, Annapure US, Singhal RS. Effect of formulation and processing parameters on acrylamide formation: a case study on extrusion of blends of potato flour and semolina. LWT-Food Sci Technol. 2011;44(7):1643–8. https://doi.org/10.1016/j.lwt.2010.11.019.
73. Guy R. 8 - snack foods. In: Guy R, editor. Extrusion cooking. Sawston: Woodhead Publishing; 2001. p. 161–81. https://doi.org/10.1533/9781855736313.2.161.
74. Bawa AS, Sidhu JS. SNACK FOODS | range on the market. In: Caballero B, editor. Encyclopedia of food sciences and nutrition. 2nd ed. Oxford: Academic; 2003. p. 5322–32. https://doi.org/10.1016/B0-12-227055-X/01096-8.
75. Cuj-Laines R, Hernández-Santos B, Herman-Lara E, Martínez-Sánchez CE, Juárez-Barrientos JM, Torruco Uco JG, Rodríguez-Miranda J. Chapter 5 - relevant aspects of the development of extruded high-protein snacks: an alternative to reduce global undernourishment. In: Holban

AM, Grumezescu AM, editors. Alternative and Replacement Foods. Cambridge: Academic; 2018. p. 141–66. https://doi.org/10.1016/B978-0-12-811446-9.00005-8.
76. Guy R. 2 - raw materials for extrusion cooking. In: Guy R, editor. Extrusion cooking. Sawston: Woodhead Publishing; 2001. p. 5–28. https://doi.org/10.1533/9781855736313.1.5.
77. Milani TMG, Menis MEC, Jordano A, Boscolo M, Conti-Silva AC. Pre-extrusion aromatization of a soy protein isolate using volatile compounds and flavor enhancers: effects on physical characteristics, volatile retention and sensory characteristics of extrudates. Food Res Int. 2014;62:375–81. https://doi.org/10.1016/j.foodres.2014.03.018.
78. Davidek T, Festring D, Dufossé T, Novotny O, Blank I. Study to elucidate formation pathways of selected roast-smelling odorants upon extrusion cooking. J Agric Food Chem. 2013;61(43):10215–9. https://doi.org/10.1021/jf4004237.
79. Cindio BD, Gabriele D, Pollini CM, Peressini D, Sensidoni A. Filled snack production by coextrusion-cooking: 1. Rheological modelling of the process. J Food Eng. 2002;52(1):67–74. https://doi.org/10.1016/S0260-8774(01)00087-5.
80. Peressini D, Sensidoni A, Pollini CM, Gabriele D, Migliori M, de Cindio B. Filled-snacks production by co-extrusion-cooking. Part 3. A rheological-based method to compare filler processing properties. J Food Eng. 2002;54(3):227–40. https://doi.org/10.1016/S0260-8774(01)00208-4.
81. Betoret E, Rosell CM. Enrichment of bread with fruits and vegetables: trends and strategies to increase functionality. Cereal Chem. 2020;97(1):9–19. https://doi.org/10.1002/cche.10204.
82. Maetens E, Hettiarachchy N, Dewettinck K, Horax R, Moens K, Moseley DO. Physicochemical and nutritional properties of a healthy snack chip developed from germinated soybeans. LWT-Food Sci Technol. 2017;84:505–10. https://doi.org/10.1016/j.lwt.2017.06.020.
83. Ktenioudaki A, Chaurin V, Reis SF, Gallagher E. Brewer's spent grain as a functional ingredient for breadsticks. Int J Food Sci Technol. 2012;47(8):1765–71. https://doi.org/10.1111/j.1365-2621.2012.03032.x.
84. Ramírez-Jiménez AK, Gaytán-Martínez M, Morales-Sánchez E, Loarca-Piña G. Functional properties and sensory value of snack bars added with common bean flour as a source of bioactive compounds. LWT. 2018;89:674–80. https://doi.org/10.1016/j.lwt.2017.11.043.
85. Singh JP, Kaur A, Singh B, Singh N, Singh B. Physicochemical evaluation of corn extrudates containing varying buckwheat flour levels prepared at various extrusion temperatures. J Food Sci Technol Mysore. 2019;56(4):2205–12. https://doi.org/10.1007/s13197-019-03703-y.
86. Cueto M, Porras-Saavedra J, Farroni A, Alamilla-Beltran L, Schoenlechner R, Schleining G, Buera P. Physical and mechanical properties of maize extrudates as affected by the addition of chia and quinoa seeds and antioxidants. J Food Eng. 2015;167:139–46. https://doi.org/10.1016/j.jfoodeng.2015.07.027.
87. Morales P, Berrios JDJ, Varela A, Burbano C, Cuadrado C, Muzquiz M, Pedrosa MM. Novel fiber-rich lentil flours as snack-type functional foods: an extrusion cooking effect on bioactive compounds. Food Funct. 2015;6(9):3135–43. https://doi.org/10.1039/c5fo00729a.
88. Arribas C, Pereira E, Barros L, Alves MJ, Calhelha RC, Guillamon E, Pedrosa MM, Ferreira I. Healthy novel gluten-free formulations based on beans, carob fruit and rice: extrusion effect on organic acids, tocopherols, phenolic compounds and bioactivity. Food Chem. 2019;292:304–13. https://doi.org/10.1016/j.foodchem.2019.04.074.
89. Hall C, Ndsu. Flaxseed as a functional food. Proceedings of the 59th flax Institute of the United States. North Dakota state Univ, flax Inst, Fargo; 2002.
90. Khouryieh H, Aramouni F. Effect of flaxseed flour incorporation on the physical properties and consumer acceptability of cereal bars. Food Sci Technol Int. 2013;19(6):549–56. https://doi.org/10.1177/1082013212462231.
91. Batista AP, Niccolai A, Bursic I, Sousa I, Raymundo A, Rodolfi L, Biondi N, Tredici MR. Microalgae as functional ingredients in savory food products: application to wheat crackers. Foods. 2019;8(12):22. https://doi.org/10.3390/foods8120611.
92. Uribe-Wandurraga ZN, Igual M, Garcia-Segovia P, Martinez-Monzo J. Effect of microalgae addition on mineral content, colour and mechanical properties of breadsticks. Food Funct. 2019;10(8):4685–92. https://doi.org/10.1039/c9fo00286c.

93. Van Drunen K. Using dried fruits to add essential nutrients to cereals, bars, and breads. Cereal Foods World. 2002;47(7):311–3.
94. Smith TJ, Karl JP, Wilson MA, Whitney CC, Barrett A, Farhadi NF, Chen CYO, Montain SJ. Glycaemic regulation, appetite and ex vivo oxidative stress in young adults following consumption of high-carbohydrate cereal bars fortified with polyphenol-rich berries. Br J Nutr. 2019;121(9):1026–38. https://doi.org/10.1017/s0007114519000394.
95. Yilmaz E, Karaman E. Functional crackers: incorporation of the dietary fibers extracted from citrus seeds. J Food Sci Technol Mysore. 2017;54(10):3208–17. https://doi.org/10.1007/s13197-017-2763-9.
96. Bchir B, Jean-Francois T, Rabetafika HN, Blecker C. Effect of pear apple and date fibres incorporation on the physico-chemical, sensory, nutritional characteristics and the acceptability of cereal bars. Food Sci Technol Int. 2018;24(3):198–208. https://doi.org/10.1177/1082013217742752.
97. Sun-Waterhouse D, Teoh A, Massarotto C, Wibisono R, Wadhwa S. Comparative analysis of fruit-based functional snack bars. Food Chem. 2010;119(4):1369–79. https://doi.org/10.1016/j.foodchem.2009.09.016.
98. Ahmed ZS, Abozed SS. Functional and antioxidant properties of novel snack crackers incorporated with Hibiscus sabdariffa by-product. J Adv Res. 2015;6(1):79–87. https://doi.org/10.1016/j.jare.2014.07.002.
99. Brown IJ, Tzoulaki I, Candeias V, Elliott P. Salt intakes around the world: implications for public health. Int J Epidemiol. 2009;38(3):791–813. https://doi.org/10.1093/ije/dyp139.
100. Ainsworth P, Plunkett A. Reducing salt in snack products; 2007. pp 296–315. doi:https://doi.org/10.1533/9781845693046.3.296.
101. Kita AM. Reducing saturated fat in savoury snacks and fried foods. Reduc Saturat Fats Foods. 2011:266–282. https://doi.org/10.1016/B978-1-84569-740-2.50013-0.
102. Parisi S, Luo WH. Maillard reaction in processed foods-reaction mechanisms. In: Chemistry of Maillard reactions in processed foods. Cham: SpringerBriefs in Molecular Science. Springer International Publishing Ag; 2018. p. 39–51. https://doi.org/10.1007/978-3-319-95463-9_2.
103. Rathod R, Annapure U. Effect of extrusion process on antinutritional factors and protein and starch digestibility of lentil splits. LWT Food Sci Technol. 2016;66:114–23. https://doi.org/10.1016/j.lwt.2015.10.028.
104. Sharif MK, Rizvi SSH, Paraman I. Characterization of supercritical fluid extrusion processed rice–soy crisps fortified with micronutrients and soy protein. LWT Food Sci Technol. 2014;56(2):414–20. https://doi.org/10.1016/j.lwt.2013.10.042.
105. Balentic JP, Ackar D, Jozinovic A, Babic J, Milicevic B, Jokic S, Pajin B, Subaric D. Application of supercritical carbon dioxide extrusion in food processing technology. Hem Ind. 2017;71(2):127–34. https://doi.org/10.2298/hemind150629024p.
106. Rozzi NL, Singh RK. Supercritical fluids and the food industry. Compr Rev Food Sci Food Saf. 2002;1(1):33–44. https://doi.org/10.1111/j.1541-4337.2002.tb00005.x.
107. Lv Y, Glahn RP, Hebb RL, Rizvi SSH. Physico-chemical properties, phytochemicals and DPPH radical scavenging activity of supercritical fluid extruded lentils. LWT. 2018;89:315–21. https://doi.org/10.1016/j.lwt.2017.10.063.
108. Bilgi Boyaci B, Han J-Y, Masatcioglu MT, Yalcin E, Celik S, Ryu G-H, Koksel H. Effects of cold extrusion process on thiamine and riboflavin contents of fortified corn extrudates. Food Chem. 2012;132(4):2165–70. https://doi.org/10.1016/j.foodchem.2011.12.013.
109. Paraman I, Wagner ME, Rizvi SSH. Micronutrient and protein-fortified whole grain puffed Rice made by supercritical fluid extrusion. J Agric Food Chem. 2012;60(44):11188–94. https://doi.org/10.1021/jf3034804.
110. Wegrzyn TF, Golding M, Archer RH. Food layered manufacture: a new process for constructing solid foods. Trends Food Sci Technol. 2012;27(2):66–72.
111. Ngo TD, Kashani A, Imbalzano G, Nguyen KTQ, Hui D. Additive manufacturing (3D printing): a review of materials, methods, applications and challenges. Compos Part B. 2018;143:172–96. https://doi.org/10.1016/j.compositesb.2018.02.012.

112. Pallottino F, Hakola L, Costa C, Antonucci F, Figorilli S, Seisto A, Menesatti P. Printing on food or food printing: a review. Food Bioprocess Technol. 2016;9(5):725–33.
113. Nachal N, Moses JA, Karthick P, Chinnaswamy A. Applications of 3D printing in food processing. Food Eng Rev. 2019;11:123–41. https://doi.org/10.1007/s12393-019-09199-8.
114. Severini C, Azzollini D, Albenzio M, Derossi A. On printability, quality and nutritional properties of 3D printed cereal based snacks enriched with edible insects. Food Res Int. 2018;106:666–76. https://doi.org/10.1016/j.foodres.2018.01.034.
115. Utela B, Storti D, Anderson R, Ganter M. A review of process development steps for new material systems in three dimensional printing (3DP). J Manuf Process. 2008;10(2):96–104.
116. Lille M, Nurmela A, Nordlund E, Metsä-Kortelainen S, Sozer N. Applicability of protein and fiber-rich food materials in extrusion-based 3D printing. J Food Eng. 2018;220:20–7. https://doi.org/10.1016/j.jfoodeng.2017.04.034.
117. Dankar I, Haddarah A, Omar FE, Sepulcre F, Pujolà M. 3D printing technology: the new era for food customization and elaboration. Trends Food Sci Technol. 2018;75:231–42.
118. Krishnaraj P, Anukiruthika T, Choudhary P, Moses JA, Anandharamakrishnan C. 3D extrusion printing and post-processing of fibre-rich snack from indigenous composite flour. Food Bioprocess Technol. 2019;12(10):1776–86. https://doi.org/10.1007/s11947-019-02336-5.
119. Derossi A, Caporizzi R, Azzollini D, Severini C. Application of 3D printing for customized food. A case on the development of a fruit-based snack for children. J Food Eng. 2018;220:65–75. https://doi.org/10.1016/j.jfoodeng.2017.05.015.
120. Vancauwenberghe V, Katalagarianakis L, Wang Z, Meerts M, Hertog M, Verboven P, Moldenaers P, Hendrickx ME, Lammertyn J, Nicolaï B. Pectin based food-ink formulations for 3-D printing of customizable porous food simulants. Innovative Food Sci Emerg Technol. 2017;42:138–50. https://doi.org/10.1016/j.ifset.2017.06.011.
121. González-Amaro RM, Figueroa-Cárdenas JD, Perales H, Velés-Medina JJ. Physicochemical and nutritional properties of different maize races on toasted tortillas. Cereal Chem. 2017;94(3):451–7. https://doi.org/10.1094/cchem-05-16-0138-r.
122. GrandViewResearch. Tortilla chips market size, share & trends analysis report by product (conventional, organic), by distribution channel (online, offline), by region (North America, Europe, Asia Pacific), and Segment Forecasts, 2019–2025. 2020. https://www.grandviewresearch.com/industry-analysis/tortilla-chips-market. Accessed 25 Mar 2020.
123. Santos EM, Quintanar-Guzman A, Solorza-Feria J, Sanchez-Ortega I, Rodriguez JA, Wang Y-J. Thermal and rheological properties of masa from nixtamalized corn subjected to a sequential protein extraction. J Cereal Sci. 2014;60(3):490–6.
124. Serna-Saldivar SO. Maize: foods from maize. Ref Module Food Sci Elsevier. 2016. https://doi.org/10.1016/B978-0-08-100596-5.00126-8.
125. Seetharaman K, Yao N, Groff ET. Quality assurance for hard pretzel production. Cereal Foods World. 2002;47(8):361–4.
126. Yao N, Floros JD, Seetharaman K. Identification of important production variables affecting hard pretzel quality. J Food Qual. 2005;28(3):222–44. https://doi.org/10.1111/j.1745-4557.2005.00032.x.
127. Lusas EW, Rooney LW. Snack foods processing; 2001
128. Mościcki L. Crispbread, bread crumbs and baby food. In: Extrusion-cooking techniques; 2011. pp. 91–97. https://doi.org/10.1002/9783527634088.ch6.
129. Ang CYW, Liu K, Huang Y-W. Asian foods: science & technology; 1999.
130. Orts W, Glenn G, Nobes G, Wood D. Wheat starch effects on the textural characteristics of puffed Brown Rice cakes. Cereal Chem. 2000;77:18–23. https://doi.org/10.1094/CCHEM.2000.77.1.18.
131. Holmes GS (2020). http://www.madehow.com/Volume-4/Rice-Cake.html. http://www.madehow.com/Volume-4/Rice-Cake.html. Accessed 29 Mar 2020.

Chapter 8
Rice: A Versatile Food at the Heart of the Mediterranean Diet

Andrea Bresciani, Maria Ambrogina Pagani, and Alessandra Marti

Abstract Although being the most consumed food in Asia, rice plays a key role also in the diet of many countries, including those of the Mediterranean area. The availability of thousands of varieties — different in pedo-climatic adaptability, size and texture of the grain, bioactive and aromatic compounds, protein and starch features and, consequently, cooking behavior — has made it possible to satisfy tastes and nutritional needs specific for each population.

The most common way of eating rice is as grain and its cooking can take place in different ways (in excess of water, pilaf method, parboiled rice or *risotto*) to obtain the desired texture according to the recipe. Various technologies have been developed to encourage the consumption of rice but satisfying the current market demands (e.g., quick cooking rice and frozen rice).

Alongside traditional consumption, in recent decades rice has been increasingly sought after as an ingredient of various food formulations, including gluten-free products (i.e., pasta, bread and bakery products) and other products that are now widespread in Mediterranean countries, such as couscous, breakfast cereals, extruded products and baby foods.

Keywords Rice · Cooking · Gluten-free · Quick-cooking rice · Steaming · Paraboiling

A. Bresciani · M. A. Pagani · A. Marti (✉)
Department of Food, Environmental and Nutritional Sciences (DeFENS), Università degli Studi di Milano, Milan, Italy
e-mail: alessandra.marti@unimi.it

F. Boukid (ed.), *Cereal-Based Foodstuffs: The Backbone of Mediterranean Cuisine*, https://doi.org/10.1007/978-3-030-69228-5_8

1 Introduction

Rice is one of the main crops cultivated across the world. The genus *Oryza* comprises two distinct types of domesticated rice: *Oryza sativa* (Asian rice) which is globally consumed and *Oryza glaberrima* (African rice), peculiar to the West Africa sub-region. Based on the different cultivation environments, the *Oryza sativa* species has differentiated into two subspecies: the *indica* subspecies, characterized by an elongated grain, and the *japonica* subspecies, which differs from having a more rounded grain. *Indicas* are adapted to the warm and humid climates, thus they are grown throughout the tropics and subtropics, whereas *japonicas* are limited to the temperate zones, i.e. northern environments, characterized by shorter and colder growing seasons. A third subspecies, which is broad grained and thrives under tropical conditions, was identified based on morphology and initially called *javanica*, but is now known as tropical *japonica*.

Rice is classified and marketed based on kernel size and shape. However, the naming system is different around the world. For example, in the United States, the main rice types are long-, medium-, and short grains. Each type has specific starch properties, that affect the cooking behavior and processing attitude: starch of medium grain rice has an amylose content about 15–20% with an average gelatinization temperature of 70 °C; long grain rice has an amylose content of 23% and an average gelatinization temperature of 60 °C. Consequently, compared with long grain rice, medium grain rice absorbs less water (1–2 times) during cooking, giving a moist, soft, and sticky product. The cereals industry in Europe classifies rice in round, medium and long grain rice according to the length and length/width ratio of the grain [1]. Specifically, grains with a length of 5.2 mm or less, and a length/width ratio less than 2 are classified as round grains; medium grains are longer than 5.2 mm and equal to or less than 6.0 mm, with a length/width ratio of less than 3. Finally, long grains are classified in two categories: long A (rice longer than 6.0 mm and with a length/width ratio greater than 2 and less than 3) and long B (rice longer than 6.0 mm and with a length/width ratio equal to or greater than 3).

Rice is grown in over 100 countries and is a primary source of food for more than 2/3 of the world's population. In 2018, 167 million hectares were dedicated to this crop with the production of 782 million tonnes of paddy rice [2]. About 90% of world production is concentrated in Asia. At the European level, the primacy belongs to Italy which produces about 40% of total production, followed by Russia (about 25%) and Spain (about 20%). Considering the production of approximately 500 million tonnes of milled rice in 2018, FAO's global milled rice production forecast for 2020 is now pegged at 509.2 million tonnes, +1.7% from 2019. World rice utilization is also predicted to reach a peak of 510.4 million tonnes in 2020/21, + 1.6% from 2019/20 based on expanding food use [2].

Several factors accounted for the increase in rice production and consumption, not only as a staple food but also as a side dish. The main drivers behind the expansion of rice-based products are: i) the nutritional aspects and the growing consumer awareness towards healthy, simple (and affordable) foods, and ii), its high versatility to be used as a suitable raw material for the production of various cereal-based products, including baby-foods, gluten-free pasta and bakery goods, ready-to-eat products such as snacks and breakfast cereals because of its color, flavour, hypo-allergenicity, and bland taste. Thus, the objective of this chapter is to provide an overview on the most common ways of using rice — as grain or as ingredients in various formulations — in the Mediterranean diet. The role of the main biopolymers present in rice (i.e., starch and proteins), as well as processing conditions, will be addressed in relation to the quality of the final products.

2 Main Components

Rice represents an important source of energy: indeed, a fifth of the worldwide total calories comes from rice. Like other cereals, starch, proteins, and lipids are the main components of the rice grain, and they greatly affect the cooking and eating quality. Starch represents more than 70% of the kernel, followed by proteins (6–8%). Moreover, rice contains most of minerals such as Fe, Zn, Cu and Se, and vitamins, which are beneficial for human health. However, the composition in macronutrients might differ according to the milling process, i.e. based on the level of bran removal (Table 8.1). Indeed, rice is mainly consumed in the form of refined (or milled) grains, whose nutritional value is lower than brown rice (Table 8.1). Indeed, most of the nutrients — especially fiber, vitamin and mineral — are lost during the removal of the outer layers, which are necessary to ensure adequate flavor and shelf life.

Table 8.1 Chemical composition of milled rice, brown rice and rice bran

Nutrients	Milled rice	Brown rice	Rice bran
Carbohydrates	77–78	73–76	34–62
Proteins	6.3–7.1	7.1–8.3	11.3–14.9
Lipids	0.3–0.5	1.6–2.8	15.0–19.7
Fiber	0.2–0.5	0.6–1.0	7.0–11.4
Ash	0.3–0.8	1.0–1.5	6.6–9.9
K	14–120	120–340	–
Ca	10–30	10–50	30–120
Na	2.2–8.5	3.1–17.6	–
Mn	1.0–3.3	1.3–4.2	–
Fe	0.2–2.7	0.7–5.4	8.6–43.0
Zn	0.3–2.1	1.5–2.2	4.3–25.8
P	0.08–0.15	0.17–0.43	1.1–2.5

Carbohydrates, proteins, lipids, fiber and ash are expressed as g/100 g; minerals as mg/100 g
Adapted from Saleh et al. [3]

2.1 Starch

Rice represents an important source of carbohydrates, which are present in the grain in the form of polysaccharides (i.e., cellulose and starch). Starch is a polymer of glucose that in nature is organized in small granules containing amylose and amylopectin. Starch granules in rice are very small (2–10 μm), up to 10 times smaller than wheat and 70 times smaller than potatoes. The small size of rice starch granules accounts for high starch digestibility, making rice an important source of energy in the diet of infants and babies.

Starch granules have alternating crystalline areas (30%) and amorphous regions (70%), which give semi-crystalline properties due to the arrangement of the two main components of starch, amylose and amylopectin. Since the amylose:amylopectin ratio greatly affects the gelatinization and retrogradation properties of rice, the amylose content should be taken into consideration in the selection of the type of rice for a specific process and, thus, end-use (Table 8.2).

Based on amylose content, rice varieties can be classified as waxy (<2%), low (10–18%), intermediate (20–25%) and high (>25%) amylose type. Considering the typical products of the Mediterranean diet, waxy or low amylose starches could be exploited for their low retrogradation tendency, thus they are suitable ingredients in bread formulation to retard the product staling. Moreover, by preventing syneresis, waxy starch is an ideal thickening agent in sauces, gravies, and puddings, as well as in frozen foods. On the other hand, intermediate-high amylose starches lend themselves well to the production of pasta because of the high retrogradation tendency.

Table 8.2 Role of amylose in rice products

	Amylose content type			
	Waxy (<2%)	Low (10–18%)	Intermediate (20–25%)	High (>25%)
Boiled rice	✗	✓	✓✓	✗
Parboiled rice	✓	✓	✓✓	✓✓
Risotto	✗	✓	✓	✗
Pasta	✗	✗	✓✓	✓
Bread	✓	✓	✗	✗
Biscuits	✓✓	✓✓	✓	✓
Snacks	✗	✓	✓	✓
Baby-foods	✗	✓	✓	✗

✗ not suitable; ✓ suitable; ✓✓ very suitable

2.2 *Proteins*

Although the main component is starch, rice is also a fair source of proteins. These can be classified according to their solubility. The main type is glutelines, which are soluble in alkaline solutions and represent 60% of total proteins, followed by prolamine (25%, soluble in alcohol), globulin (10%, soluble in salt water) and albumin (5%, water-soluble). Proteins are more abundant in sub-aleuron layers but are also present in aleuron cells. Endosperm proteins include 7–18% albumin plus globulin, 5–12% prolamine and the rest is glutelin [4]. Although rice proteins are not very represented from a quantitative point of view, they have a good biological value in relation to the optimal amino acid ratio; indeed, rice contains essential amino acids such as lysine, tryptophan and methionine in higher amount than in wheat and corn. Rice proteins are therefore qualitatively superior to those of any other cereal. Finally, rice contains non-gluten proteins, making it ideal for developing products for people affected by celiac disease.

2.3 *Lipids*

Rice contains a small amount of lipids (less than 5%), unevenly located in the grain regions. Unlike milled rice, rice bran is rich in fatty acids (about 15%) [5]. Based on cellular distribution, rice lipids are generally classified as starch lipids, which are associated with starch granules, and non-starch lipids, that are distributed throughout the grain but concentrated in the bran [4]. The main fatty acids of non-starchy lipids are linoleic, oleic and palmitic acids [4]. Starch lipids are mainly fatty acids and phospholipids, complexed with amylose. The main free bound fatty acids in rice starch are palmitic (C16: 0) and linoleic (C18: 2) acids. These essential fatty acids are fundamental constituents of phospholipids of all cell membranes and are therefore indispensable for the functions of all tissues and organs. A diet based on brown rice (with bran and germ) therefore provides an adequate amount of essential fatty acid, allowing the physiological performance of all cellular metabolic reactions. However, these unsaturated lipids can degrade during storage, resulting in deterioration of taste, palatability and food quality and safety.

3 Rice Consumption as Grain

Rice has always occupied an honorable place in the Mediterranean culinary tradition. The most common way of eating rice is as grain with meat and/or vegetables. In each Mediterranean country, several interesting recipes have been proposed, resulting in nutritionally balanced meals. Rice is cooked in a variety of ways depending on the cultural background, cuisine and rice variety. In the following section, the most commonly used cooking methods are discussed. For an in-depth analysis of the topic, the readers might refer to the recent review of Yu et al. [6].

3.1 Cooking Methods

Rice cooking is mainly carried out by water absorption (Pilaf method) or in excess of water. Consumer preference for cooking method is dependent on the rice variety. For example, the acceptance of long grain and Thai Jasmine rice was higher when cooked by Pilaf method, producing stickier, firmer, drier rice with a more acceptable flavor and appearance; while Basmati rice was preferred when cooked in excess water [7]. It follows that Western consumers — who look for firm and unsticky cooked rice — prefer cooking rice in excess of water, while Asian consumers — who look for glutinous properties — prefer the Pilaf method [8].

Pilaf method consists in cooking rice with a fixed amount of water until the water is fully absorbed. Cooking is generally carried out in electronic cookers, with a water/rice ratio depending on the amylose content of rice: high amylose rice requires more water than waxy starch (i.e., for each unit of milled rice, 1.3 and 2.1 times as much water for waxy and high amylose rice, respectively) [9]. Maintaining the right water/rice ratio is important since it affects starch gelatinization and thus product texture and quality. Indeed, a low water/rice ratio might create a thin layer of gelatinized starch on the surface of the grain, without leaving enough water for assuring starch gelatinization at the center of the grain [10]. Conversely, a higher water/rice ratio — favoring starch gelatinization — leads to a greater amount of starch leaching from rice grains and forming a gel that coats the surface of the rice kernel [11]. This phenomenon results in hardness decreasing and stickiness increasing [12].

Cooking by absorption does not ensure a uniform treatment throughout the bulk sample because the moisture content of individual rice kernels varies with their location due to non-uniformity of heat distribution [13]. Thus, cooking in excess of water is preferred at industrial level as it can be conducted as a continuous process and allows an even distribution of moisture within rice grains [6]. In this approach, rice is cooked in boiling water for a fixed time and using water-to-rice ratio generally ranging from 10:1 to 20:1. After the rice is cooked, the excess water is discarded with all the leached components.

Regardless the rice:water ratio, the cooking process comprises two mechanisms: the gradual absorption of water from the surface to the core of the grain, and the structural changes of the rice components by heating [14]. Since cooking in excess water does not limit the diffusion of water, excessive cooking time leads to grain disintegration. Rice grains are periodically sampled during cooking and pressed between two parallel glass plates to determine the cooking time. The optimal cooking time is achieved when the grains do not show a white ungelatinized core anymore [15]. Though the rice is cooked through completely, this stage of cooking does not necessarily represent the most desirable texture. Indeed, firmness decreased, while stickiness increased with an increase in the cooking time [16]. Apart the cooking time, rice texture is strongly affected by amylose content. Varieties characterized by a low amylose content (10–18%) show, after cooking, a sticky and not very consistent cooked grain. Conversely, rice cultivars with an intermediate (20–25%) and high amylose content (>25%) have a little sticky grain with a good consistency.

In Italy, rice is mainly consumed as *risotto*, whereas in Spain as *paella*; both the products are becoming more and more appreciated all over the world. This peculiar cooking style of both *risotto* and *paella* normally involves a slow cooking (16–20 min), under a careful mixing, covering the rice grains with hot broth, with a broth/rice ratio of about 3.5:1 (w/w). Broth, which is slowing added to rice during the dish preparation, is completely absorbed at the end of the cooking. Specialty cultivars such as Carnaroli, Arborio and Baldo, classified as a medium grain (according to the biometric characteristics of Reg. UE n.1308/2013 [1]), have medium-high amylose content (ranging from 18 to 24%) and cooking kinetics considered most suited for *risotto* preparation. These varieties often have a white core (e.g., chalky center) that is thought to be responsible for their ability to take up the flavor of a cooking sauce. After cooking, these types of rice have a firm internal texture and a creamy exterior. Since the *risotto* cooking allows a complete absorption of water and no losses in the cooking water, it would be a good cooking method to retain phenolic compounds in pigmented and non-pigmented whole meal rice [17–19].

In North African countries such as Tunisia, rice is cooked by steaming using a traditional double-chambered (with two pans) steamer like the one used for cooking couscous, also called "couscoussier". After a quick soaking step, rice is placed in the top pan of the steamer having perforated bottom, while the mixture of meat and vegetables are placed in the bottom pan. During steaming, hot water is gradually added to soften the rice kernels with mixing to homogenize the cooking. After cooking, rice and the blend of vegetables are mixed together.

Finally, *rice salad* is commonly consumed in the Mediterranean area. Product quality is assured if rice kernels show high firmness and absence of stickiness, so that rice grains are kept separated from each others. Such characteristics are achieved by selecting varieties with intermediate amylose content and/or using parboiled

Table 8.3 Main changes on rice features promoted by milling or parboiling

	Milled rice	Parboiled rice
Milling yield	✗	✓
Broken kernels	✗	✓
Nutritional properties		✓
Color	✓	✗
Cooking time		✓
Texture		✓

✗ negative effect; ✓ positive effect; / no effect

rice. Parboiling process and its main effect on the physicochemical properties of rice are discussed in the following section.

3.2 Parboiled Rice

Parboiled rice is 'par'-tially 'boiled' (i.e. partially cooked) rice. In other words, rice grains have been subjected to a hydrothermal treatment (i.e., parboiling) that promotes important changes in physicochemical, nutritional, and sensory properties of paddy rice (Table 8.3). The main steps of parboiling process consist in soaking, steaming, and drying. This process was introduced in Asia to make the husks easier to be removed, thus assuring a high milling yield as well as a higher resistance to pests. In fact, the cracks that are frequently present in rice grains, are healed during steaming, so that breakages during milling are highly reduced.

During soaking, paddy rice is sufficiently hydrated to enable starch gelatinization during steaming. As well known, starch is insoluble in cold water but the disruption of hydrogen bonds during hot soaking weakens the granule structure, and therefore enhances surface water absorption by starch granules [20]. Moreover, hydration rate increases as temperature increases. Indeed, below the gelatinization temperature (about 70 °C), the paddy rice absorbs water slowly, eventually reaching an equilibrium about 30% moisture. If the temperature is below 60–65 °C, more time to reach the equilibrium can be required with the potential occurrence of fermentation. On the contrary, below the gelatinization temperature, the hydration rate, after an initial lag period, increases exponentially, because starch begins its gelatinization. Then, when the grain moisture exceeds about 30–32%, the husk splits open, causing leaching and grain deformation [21]. Such evidence suggests to monitoring the soaking time when high temperatures (>70 °C) are used and draining the water out when the grain moisture reaches 30–32%, after which the paddy is tempered to equalize its moisture [22].

After soaking, the paddy rice is subjected to steaming which promotes starch gelatinization. The severity of this step affects the product quality. The more severe the steaming (high steam pressure or longer time), the greater the degree of gelatinization as well as the ability of the parboiled grain to withstand adverse conditions of drying without giving rise to cracks. Various combination of time and pressure can be adopted based on final product quality desired. Indeed, more severe steaming makes the kernels harder, reducing milling breakage but increasing cooking time. At the same time, kernel discoloration increases with increasing degree of steaming.

At the end of steaming, the moisture content of kernels (about 35–40%) needs to be reduced at storage level (approximately 13%). Maillard reaction occurring during drying is responsible for slight changes in kernel color, from a light yellow to amber. In addition, gelatinized starch undergoes to retrogradation, accounting for the higher resistant starch content in parboiled rice compared to non-parboiled one [23]. Starch structural rearrangement upon parboiling result in increased kernel

Table 8.4 Effect of parboiling process on the nutritional composition of milled rice [24]

Nutrients	Milled rice	Parboiled rice
Starch	72.9	73.6
Proteins	6.7	7.4
Lipids	0.4	0.3
Fiber	1	0.5
Cu	0.18	0.34
Fe	0.8	2.9
Zn	1.3	2
Na	5	9
Se	10	14
Ca	24	60
K	92	150
P	94	200
Thiamine	0.11	0.34

Starch, proteins, lipids, and fiber are expressed as g/100 g; minerals and thiamine in mg/100 g

hardness with the advantage of reducing grain breakage during milling. In addition, after cooking parboiled rice exhibits a higher firmness and lower stickiness than unparboiled rice. Moreover, parboiled rice would not easily overcook, so timing of its cooking is not so critical. These changes in kernel texture have been associated with starch retrogradation and amylose-lipid complex formation.

From a nutritional standpoint, besides the high resistant starch and low glycaemic index [23], micronutrients increase upon parboiling (Table 8.4). Their migration from the bran layers (where they are placed) towards the endosperm is one of the possible explanations for the reduced loss of water-soluble constituents during milling of brown rice after parboiling. However, their adhesion to the endosperm as a consequence of starch gelatinization should be also considered [22]. Lipophilic antioxidants such as tocopherols and γ-oryzanol in both pigmented and non-pigmented cultivars increase by parboiling [25]. As regards phenolic compounds, the increased release of bound phenolics was not enough to compensate for the loss of soluble phenolics [25]. Possible explanations for the decrease in soluble phenolic compounds are: loss in the processing water, thermal decomposition and interaction with other chemical components [26]. On the other hand, parboiling seems to partially preserve free phenolics content in polished rice [27].

3.3 *Quick-Cooking Rice (QCR)*

Recently, as a result of consumer demand for instant cooking foods, quick rice cooking (QCR) is becoming more popular. Indeed, white or milled rice normally takes 15–25 min of boiling in water to cook, while QCR can cook in 5–10 min. The reduced cooking time in QCR is even more attractive in the case of brown rice, whose cooking time is at least 40 min. Thus, the modern consumer is struggling between the awareness of eating brown rice as source of fiber and other bioactive compounds and the lifestyle changes in a modern society. This brought to an increase in the worldwide consumption of instant, fast or quick cooking foods.

The procedure for producing QCR involves the basic steps of soaking in water, cooking and drying. Actually, QCR cannot be considered as an innovation of the food industry, since the QCR concept arose from around the middle of the last century, with the first patent granted in 1948. Since then, newer patents have been granted off and on and all along further improvements have been regularly made. Searching "quick-cooking rice" in the Food Science and Technology Abstracts Database, 43 patents have been found from 1968 to 2019. The most recent one [28] proposes the treatment of brown rice with super-heated steam (at 100 °C or higher), without requiring soaking treatments. The brown rice is subsequently cooled and dried, as most of QRC production process.

The basic principle of QRC processes lies in precooking the rice in water, followed by drying in such a way as to leave a large number of microscopic cracks or openings in it. The porous structure allows the hot water uptake that enables the product to be cooked quickly. Alternatively to the "soak-cook-dry method", the grain is micro-damaged by suitable dry heat or freezing. The resulting fissured structure can enable the rice to hydrate rapidly by allowing entry of water and thus to cook quickly [29]. Fissured structure can be formed also mechanically during the milling process; however, no details are reported.

Variations in the procedures have been shown to have a significant effect on the cooking time and sensory quality of the finished products [30]. Specifically, dry grains of QCR should be separate, with a low percentage of broken kernels, and should resemble in color, flavor and texture to milled rice. As regards processing conditions, Prasert and Suwannaporn [31] investigated the effects of moisture content, pressure and drying temperature on physicochemical properties of QCR and its eating quality. Results showed that only pressure and moisture content affected density, rehydration ratio, and volume increase as consequence of the increased the kernel's porosity.

At molecular level, heating promotes starch gelatinization and the loss of molecular order, while during the cooling and drying steps the formation of amylose-lipid complex occurs, as well as the increased extent of interaction between partially gelatinized starch and partially denatured proteins [32, 33]. As regards proteins, a stronger protein-protein and protein-starch matrix is present in QCR when compared to the untreated counterparts, which lead to a decreased protein extractability in denaturant agent (i.e., SDS) in presence or absence of a reducing agent (i.e., 2-mercaptoethanol) [33]. Indeed, the denaturation of rice glutelins occurs when rice is subjected to hydration and heating, thus favoring the occurrence of protein-protein and/or protein-starch matrix with greater hydrophobicity while cooling and drying afterwards [33]. Alterations in structure and composition of rice kernel during the cooking procedure of QCR preparation explain its greater glucose release in the gastric digestion (the first 30 min) compared to the untreated rice. However, during intestinal digestion, the digestive enzymes were more able to digest the constituents of untreated rice compared to the QCR, due to starch retrogradation occurring just after the cooking step [33]. In fact, retrogradation promotes the amylose molecules re-association. This limits the hydration of amorphous regions in the starch granules, decreasing the starch accessibility to enzyme digestion.

Starch-protein and protein-protein interactions are also favored during retrogradation, which may contribute into the low QCR starch digestibility.

3.4 *Rice in Pouches*

Another form of distribution of cooked rice is in the form of rice in pouches. Rice is either cooked then put in flexible heat-sealable pouches alone or with all the other ingredients (sauces, spices, vegetables or meats or a combination thereof). The pouch is then sealed and subjected to sterilization to guarantee the storage at room temperature more than one year, with no need of the cold chain.

Use of this kind of product has been increasing in recent times, since it is extremely versatile and convenient for use of single persons or even by small families. Final cooking can be carried out in pan with hot water or in the microwave for a couple of minutes.

The product quality depends as much on the production process as on the type of rice. As regards the raw material, high-amylose rice is more likely to yield a better product, with lesser grain distortion or fraying than low-amylose variety. Similarly, an appropriate level of parboiling is essential to make the rice tolerate the stress of the treatment.

3.5 *Frozen Rice*

Use of frozen rice is another way to distribute cooked rice ready for consumption as table rice or in mixed cooked form, with various combinations of sauces, vegetables, meat or fish. In this system, rice is cooked and frozen. It is then preserved and distributed in frozen condition and heated in the microwave oven before serving. Although frozen rice was the first product in the category of ready-to-eat meals to be launched on the market, its consumption has never really taken off because of the longer cooking time (at least 8 min) compared to rice in pouches as well as the high price due to the high costs of the cold chain. The quality of the product greatly depends on the process more than of the variety of rice used.

4 Rice-Based Products in the Gluten-Free Diet

Rice is not solely consumed in the form of cooked table rice but in many other forms, that may be called "rice products", mainly driven by the growing interest in the gluten-free (GF) products.

The market of GF products represents one of the most prosperous markets in the field of food and beverages soon. Indeed, it is projected to reach a value of USD 6.47 billion by 2023, growing at a compounded annual growth rate of 7.6% [34].

Rise in awareness about celiac disease and adoption of special dietary lifestyles are the major factors driving this market. Indeed, besides people affected by celiac disease or gluten sensitivity, their relatives that might be genetically predisposed to celiac disease are occasional consumers of GF products. More worrying is the adherence to a GF diet from those who want to exclude gluten from their diet for health reasons since they think it may be dangerous for their health. Currently, there are contrasting data about the nutritional adequacy of a GF diet in terms of nutrients and food categories, mainly due to the different dietary habits among the countries. However, researchers agree that the celiac population on the whole does not introduce the recommended amount of complex carbohydrates or fiber and they also have a low consumption of several minerals and vitamins. Conversely, this population tends to have a high intake of energy from protein, saturated and total fats [35].

Recently, Morreale et al. [36] recorded the dietary habits of a group of individuals with celiac disease and a group of healthy subjects (i.e., not following a GF diet) to determine the adherence of the two groups to the Mediterranean diet. Typical food consumption was not significantly different between individuals with celiac disease and the healthy participants, except for the lower fruit consumption and the higher amounts of potatoes and red and processed meat. Thus, the resulting average Italian Mediterranean Index (an index introduced to evaluate the adherence to a Mediterranean diet) was significantly higher in healthy participants than in individuals with celiac disease. However, considering that the GF diet composition is based on personal food choices, the authors suggest the need of encouraging people to make better food choices in line with the Mediterranean diet in order to improve their nutritional status and protect them from a number of other disease and/or health complications linked to celiac disease since most of GF products are mainly formulated with starchy ingredients thereby having a high glycemic index. At the same time, continuous efforts from the food industries should be devoted to the reformulation of GF products to improve their nutritional quality.

As a matter of fact, in response to the continuous increase in the demand for GF products and the recommendation of nutritionists, the label of such products have been extensively changed in the last decade. The overall nutritional improvement has been attributed to the incorporation of wholegrain cereals (such as brown rice) and pseudocereals, at the expense of starches or refined flours with a low fiber, protein, mineral and vitamin content [35]. In the case of pasta, the number of references on the Italian market has doubled in just 5 years (2010–2015), with a preference for formulations that include multigrain and / or wholemeal flours (Table 8.5).

Besides the enhancement of the nutritional profile, important advances have been concerned the improvement of the sensory and texture characteristics of GF products that, however, are still far from those of wheat-based products. Although gluten plays an important role in determining the structure of the product and thus its quality, its absence weighs in a different way considering the various cereal based products. Indeed, the production of GF bread or pasta is much more challenging than producing GF snacks or biscuits. Regardless the type of product, rice flour is one of the main ingredients. As an example, rice (mainly as refined flour) represents the 78% of the flours used in GF biscuits [37] and used in more than the 70% of the GF pasta [38] available on Italian market.

Table 8.5 Formulation of gluten-free pasta in the Italian market

Sample	Ingredients (as reported in the label)	Year	References
1	Rice flour (100%)	2010	[47]
2	Rice flour, E-471		
3	Rice flour, 2% rice germ, E-471		
4	70% rice flour, 29% corn flour, E-471		
5	Corn flour (100%)		
6	Corn flour, rice flour, pea protein isolate, E-471		
7	Corn flour, rice flour, pea protein isolate, E-471		
8	Corn starch, corn flour, rice flour, pea protein isolate, E-471		
9	Corn starch, potato starch, inulin, E-471, aroma cardamom		
10	Corn starch, potato starch, lupin flour, lupin proteins, E-471		
11	Rice flour	2015	[38]
12	Rice flour (99.5%), emulsifiers: E471		
13	Whole rice flour		
14	Rice flour, corn flour		
15	Rice flour (75%), quinoa flour (25%)		
16	Rice flour (59.5%), corn flour (30%), buckwheat flour (10%), emulsifiers: E471		
17	Whole rice flour (51.2%), whole corn flour (41.8%), corn bran (6.5%), emulsifiers: E471		
18	Corn flour, rice flour, emulsifiers: E471		
19	Yellow corn flour, rice flour, emulsifiers: E471		
20	Corn flour (90%), rice flour (10%)		
21	Corn flour (80%), rice flour (20%)		
22	Corn flour (70%), rice flour (29.5%), emulsifiers: E471		
23	Corn flour (59%), rice flour (45%), emulsifiers: E471		
24	Corn flour (33%), rice flour (67%)		
25	Corn and rice flours, corn starch		
26	Corn flour (70%), rice flour (18%), quinoa flour (3%), corn starch, emulsifiers: E471		
27	Corn flour, rice flour, millet flour, sugar cane syrup		
28	White corn flour (65%), yellow corn flour (29,5%), rice flour (5%), emulsifiers: E471		
29	Corn starch and flour, rice starch and flour, pea proteins, lupin flour, emulsifiers: E471		
30	Corn starch, potato starch, lupin flour, emulsifiers: E471		
31	Corn flour (97%), vegetable fibers (1.5%), emulsifiers: E471		
32	Corn flour (100%)		
33	Corn starch, potato starch, vegetable fibers, emulsifiers: E471, flavors agent, safflower		
34	Corn flour, emulsifiers: E471		

E471 mono- and di-glycerides of fatty acids

4.1 Pasta

Pasta is a typical Mediterranean food, widely consumed around the world thanks to several factors, including its easy preparation, palatability, and low glycemic index. The possibility of using grains other than wheat — including rice, corn and more recently pulses — as raw materials is an additional reason for pasta popularity.

Durum wheat semolina is recognized as the most suitable raw material for pasta production mainly due to the quality of its gluten proteins, which are unique in creating a strong and elastic network able to withstand to physical stress occurring during the pasta-making process and the final cooking in boiling water [39]. Despite the importance of protein quantity and quality, starch — that represents up to 70% of dry matter in semolina — cannot be considered as an inert filler, indeed it contributes into pasta structure and quality. Generally, starch that is good for pasta-making is characterized by a high gelatinization temperature, which delays starch swelling and solubilization, thus reducing interference with protein reticulation [40]. The role of starch in pasta quality is even more important in the case of GF pasta, as discussed below.

The role of rice in the GF pasta sector has changed over the last few decades. Initially, rice flour from broken kernels (i.e., rice milling by-products) was used for making GF pasta. Nowadays, due to the increased of volumes and quality requirements, great attention is paid to the selection of the ideal raw material. Moreover, if about ten years ago, pasta samples from 100% rice were present on the market, today the food industry prefers to propose formulations in which the rice is mixed with yellow corn and/or pseudocereal flours (Table 8.5). Indeed, the corn gives better taste to the product and above all a yellow color that recalls that of semolina pasta, which is continuously used as reference (in terms of texture, appearance, and sensory attributes), especially for those consumers with a tradition of wheat-based products. Conversely, the choice of including pseudocereals or using brown rice instead of milled rice is driven by the need for a nutritionally balanced product. At the same time, the presence of multiple ingredients and/or fiber might weaken the network [41, 42], accounting for the faster amylase hydrolysis of starch [43] and lower cooking quality compared to pasta from milled rice. However, brown pasta quality can be improved by optimizing the pasta-making process [44, 45]. As an example, brown rice pasta produced with an extrusion temperature of 120 °C and screw speed at 120 rpm had a similar quality to that made from other GF flours [44]. Alternatively, rice bran can be included in the formulation up to 15% to provide pasta of an acceptable quality with suitable attributes for color, cooking loss, hardness and adhesiveness [46]. Although this aspect has not been addressed yet, using of rice bran might overcome the use of mono and diglycerides of fatty acids, exploiting their presence already in the bran, potentially facilitating the adoption of clean label and having a positive impact on consumers' acceptance.

Oriental rice noodles can be considered the "ancestor" of GF pasta. Although similar in several aspects, noodles and pasta are produced in a different way resulting in products with different shape and texture. At the base of the traditional process for noodle production, there is the treatment of flour/starch with steam to

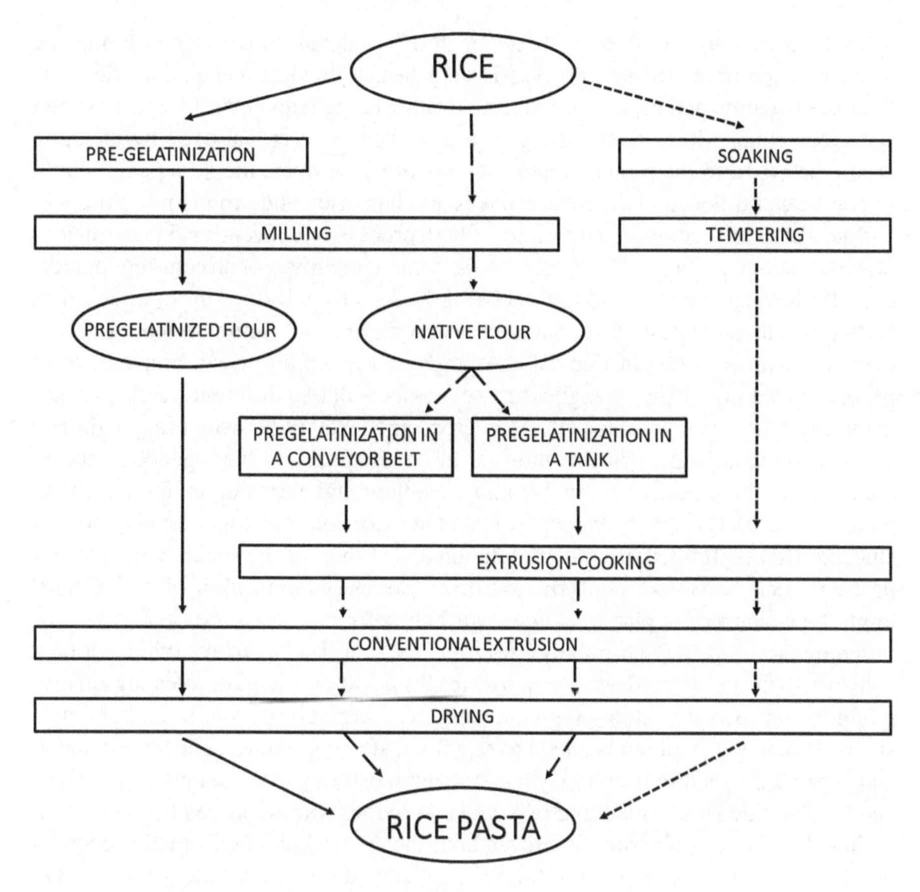

Fig. 8.1 Approaches for producing gluten-free pasta

promote complete starch gelatinization followed by cooling of the shaped-product to promote starch retrogradation, thus the formation of an extensive reticular and fibrillar network [48, 49]. Extensive research on rice noodles suggests that the ideal starch for GF pasta products should have an adequate amylose content (22–25 g/100 g) leading to a marked retrogradation tendency. These physico-chemical properties are closely correlated to cooking quality in terms of texture and low cooking loss [21].

Nowadays, pasta-makers can choose between two different approaches for producing GF pasta (Fig. 8.1): a conventional extrusion of flour in which starch is already mostly gelatinized (known as pregelatinized flours) *vs* extrusion-cooking of native flour. In the latter approach, steam (>100 °C) is insufflated in a tank where flour and water are mixed and extruded using an heated screw, applying high shear-stress and high temperature (120–160 °C) to obtain pellets (i.e., pre-gelatinized dough) that are formed into pasta in a conventional automatic press and finally dried [50]. Currently, using pre-gelatinized flours is the favorite solution for companies owing to the use of conventional continuous press used for semolina pasta [50]. Pre-gelatinized flour can be produced from both flour and grits. The first approach consists in treating the flour with steam in a horizontal cooker. Considering the small flour particle size this process is very fast, following pregelatinization, the flour steam dampened must be dried to a moisture content less than 15% to be stored for a long time. Alternatively, the grits are cooked in a vertical oven by injecting steam directly into the product mass. At the exit of the oven, the grits pass thought of two heated rollers and the related flakes are then dried and ground into flour.

Marti et al. [51] compared the effect of both processes (conventional *vs* extrusion-process) on the quality of 100% rice pasta, without additives or structuring ingredients. Both pasta samples had high cooking losses (10 g/100 g), up to three times higher than those of pasta from durum wheat semolina and similar to those of commercial pasta from brown rice [38]. As regards the texture, pasta prepared from pregelatinized flour exhibits higher firmness as a result of a different starch arrangement inside the two products [51]. At microscopic level, pasta from pre-gelatinized rice and conventional extrusion shows a more compact and homogeneous matrix compared to the product obtained from native flour and extrusion-cooking [50]. At molecular level, the structure derived from the extrusion- cooking of native flour is characterized by limited protein reticulation and by part of the starch being present in a compact native-like conformation [52]. Starch gelatinization of native flour might be enhanced by placing a conveyor belt before the cooker-extruder, as proposed by Fava S.p.A. (Cento, Italy). Specifically, flour is placed on a thin layer on a conveyor belt and treated with steam (Fig. 8.1). Moreover, pasta cooking quality might be improved by adding ingredients such as structuring proteins and/or emulsifiers. When egg albumen is added to rice flour, starch granules appear surrounded and separated by a thin protein layer stabilized mostly by hydrophobic interactions and by disulfide bonds, resulting in a cooked product with improved firmness [53]. Although using egg albumen improves both the textural and nutritional properties of GF pasta, this approach is not followed by companies due to allergen issue. On the other hand, emulsifiers (i.e., mono- and di-glycerides of fatty acids) are present in almost all GF pasta formulation (Table 8.5). Indeed, they play a dual role: they improve the smoothness of the extrusion process and the quality of the product, by decreasing the cooking loss as result of the formation of an emulsifier-amylose complex during processing and cooking [54].

A pasta formulation without the inclusion of additives has been proposed by using parboiled rice as alternative to either native or pregelatinized flour [55]. The rationale behind the idea is that parboiled rice is less sticky and firmer than native one, as result of starch modification promoted during the steaming and drying steps [22], as already discussed. Combining parboiling to extrusion-cooking results in a product with yellow color and low cooking loss, but to high firmness [41, 51]. This seems to be consequent of the presence in the uncooked pasta of starch matrix with an external region characterized by an amorphous structure and a core characterized by crystalline structure [41, 56, 57]. This peculiar starch organization, which results from the combination of parboiling and extrusion-cooking, would account for the increased resistant starch [52]. Last but not least, besides the enhanced cooking behavior and the decreased glycemic index [52], the advantage of using parboiled rice flour in GF pasta formulation lies in obtaining a product with an increased yellowness [52], an attribute appreciated by the pasta consumers in the Mediterranean area [58].

More recently, a new method for producing pasta directly from rice kernels, skyping the grinding step has been proposed [59, 60]. The process involves few steps prior to the pasta-making process: soaking of kernels (20 min at 50 °C; kernel:water ratio = 1:1.5), draining the excess of water and tempering (up to 3 h) (Fig. 8.1). The cooking behaviour was not negatively affected by the process, whose conditions need to be optimized to make the process more suitable at industrial level. The implementation of this process would be of great interest especially in developing countries, since no milling plants are necessary avoiding flour attacks from pest.

From the studies cited above, it is well accepted that, while gluten plays a key role in durum wheat pasta, starch acts as "structuring" ingredient in GF pasta. However, rice proteins – although they are not able to aggregate as gluten – are relevant to the physical features of the products. In rice pasta almost all proteins are linked through hydrophobic interactions [52]. The effects of extrusion-cooking on protein structure rearrangements are more dramatic than those of conventional extrusion. A previous parboiling step on rice flour provides further stabilization of these interactions by intramolecular disulfide bonds, whose formation seems insensitive to the conditions used for pasta making and apparently involving specific proteins in a preferential way [52]. Finally, starch-protein interactions are important. As an example, in a rice-based formulation containing 25% amaranth, the best cooking quality is obtained when starch in rice flour is allowed to interact during gelatinization with amaranth proteins that are simultaneously undergoing denaturation in the cooker-extruder [42]. In other words, the simultaneously cooking and extrusion of both rice and amaranth helps in creating a network where proteins are entrapped in the crystallized starch network. Formation of inter-protein network by the combination of covalent (disulfide) and hydrophobic interactions also may play an independent role in pasta quality. Finally, in the case of high-protein ingredients, soybean strengthens the inter-protein network generated during the pasta-making process, forming a network capable of retaining the swollen starch granules upon cooking [61].

4.2 Bread

By analyzing the label of 228 commercial GF breads from 32 brands produced and consumed worldwide — including the Mediterranean area — Roman et al. [60] found that 30% of the commercially available products has white rice flour as the first or second ingredient. Beside its high starch content, the reasons for the rice to being widely used include its wide availability, low cost, bland taste, and hypoallergenic properties. Actually, rice flour is often used in combination with corn starch: despite giving breads of lower volume than those made with only corn starch, rice flour improves their sensory acceptance, especially because they result in a less dry texture [63]. Interestingly, rice flour and corn starch are the preferred ingredients also in the preparation of experimental GF breads [64].

From a nutritional standpoint, the low fiber intake in GF diet can be overcome by either using wholegrains or adding isolated dietary fiber sources (i.e., bran) to refined flour. Currently, the first approach seems to be preferred at industrial level. Indeed, brown rice flour appeared in 8% of the breads studied as the main or the second ingredient [62]. However, including insoluble fiber negatively affects both sensory and technological properties, such as dough or batter hydration [65].

Different bio-technological approaches may help in delivering GF breads high in fiber and with good performance. Proteases have been applied to support structure formation in GF batter from brown rice flour [66]. The enzymatic treatment yields to a batter with decreased consistency, which might favor its expansion during proofing and early stage of baking, thus resulting in bread with increased specific volume and decreased crumb harness and chewiness [66]. Besides the exogenous enzymes, the role of endogenous enzymes has been tested by germinating brown rice [67]. Considering the nutritional benefits of germinated brown rice — i.e., formation of new bioactive compounds, such as gamma-aminobutyric acid (GABA) [68] — germinated brown rice was used as functional ingredient in bread-making. The enzymatic pattern developed during germination leads to an improvement in bread volume, texture, color and shelf-life [67, 69]. The enhancement of bread quality can be attributed to the greater protease and amylase activities which significantly increased foaming capacity and reduced starch molecule size, respectively, compared to those of ungerminated rice [69]. Monitoring the accumulation of enzymes during germination is important since extensive enzymatic activity compromises the quality of the product. In this context, it is worthy to mention that both rice varieties and sprouting conditions greatly affect the development of hydrolytic enzymes, thus affecting the quality of the end-product.

Moving to the use of isolated dietary fiber sources (e.g., bran fractions), this approach allows to eventually select an appropriate milling fraction (i.e., richer of soluble or insoluble dietary fiber), as well as to control the amount of its addition in order to balance between high fiber content and bread-making performance. The addition of 10% of rice bran, in particular, when containing a high amount of soluble dietary fiber, produces better bread color, higher specific volume, softer crumb firmness and improved porosity profile [70].

Rice bran has been included in the formulation of GF breads also as an economical and hypoallergenic source of proteins, as an alternative to the use of egg albumin. As rice proteins have interesting foaming and emulsifying ability, they can help the development of the typical bubble structure of bread [71]. Specifically, enrichment of 2% rice bran protein concentrate results in bread with high specific volume and improved gas retention, crumb porosity, shelf-life and sensory attributes [72].

From a technological standpoint, processing conditions, from milling to bread-making, play a role in determining rice bread quality. In this context, a larger bread specific volume was obtained when coarser fraction (>180 μm) and great dough hydration (90–110%) were combined [73, 74] On the other hand, flour with the finest particle size leads to poor gas retention during leavening and low bread volume [73].

All the researchers and all the GF baking companies agree on the use of hydrocolloids being irreplaceable to (partially) mimic gluten functionality. Among the hydrocolloids used, hydroxypropyl methylcellulose (HPMC) was the most widely used [62], since it improves both volume and texture of rice-based GF breads, as extensively reviewed [75, 76]. In respect to rice bread, HPMC helps in preparing a dough with enough consistency and a bread with higher height/width ratio of the slices [77].

Regarding variety, there are contrasting results about the influence of rice grain length on bread quality. In general, the size of the grain is related to its amylose content, with short-grain rice having lower amylose content and thus lower retrogradation tendency than long-grain rice which is a positive feature in baking. However, Cornejo and Rosell [78] showed the length of the rice is not a determining factor for breadmaking, and thus long-grain rice could be used for bakery. The Authors pointed out a synergistic effect of several factors, including particle size, protein conformation, lipid and protein content, as well as lipid-amylose complex and starch structure that could affect the bread-making performance of rice flour [78]. Recently, Roman et al. [79] showed that differences in amylose chain length among rice flours affect the starch retrogradation extent and, ultimately, the texture of GF breads. Selecting varieties with intermediate length amylose might help industries to improve the textural quality of GF baked goods by reducing the staling phenomena without the use of additives. As regards starch shape, Kang et al. [80] assessed the bread-making performance of the variety Seolgaeng — a new *Japonica* rice variety developed in Korea — that possesses an opaque endosperm with round starch granules and larger void spaces between the components compared to the majority of rice varieties. Such characteristics account for the low grain hardness and starch damage upon milling info flour, resulting in a product with enhanced dough expansion during proofing and volume during baking [80].

Overall, it seems important to select specific varieties to guarantee consistent quality and avoid problems during bread-making. In this context, breeding could also be a good tool to obtain new varieties with specific features and specific applications.

4.3 Biscuits

Looking at the labels of more than 280 GF biscuits available on the Italian market highlighted that the 22% of the starches and the 78% of the flours used in GF biscuits are from rice [37]. The use of rice in GF biscuits has been extensively reviewed by Di Cairano et al. [37]. The addition of rice milling by-products (i.e., rice bran and flour from rice broken kernels) have been explored, although obtaining biscuits with low quality (i.e., low specific volume and diameter) compared to the commercially available samples [81]. On the contrary, biscuits with similar quality of wheat biscuits were obtained using brown rice flour (70%) in a formulation containing soya flour (10%), corn (10%) and potato starch (10%) [82].

The nutritional profile or rice biscuits can be achieved by incorporating high fiber/protein sources in the formulation. In this context, recent findings suggested that alfalfa seed flour could be a valuable ingredient for improving health-promoting properties of rice biscuits [83, 84]. Adding alfalfa seed flour to rice results in higher protein and fiber content, a better fatty acid composition (i.e., high total polyunsaturated, total n-3 and n-6 fatty acid contents), and higher total phenolic content and *in vitro* antioxidant capacity compared to control (100% rice). Specifically, such changes are a function of the gradual substitution of rice flour with alfalfa seed flour (15, 30 and 45% w/w). As regards the starch fraction, the inclusion of alfalfa seed flour contributed into formulate GF cookies with slowly digestible starch and high resistant starch. Taking into consideration also the technological quality and sensory attributes of biscuits, the 30% substitution level can be considered the threshold for obtaining rice-based biscuits with improved nutritional quality and acceptable sensory quality.

Certainly, the downside of using rice as main ingredient in biscuits is the higher glycemic index compared to biscuits from other grains, including buckwheat or bean [37]. Given that the glycemic index depends on the whole recipe, ingredient composition, processing conditions, researchers and producers should focus on the improvement of this aspect of GF biscuits. One of the most successfully approaches is based on the partial replacement of native starch with high amylose starch, either in the native form or after modification through hydrothermal, enzymatic and/or chemical treatments [85]. Considering the interest in using waxy starch in many food applications due to its low retrogradation tendency [86], the attention shifted to rice waxy starch. Several studies reveal that higher amount of resistant starch could be obtained from waxy rice starch after amylopectin debranching by enzymatic treatment [87], annealing [88] or a combination of acid and heat-moisture treatments [89]. Giuberti et al. [90] closed the loop by assessing the resistant starch content in biscuits. Although all the tested treatments are effective in increasing the proportion of resistant starch compared to native starch, the higher resistant starch content, along with the lower in vitro glycemic index, is obtained for biscuits enriched in debranched waxy starch. However, in order to confirm the *in vitro* results of Giuberti et al. [90], *in vivo* trials are strongly warranted.

4.4 Snacks

The snack food market is expanding rapidly and is predicted to continue growing in the future. Indeed, this product category completely meets the requests of the "modern" consumer whose eating habits have been changed together with the changes in lifestyle. Specifically, snacks satisfy the demand for healthy, minimally processed, versatile and convenient ready-to-eat foods. Snacks can be designed using a wide variety of starch and/or grains, including rice, that play a key role in providing all the features desired for highly acceptable snack products, such as structure, texture, and mouth feel. Moreover, rice is known for its blandness in flavor and desirable white color and, therefore, does not affect either color or flavor of the formulation. Various ingredients — including rice bran [91] or pulses [92] — can be then included to improve the nutritional profile of the products. It is no surprise that a wide range of snacks are available on supermarket shelves with a wide selection of sizes, shapes, colors and flavors designed to appeal the consumer. Among the various products, extruded snacks are the most popular.

4.4.1 Extruded Snacks

Although there are many ways to classify extruded snacks, manufacturers use three main terms to identify snacks [93]:

- Direct expansion snacks: they represent the most present and consumed category on the market. They are defined "light" because they have a low density; they are seasoned with a wide range of aromas.
- Pellets: they are semi-finished products formed at low pressure to prevent expansion; they have to be cooked before final consumption.
- Co-extruded snacks: they represent relatively the latest technology for the snack industry. In this process, two different materials are extruded from the extruder die; the raw materials to be extruded can come from two extruders or from a single extruder and a pump. The most common types of co-extruded snacks consist of a solid outer cylinder usually based on cereals and an inner part consisting in a filling cream (jams, sweet or savory creams).

Snack structure (i.e., expansion, crunchiness, mouthfeel, etc.) is mainly given by the versatility of the extrusion-cooking technology. This technology provides the opportunity to process a wide range of unique food products by simply changing the raw material and processing conditions (head and barrel temperature, extruder screw and feeder speed). In general, the extrusion process is a system consisting of multiple mixing, kneading, cooking, shear and forming operations in a single plant. The ingredients in the form of flour are mixed with water (about 10%) to create a homogeneous hydrated dough. The dough is mixed by the screw along the longitudinal axis of the extruder; inside the extruder the ingredients are subjected to a high shear stress, high temperatures and high pressures for a short time, after which the product is cooked keeping the water in a liquid phase. The sudden drop in pressure at the end of the extruder causes an immediate product expansion, thanks to the rapid evaporation of the water. The product is finally shaped by the head of the extruder and cut continuously to the desired length by means of a series of blades placed downstream of the die. In each phase of the process the time-temperature combination is important. To produce extruded snacks, the food industry adopts HTST (high temperature/short time) processes that allow the materials to be subjected to a predetermined maximum temperature for the minimum time necessary. The HTST process reduces microbial contamination and inactivates enzymes, it also guarantees a high shelf life thanks to the low moisture content and water activity. The applied thermal and shear stress cause structural, chemical and nutritional changes such as starch gelatinization, protein denaturation, lipid oxidation, vitamin degradation, loss of nutrients and phytochemicals [94]. These component changes modify the functional properties of flours and the quality of extruded products [95]. The desired characteristics of the extruded product are provided by the correct control of primary process parameters (barrel temperature, feed humidity, feed rate and screw speed) and secondary parameters (dough temperature, die pressure, torque and specific mechanical energy).

The effect of the extrusion process on starch characteristics is the most relevant and influential on the characteristics of the product. Starch plays the role of binder and offers the expansion of snacks [96]. During extrusion, the hydrated flour undergoes a thermo-mechanical treatment that transforms it into a dough. Such conditions induce starch gelatinization and the loss of its crystallinity [95]; therefore, it is not surprising that extruded snack products tend to be considered with a high glycemic index value [97]. However, the formation of amylose-lipid complexes, starch-protein complexes or resistant starch upon extrusion might reduce starch digestibility. Physical modifications induced by extrusion cooking can be an alternative way to improve the properties of native starches without chemical modifications. For example, extruded flour from low-amylose (5%) rice — exhibiting low pasting viscosity — has been suggested as a base material in food formulations where a high solid density per unit volume is required. The extrudate from high-amylose rice (about 30%), on the other hand, could be used as a thickener in convenience foods such as soup mixes [98].

In addition to starch, proteins, lipids and phenolic compounds can also be significantly affected by extrusion-cooking. Proteins can suffer a loss of solubility combined with polymerization through the creation of disulfide bonds. Therefore, they undergo changes in their molecular interactions involving covalent, non-covalent bonds and protein-starch and protein-lipid interactions [99]. As far as proteins and amino acids are concerned, there is a lysine loss due to Maillard reaction.

Lipids play an important role during extrusion because they act as emulsifiers and provide a consistency and viscosity appropriate for the extrusion process [96]. The lipid content, however, is reduced due to the formation of lipid-protein complexes; moreover, the high temperature causes the oxidation of unsaturated fatty acids into lipid hydroperoxides and therefore a decrease in lipids in the final product [100]. In addition, lipids can form a complex with amylose that alters the structure and consistency of extruded products. In fact, raw materials with adequate amylose content (>25%) could reduce the degree of gelatinization of starch and consequently the characteristics of snacks by reducing their explanation [101].

Finally, due to high temperature and high shear, most of the bioactive compounds are deteriorated during extrusion cooking. Specifically, extrusion-cooking results in a significant decrease in soluble phenolic acids in brown and pigmented rice [102, 103]. On the other hand, an increase of in the insoluble phenolic compounds occurs due to the polymerization of free phenolic acids [103]. The vacuum-assisted extrusion has been proposed to expand the extrudate at low temperatures and enhance the retention of bioactive compounds [104].

4.4.2 Cakes

Rice cakes (also known as puffed rice cakes) are a popular snack that are also consumed as a bread substitute. Thanks to their 35–40 kcal per cake (about 9–10 g per cake), rice cakes meet the consumer demand for low calorie and healthy foods.

They are traditionally made by heating whole grain rice or fractured kernels (with a typical moisture content of 16–20%, w/w) in a confined mold at temperatures of 190–250 °C. During heating, steam pressure builds within the mold. Upon opening the mold, pressure is released rapidly, causing flash vaporization of the superheated water and steam. This process causes the rice kernels to melt and fuse during their expansion into a disc. Minor ingredient such as salt and other flavoring agents may be added.

From a structural standpoint, rice cakes differ from classical food and polymeric solid foams in that the solid matrix is not continuous but made of aggregated individual puffed particles. Among the drawbacks that have limited the market growth of the rice cakes there is their tendency to crumble during packaging, transportation, and storage. Since over- or under-puffed rice cake is difficult for packaging and transportation, the specific volume of the product is one of the major quality parameters to be checked during processing. Besides specific volume and integrity, other important quality indices of rice cake are appearance, color, and texture.

Both raw material characteristics (i.e., amylose content, type of rice) and processing conditions (i.e., time, and temperature) are of foremost importance to guarantee the desired expansion and to avoid the breakage of rice cakes during production that would result in the decrease in their value. In this context, medium or short-grain *japonica* (low amylose) rice varieties are preferred, while the long grain varieties, as well as waxy varieties, do not expand much during cooking [105]. Indeed, the branched chains of amylopectin produce desirable, lighter, expanded texture in products. Using long-grain rice would require process optimization. In general, a lower tempering moisture, higher heating temperature and longer heating time leads to cakes from long-grain brown rice with increased specific volume [106, 107]. On the contrary, for medium grain rice, higher tempering moisture yielded a larger specific volume. Medium-grain brown rice also produced cakes that are much more fragile than those produced from long-grain brown rice [107]. As regards product formulation, adding black rice to brown rice negatively affects the integrity of the product [108]. The specific volume of black rice cakes increases with increasing tempering moisture, heating temperature and heating time [108, 109]. The processing conditions do not significantly affect the color of black rice cakes, while its hardness decreases as tempering moisture and heating time increases [109]. Maximum values for cake hardness occur at 250 and 260 °C for 6 s at 16 and 20% moisture contents [108].

4.5 Breakfast Cereals

Breakfast cereals are convenient and relatively shelf stable foods made from corn, wheat, rice, or oats, and typically consumed as the first meal of the day. They are classified into ready-to-eat cereals (such as flakes, puffed, and shredded grains) and hot cereals that include oatmeal, hot wheat, and other grain products.

Flakes and puffed cereals are appreciated mainly for their lightness, friability and crispness; such qualities are related to their cellular structure and degree of expansion/bulk density, as a result of changes — occurring during processing — to the grain at the molecular, microstructural, and macrostructural levels. In addition to flakes and puffed cereals, a wide range of extruded-expanded have been developed over the last two to three decades by extrusion-cooking, whose features have been described in Sect. 4.4.1.

4.5.1 Rice Flakes

Although corn and wheat are the most common raw materials for flakes production, flakes from rice are also produced. Using the large broken kernels (the small ones are generally used for brewing) is usually preferred from an economic standpoint due to the fact that broken kernels could not be consumed as table rice, whereas the whole grains are used for the production of puffed cereals. The process of rice flakes production begins with the cooking of rice grits (generally 60 min at a steam pressure of 1–1.5 bar) [110], with the aim of achieving the proper starch gelatinization and the development of flavor and color. Appearance of stickiness during cooking would make the product difficult to handle through the cooling, lump breaking, and drying phases. This can be avoided by keeping the moisture content below the 28% and selecting rice varieties with intermediate amylose content (18–24%). The cooked grits tend to aggregate in big lumps, that are conveyed to a lump-breaking machine that separates agglomerated grits into unit of lower size. Cooked rice clusters must be handled gently to avoid the production of useless fines, which have to be reworked or used in other products. The cooked grits are partially dried to about 17% moisture content. If the moisture content is too low, the cluster shatters during flaking, and small flakes and fines are produced. If the moisture content is too high, the flaking rolls gum up [110].

After drying, the grits are cooled to about 40–45 °C and tempered typically for 24 h to allow the equilibration of the temperature and moisture content of the grains, to facilitate further processing. After tempering, the grits are ready for flaking in large, smooth rolls and toasting with hot air for about 50 s at 300 °C in a cylindrical rotary oven. The toasting not only dehydrates the flakes to less than 3% moisture but is also responsible for the crispy, blistered texture and characteristic flavor and color. The moisture content of the incoming flakes and the oven temperature must be designed such that the flakes blow up or puff up during roasting. If they do not, they become excessively hard and fragile [110].

4.5.2 Puffed Rice

Puffed rice is widely used as ready-to-eat breakfast foods or as ingredients in snack formulations. Either whole grain kernels or flour are cooked by traditional cooking or extrusion-cooking process that leads to the production of pellets. Either cooked kernels or extruded pellets are then puffed in a toaster oven (oven-puffed cereals) or in a gun puffing machine (gun-puffed cereals). Oven puffing is based on a sudden application of heat at atmospheric pressure to a prewetted cereal so that the water in the grain is vaporized *in situ*, before it has time to diffuse to the surface of the kernel. The internal vaporization expands (or puffs) the product from 2 to 5 times its original size. However, these products have less expansion rate compared to products expanded with gun-puffing. Gun puffing is based on a sudden transfer of grain containing superheated steam from a high pressure to a low pressure, thereby allowing the water to suddenly vaporize and cause expansion.

During puffing, the size of rice kernels increases, while the bulk density decreases, and a fully heat-treated crispy and porous structure is created. Puffed rice is a matrix made up of numerous cavities of different sizes and separated by a very thin "wall" [111–113]. The expansion during puffing progresses from the outer endosperm portion towards the center gaining more freedom for expansion which results in bigger pores. The treatment also radically affects the water absorption capacity of both the grains and the related flours, due to the new organization of the outer layers and the high porosity of the matrix, that allow a faster hydration [112]. Taking into consideration the changes induced by puffing on starch structure and its physical features, flour from puffed rice might be successfully used not only as ingredients in snack formulations, but also in baked goods as a means to control water migration, thus slowing staling phenomenon and keeping the product softer for a longer period, as already successfully shown for other puffed grains [114].

At molecular level, the puffing process degrades the starch integrity, which significantly affects the physical properties of grain. The degraded starch during puffing treatments leads to changes in X-ray diffraction pattern. Specifically, the A-type of diffraction pattern, typical of raw rice, is altered in favor of the formation of B- and V-type patterns [115]. Moreover, since most of the granules are gelatinized during puffing, flours from puffed grains present a higher initial viscosity and a less pronounced peak viscosity in a viscoamylograph test [112]. Parboiling process prior to puffing enhances the expansion rate of rice [115], likely due to the formation of retrograded starch in parboiled rice. Indeed, a positive correlation was found between crystallinity and expansion ratio of puffed rice, which positively affects the hardness, fracturability, and bulk density of the products [116]. In this context, the amylose-amylopectin ratio has been known to be an important parameter affecting the physical properties of starch and thus the characteristics of puffed products. By screening twelve varieties of *Indica* rice, with amylose content ranging from 21.5 and 28.5%, Joshi et al. [117] found that higher-amylose rice would have better puffed volume than samples with lower amylose content. Overall, oven-puffing requires the use of medium or short white rice with intermediate amylose content (15–20%), while the rice used for gun-puffing is either long-grain white rice or parboiled medium-grain rice. Conversely, neither waxy rice or rice with a high incidence of microfissures or stress cracks are recommended because they expand less, thus resulting in products with low quality.

Besides amylose content, other characteristics of the material to be puffed affect the quality of the final product. Protein content was found to negatively affect the expansion ratio of puffed rice [117], as well as the presence of fiber in brown rice [118]. Considering the growing interest in using brown rice due to nutritional properties, some studies focused on the process optimization to get a product with high expansion ratio [115]. Nevertheless, puffing causes a drastic decrease in total phenolic, flavonoid and anthocyanin content, as well as in reducing power and DPPH radical scavenging activity in both black [118] and brown [119] rice.

5 Other products

5.1 *Couscous*

Couscous is one of the most ancient dishes prepared and consumed in the Mediterranean area. Originally of North Africa, nowadays, thanks to the globalization, couscous is appreciated and widespread in Western countries.

Generally, it is produced starting from durum wheat semolina and water and the end-product is presented in the form of agglomerates with size between 1 and 2 mm in diameter [120]. Nowadays, many other cereals, including rice, can serve as raw materials, helping to solve the issue of low diversity and availability of GF products [121].

The artisanal method, still applied today at a household level in the countries of North Africa, requires the semolina to be humidified and worked manually with a prolonged stirring action [122]. The agglomerates thus obtained are then sieved in order to create a uniform product in size, dried in the sun and stored for several months at room temperature until the moment of consumption. The product can undergo steam cooking for about 10 min and, to increase its shelf life, a phase of drying in the sun is foreseen. The final stage of preparation of this typical dish involves subsequent hydration and steam cooking. Couscous is then mixed with meat and/or vegetable sauce when eaten.

The industrialization has made it possible to simplify and shorten the process. In the current industrial process, semolina and water are first mixed to promote the particle agglomeration and form the wet semolina agglomerates. These are then rolled in a rotating screen drum to improve the structure of the grains and to select by size classification the fraction with diameter ranging between 1 and 2 mm. The rolled agglomerates are then cooked using steam at 100 °C to strengthen the structure and to reduce the final preparation of the dish to a few minutes. At the industrial scale, the steam cooking stage is conducted in continuous oven, at 100 °C using steam at atmospheric pressure. The wet grains are layered on a perforated metallic surface. The thickness of the grain layer (about 20 cm) requires about 15 min to complete the cooking [123]. Finally, the product is dried to reduce the water content (<13–14%) and guaranty stability over the storage. The drying stage could contribute into the texturing mechanisms of the couscous grains through the possible compaction of the structure. At the industrial scale, the drying stage is conducted in continuous drier at high temperature (90–120 °C), with a short residence time close to 15–30 min [123].

The high quality of couscous is defined as a product with regular and homogeneous particle size. After cooking, the grains of couscous should remain individualized without disintegrating or sticking together. When GF grains are used in the manufacture of couscous, some problems can be encountered due to poor protein aggregation properties [120, 123].

Although rice couscous is commercially available, very few scientific works have focused on this kind of product. Benatallah et al. [121] investigated the technological feasibility to obtain GF couscous based on rice-leguminous supplementation. Among the rice-based couscous, the best results were obtained by the formulation containing field bean (*Vicia faba*), followed by chickpea (*Cicer arietinum*) and proteaginous pea (*Pisum arvense*) [121].

Considering the growing interest in GF couscous and the limited number of investigations, further efforts should be directed toward the study of the properties required for agglomeration and the mechanisms involved during water hydration of rice and other GF grains.

5.2 Baby Foods

The ideal baby food should be nutrient dense, easily digestible, and of suitable consistency and, most importantly, affordable to the target market [124]. In this context, cereal-based baby foods are major source of energy, protein, vitamins, and minerals for infants from 6 to nearly 36 months old [124]. An extensive review of baby foods based on cereals has been recently provided by Jeelani et al. [125].

Among cereals, rice is considered one of the most suitable raw materials for baby food production, thanks to its high digestibility, low allergenicity as well as being nutritionally balanced, available and with a bland taste. However, one of the issues related to the use of rice in this product category is the presence of inorganic arsenic [126]. In 2015, the European Commission (EC) established a limit of 100 μg/kg for inorganic arsenic in rice intended for the production of infant foods [127]. Interestingly, the European Food Safety Authority (EFSA) stated that infants in the EU are exposed to a mean of 1.6 μg/kg per day of inorganic arsenic through rice-based infant food alone [128]. More recently, by analyzing the dietary exposure to inorganic arsenic via rice and rice-based products consumption, González et al. [129] found that none of the baby products analyzed exceeded that threshold, being levels actually well below: 4.0 μg/kg for baby food with fish and rice, and 3.1 μg/kg for baby food with chicken and rice.

Cereals for infants are usually manufactured by heating cereal flour in excess of water. Other ingredients such as milk powder, flavorings, salt, sugars, vitamins and minerals, as well as other essential nutrients can be included in the formulation [124]. The heat-treatment of the cereal slurry varies in order to achieve the required sensory quality of the final product. Despite that, the complete starch gelatinization needs to be guaranteed in baby cereal-based foods before ingestion by the infant. After heating, the slurry is passed on to the drying stage. In this step, the slurry is distributed in a thin film of consistent thickness onto heated rotating drums. This results in an almost immediate evaporation of the water leaving a thin dry film of the product which is then scraped off the dryer and onto a conveyor. After drying, instant cereal formulations are packaged as dry cereal flakes and readily reconstituted by adding milk or water to produce a cereal with a "cooked" texture.

During the heat-treatment, the viscosity of the slurry increases as starch gelatinizes, making the product processing and handling more difficult, especially during drying. The reduction of starch molecular weight by using hydrolytic enzymes (e.g., malt or fungal α-amylases) might help in decreasing the viscosity of the slurry during the heat treatment. At the same time, as a consequence of its low viscosity, the dry product might tend to settle out of the milk or water, during the rehydration process. Moreover, the enzymatic treatment commonly results in high levels of glucose, that reacting with lysine (i.e., Maillard reaction), leading to a dark, brown product, whose nutritional quality is remarkably lower and commonly perceived by consumers as unattractive.

Baby foods can be also produced by extrusion-cooking. Indeed, this process increases mineral bioavailability, protein solubility, and starch accessibility to amylase activity, which are the required properties for baby foods. There are several advantages of producing pre-gelatinized flours for baby food application by extrusion-cooking. Firstly, the possibility of controlling the viscosity, hydration properties and solubility rate of the final product allows obtaining products with consistent quality. Moreover, the versatility of the process offers the opportunity to formulate a large variety of products attractive to different consumer demographics. Last but not least, the lower energy costs compared to drum dryers and the reduced floor space required make the extrusion-cooking an interesting processing for formulate baby foods from cereals.

5.3 *Rice Milk*

Plant-based milk analogues are gaining popularity in the Western countries due to several factors. From one side, the number of people facing with cow's milk intolerance or allergy is increasing, as well as the consumer's awareness to emerging topics such as sustainability and food environmental impact. Moreover, changes in lifestyle towards vegetarian and vegan diet foster the market of milk alternatives. Nowadays, milk alternatives are obtained from a variety of vegetable sources, including soy, rice, almonds. Using rice allows solving the allergy issues related to soy and almonds. From a nutritional standpoint, rice milk is higher in carbohydrates and lower in proteins and fats compared to other plant-based milks. The lack in minerals and vitamins such as calcium and vitamin in rice is overcome by fortification, as in the case of most commercial milks [130]. Interestingly, the high sugar content in rice allows to obtain a product with a sweet taste without addition of other sugars [130–132].

Rice milk is prepared by mixing the whole grain, either white or brown rice, or the related flours, with water. Because of the high starch content in rice, the product is characterized by poor emulsion stability. Thus, an enzymatic treatment with α and β-amylase or glucosidase is usually carried out to hydrolyze the starch granules. However, the reaction time should be limited to avoid the development of unwanted off-flavors [132]. After saccharification, the product is filtered or decanted, to remove the grinding waste and insoluble plant material. Addition of other ingredients like oils, flavorings, or stabilizers may be added at this point, depending on the desired product. Finally, ultra-high temperature treatment (UHT) or pasteurization is used to extend the product shelf life. After pasteurization – usually carried out at temperatures below 100 °C – the product can be stored for about one week at refrigerated temperatures. On the other hand, the UHT treatment (135–150 °C for few seconds) yields a commercially sterile product [133]. Innovative processing technologies (e.g. high-pressure processing, high-pressure homogenization, pulsed electric fields, and ultrasound) are developing fast, to improve physicochemical, nutritional, and quality properties of plant-based beverages [131].

5.4 *Rice Oil*

The rice milling process involves the removal of husks, bran and germ to obtain the milled rice, that—as discussed above—can be consumed as grain or as flour/grits in several food formulations. The woody and siliceous nature of the husks make them unsuitable even as animal feed. On the other hand, rice bran and germ contain up to 20 and 40% of lipids, respectively [28], thus they are valuable raw materials for oil extraction. In some European systems (i.e., Spain, Italy), the germ is separated from the bran by sieving and aspiration, whereas in most milling systems, the germ is not separated and remains in the bran [29]. The lipid material that is extracted from rice bran/germ to make rice oil comes primarily from the germ and the outer layers (aleurone and subaleurone) of the rice grain where it exists as lipid droplets or spherosomes.

Rice oil is a valuable cooking oil, generally consumed in Asian counties. However, its interest is growing also in Western countries, thanks to its unique health benefits, due to the high content in nutraceutical compounds like γ-oryzanol and tocotrienolen (Vitamin E) whose beneficial implications, such as lowering plasma cholesterol levels, are well known [134, 135]. Specifically, rice oil has the property of lowering low density lipoprotein cholesterol and increasing the high density lipoprotein cholesterol to some extent either by influencing absorption of dietary cholesterol or by enhancing the conversion of cholesterol to fecal bile acids and sterols [136].

Besides the health benefits, another reason accounting for the growing interest in this product could be found in the "sustainability" topic and, specifically in the exploitation of rice bran/germ, generally considered a waste, for the oil production for human consumption.

Rice oil extraction, like other vegetable oils, is carried out with supercritical carbon dioxide (SC-CO_2). CO_2 is a non-toxic solvent and can be easily and completely removed from products; moreover, it is non-corrosive, non-flammable, and readily available in large quantities with high purity. Although supercritical fluid extraction provides total yields like the ones provided by the traditional hexane extraction, the process provides a two-step extraction, according to SC-CO_2 selectivity: free fatty acids are the first compounds extracted at lower pressures, followed by the obtaining of highly purified triglycerides when higher pressures are employed [137].

Rice oil is characterized by slightly bland flavor like peanut oil or other vegetable oils. It has high levels of neutral lipids and low amounts of glycolipids and phospholipids. Compared with other vegetable oils, the fatty acid profile of rice oil tends to be higher in oleic acid and lower in linoleic acid, similar to peanut oil. Because of its relative stability during frying or other heat processing, rice oil actually is considered a lower risk oil, relative to the production of toxic heat-generated products, such as hydroperoxides and toxic polymers during frying or other heat processing [138]. Rice oil is also the unique oil that contains higher levels of wax (3–9%) and unsaponifiable components (2–5%) than other vegetable oils, which can cause processing problems but also may contribute into improve health benefits attributed to rice bran oil. Indeed, one of the most positive features of rice oil is its abundance of minor constituents, many of which are suggested to have significant health benefits such as tocols (tocopherols and tocotrienols), oryzanol, phytosterols, and policosanol [138].

6 Conclusions

In recent decades, the ways in which rice is consumed worldwide have changed radically: it is no longer just the staple food of all Asian countries and the ingredient of some Mediterranean dishes (e.g. *risotto* in Italy and *paella* in Spain). Currently, rice is one of the most important ingredients — both in terms of quality and quantity — of several food formulations, different in terms of occasions and frequency of consumption and type of consumers to whom they are intended. This chapter describes the main characteristics of the rice-based foods most present in the Mediterranean diet, and the technological processes adopted. The fundamental role of rice in these foods is undoubtedly related to the sensory characteristics, to the peculiar properties of the proteins in which the fractions responsible for the formation of gluten are absent, as well as to the characteristics of starch and its modifications during processing.

To fully exploit rice, it is now essential to monitor changes in constituents, especially starch, during the technological process and propose new classification models. The current characterization/classification of rice into different classes based on the biometric characteristics and the amylose content must be completed with indices that can, in a simple, fast and reliable way, directly describe the behavior of the flour during processing. Only by predicting the nature and intensity of the changes in the starch of a particular type of rice — that depend on the amount of water and the heat treatment — it is possible to obtain the desired texture of gluten-free bread, pasta, and snacks. The sensory demands of consumers, in fact, are just as important as the nutritional ones to guarantee the success of a food.

References

1. Regulation (EU) No 1308/2013 of the European Parliament and of the Council of 17 December 2013 establishing a common organisation of the markets in agricultural products and repealing Council Regulations (EEC) No 922/72, (EEC) No 234/79, (EC) No 1037/2001 and (EC) No 1234/2007.
2. www.fao.org.
3. Saleh AS, Wang P, Wang N, Yang L, Xiao Z. Brown rice versus white rice: nutritional quality, potential health benefits, development of food products, and preservation technologies. Compr Rev Food Sci Food Saf. 2019;18:1070–96.
4. Zhou Z, Robards K, Helliwell S, Blanchard C. Composition and functional properties of rice. Int J Food Sci Technol. 2002;37:849–68.
5. Kitta K, Ebihara M, Iizuka T, Yoshikawa R, Isshiki K, Kawamoto S. Variations in lipid content and fatty acid composition of major non-glutinous rice cultivars in Japan. J Food Compos Anal. 2005;18:269–78.
6. Yu L, Turner MS, Fitzgerald M, Stokes JR, Witt T. Review of the effects of different processing technologies on cooked and convenience rice quality. Trends Food Sci Techn. 2017;59:124–38.
7. Crowhurst DG, Creed PG. Effect of cooking method and variety on the sensory quality of rice. Food Serv Technol. 2001;1:133–40.
8. Son JS, Do VB, Kim KO, Cho MS, Suwonsichon T, Valentin D. Consumers' attitude towards rice cooking processes in Korea, Japan, Thailand and France. Food Qual Pref. 2013;29:65–75.
9. Perez CM, Juliano BO. Indicators of eating quality for non-waxy rices. Food Chem. 1979;4:185–95.
10. Kasai M, Lewis A, Marica F, Ayabe S, Hatae K, Fyfe CA. NMR imaging investigation of rice cooking. Food Res Int. 2005;38:403–10.
11. Fitzgerald MA. Starch. In: Champagne ET, editor. Rice: chemistry and technology. St Paul: American Association of Cereal Chemists International; 2004. p. 109–41.
12. Bett-Garber KL, Champagne ET, Ingram DA, McClung AM. Influence of water-to-rice ratio on cooked rice flavor and texture. Cereal Chem. 2007;84:614–9.
13. Das T, Subramanian R, Chakkaravarthi A, Singh V, Ali SZ, Bordoloi PK. Energy conservation in domestic rice cooking. J Food Eng. 2006;75:156–66.
14. Suzuki U, Kubota K, Omichi M, Hosaka H. Kinetic studies on cooking of rice. J Food Sci. 1976;41:1180–3.

15. Billiris MA, Siebenmorgen TJ, Meullenet JF, Mauromoustakos A. Rice degree of milling effects on hydration, texture, sensory and energy characteristics. Part 1. Cooking using excess water. J Food Eng. 2012;113:559–68.
16. Mestres C, Ribeyre F, Pons B, Fallet V, Matencio. Sensory texture of cooked rice is rather linked to chemical than to physical characteristics of raw grain. J Cereal Sci. 2011;53:81–9.
17. Finocchiaro F, Ferrari B, Gianinetti A, Dall'Asta C, Galaverna G, Scazzina F, Pellegrini N. Characterization of antioxidant compounds of red and white rice and changes in total antioxidant capacity during processing. Mol Nutr Food Res. 2007;51:1006–19.
18. Zaupa M, Calani L, Del Rio D, Brighenti F, Pellegrini N. Characterization of total antioxidant capacity and (poly) phenolic compounds of differently pigmented rice varieties and their changes during domestic cooking. Food Chem. 2015;187:338–47.
19. Catena S, Turrini F, Boggia R, Borriello M, Gardella M, Zunin P. Effects of different cooking conditions on the anthocyanin content of a black rice (*Oryza sativa* L.'Violet Nori'). Eur Food Res. 2019;245:2303–10.
20. Ali N, Pandya AC. Basic concept of parboiling of paddy. J Agric Eng Res. 1974;19:111–5.
21. Bhattacharya M, Zee SY, Corke H. Physicochemical properties related to quality of rice noodles. Cereal Chem. 1999;76:861–7.
22. Bhattacharya KR. Parboiling of rice. In: Champagne ET, editor. Rice: chemistry and technology. St. Paul: The American Association of Cereal Chemists; 2004. p. 329–404.
23. Casiraghi MC, Brighenti F, Pellegrini N, Leopardi E, Testolin G. Effects of processing on rice starch digestibility evaluated by in vivo and in vitro methods. J Cereal Sci. 1993;17:147–56.
24. https://www.alimentinutrizione.it/tabelle-nutrizionali/
25. Min B, McClung A, Chen MH. Effects of hydrothermal processes on antioxidants in brown, purple and red bran whole grain rice (*Oryza sativa* L.). Food Chem. 2014;159:106–15.
26. Walter M, Marchesan E, Massoni PFS, da Silva LP, Sartori GMS, Ferreira RB. Antioxidant properties of rice grains with light brown, red and black pericarp colors and the effect of processing. Food Res Int. 2013;50:698–703.
27. Paiva FF, Vanier NL, Berrios JDJ, Pinto VZ, Wood D, Williams T, et al. Polishing and parboiling effect on the nutritional and technological properties of pigmented rice. Food Chem. 2014;191:105–12.
28. Mizuno H, Kajiwara K. Method for producing quick-cooking rice. PCT International Patent Application; 2019.
29. Bhattacharya KR, Ali SZ. Rice products. In: Bhattacharya KR, Ali SZ, editors. An introduction to rice-grain technology. Boca Raton: CRC Press; 2015. p. 188–217.
30. Rizk LF, Doss HA. Preparation of improved quick cooking rice. Food Nahrung. 1995;39:124–31.
31. Prasert W, Suwannaporn P. Optimization of instant jasmine rice process and its physicochemical properties. J Food Eng. 2009;95:54–61.
32. Saleh MI. Protein-starch matrix microstructure during rice flour pastes formation. J Cereal Sci. 2017;74:183–6.
33. de Souza Batista C, dos Santos JP, Dittgen CL, Colussi R, Bassinello PZ, Elias MC, Vanier NL. Impact of cooking temperature on the quality of quick cooking brown rice. Food Chem. 2019;286:98–105.
34. www.marketsandmarkets.com
35. Pellegrini N, Agostoni C. Nutritional aspects of gluten-free products. J Sci Food Agric. 2015;95:2380–5.
36. Morreale F, Agnoli C, Roncoroni L, Sieri S, Lombardo V, Mazzeo T, et al. Are the dietary habits of treated individuals with celiac disease adherent to a Mediterranean diet? Nutr Metab Cardiovasc Dis. 2018;28:1148–54.
37. Di Cairano M, Galgano F, Tolve R, Caruso MC, Condelli N. Focus on gluten free biscuits: Ingredients and issues. Trends Food Sci Technol. 2018;81:203–12.

38. Morreale F, Boukid F, Carini E, Federici E, Vittadini E, Pellegrini N. An overview of the Italian market for 2015: cooking quality and nutritional value of gluten-free pasta. Int J Food Sci Technol. 2019;54:780–6.
39. Marti A, D'Egidio MG, Dreisoerner J, Seetharaman K, Pagani MA. Temperature-induced changes in dough elasticity as a useful tool in defining the firmness of cooked pasta. Eur Food Res Technol. 2014;238:333–6.
40. Marti A, Seetharaman K, Pagani MA. Rheological approaches suitable for investigating starch and protein properties related to cooking quality of durum wheat pasta. J Food Qual. 2013;36:133–8.
41. Marti A, Seetharaman K, Pagani MA. Rice-based pasta: a comparison between conventional pasta-making and extrusion-cooking. J Cereal Sci. 2010;52:404–9.
42. Cabrera-Chávez F, de la Barca AMC, Islas-Rubio AR, Marti A, Marengo M, Pagani MA, et al. Molecular rearrangements in extrusion processes for the production of amaranth-enriched, gluten-free rice pasta. LWT Food Sci Technol. 2012;47:421–6.
43. Marti A, Abbasi Parizad P, Marengo M, Erba D, Pagani MA, Casiraghi MC. In vitro starch digestibility of commercial gluten-free pasta: the role of ingredients and origin. J Food Sci. 2017;82:1012–9.
44. Wang L, Duan W, Zhou S, Qian H, Zhang H, Qi X. Effects of extrusion conditions on the extrusion responses and the quality of brown rice pasta. Food Chem. 2016;204:320–5.
45. da Silva EMM, Ascheri JLR, Ascheri DPR. Quality assessment of gluten-free pasta prepared with a brown rice and corn meal blend via thermoplastic extrusion. LWT Food Sci Technol. 2016;68:698–706.
46. Wang L, Duan W, Zhou S, Qian H, Zhang H, Qi X. Effect of rice bran fibre on the quality of rice pasta. Int J Food Sci Technol. 2018;53:81–7.
47. Mariotti M, Iametti S, Cappa C, Rasmussen P, Lucisano M. Characterisation of gluten-free pasta through conventional and innovative methods: evaluation of the uncooked products. J Cereal Sci. 2011;53:319–27.
48. Resmini P, Pagani MA. Ultrastructure studies of pasta. A review. Food Struct. 1983;2:1–12.
49. Tan HZ, Li ZG, Tan B. Starch noodles: history, classification, materials, processing, structure, nutrition, quality evaluating and improving. Food Res Int. 2009;42:551–76.
50. Marti A, Pagani MA. What can play the role of gluten in gluten free pasta? Trends Food Sci Technol. 2013;31:63–71.
51. Marti A, Caramanico R, Bottega G, Pagani MA. Cooking behavior of rice pasta: effect of thermal treatments and extrusion conditions. LWT Food Sci Technol. 2013;54:229–35.
52. Barbiroli A, Bonomi F, Casiraghi MC, Iametti S, Pagani MA, Marti A. Process conditions affect starch structure and its interactions with proteins in rice pasta. Carbohydr Polym. 2013;92:1865–72.
53. Marti A, Barbiroli A, Marengo M, Fongaro L, Iametti S, Pagani MA. Structuring and texturing gluten-free pasta: egg albumen or whey proteins? Eur Food Res Technol. 2014;238:217–24.
54. Lai HM. Effects of rice properties and emulsifiers on the quality of rice pasta. J Sci Food Agric. 2002;82:203–16.
55. Grugni G, Mazzini F, Viazzo G, Viazzo N. Patent EP 2110026 B1. 2009; Application number: 09000385.6.
56. Marti A, Pagani MA, Seetharaman K. Understanding starch organisation in gluten-free pasta from rice flour. Carbohydr Polym. 2011;84:1069–84.
57. Marti A, Pagani MA, Seetharaman K. Characterizing starch structure in a gluten-free pasta by using iodine vapor as a tool. Starch-Starke. 2011;63:241–4.
58. Marti A, D'Egidio MA, Pagani MA. Pasta: quality testing methods. In: Wrigley CW, Corke H, Seetharaman K, Faubion J, editors. Encyclopedia of food grains. Oxford: Academic; 2015. p. 161–5.
59. Grugni G, Mazzini F, Viazzo G, Viazzo N. Patent EP EP 2534960. 2014; Application number: 2454616.

60. Marti A, Ragg EM, Pagani MA. Effect of processing conditions on water mobility and cooking quality of gluten-free pasta. A magnetic resonance imaging study. Food Chem. 2018;266:17–23.
61. Marengo M, Amoah I, Carpen A, Benedetti S, Zanoletti M, Buratti S, et al. Enriching gluten-free rice pasta with soybean and sweet potato flours. J Food Sci Technol. 2018;55:2641–8.
62. Roman L, Belorio M, Gomez M. Gluten-free breads: the gap between research and commercial reality. Compr Rev Food Sci Food Saf. 2019;18:690–702.
63. Mancebo CM, Merino C, Martinez MM, Gomez M. Mixture design of rice flour, maize starch and wheat starch for optimization of gluten free bread quality. J Food Sci Technol. 2015;52:6323–33.
64. Masure HG, Fierens E, Delcour JA. Current and forward looking experimental approaches in gluten-free bread making research. J Cereal Sci. 2016;67:92–111.
65. Föste M, Verheyen C, Jekle M, Becker T. Fibres of milling and fruit processing by-products in gluten-free bread making: a review of hydration properties, dough formation and quality-improving strategies. Food Chem. 2020;306:125451.
66. Renzetti S, Arendt EK. Effect of protease treatment on the baking quality of brown rice bread: from textural and rheological properties to biochemistry and microstructure. J Cereal Sci. 2009;50:22–8.
67. Cornejo F, Rosell CM. Influence of germination time of brown rice in relation to flour and gluten free bread quality. J Food Sci Technol. 2015;52:6591–8.
68. Cho DH, Lim ST. Germinated brown rice and its bio-functional compounds. Food Chem. 2016;196:259–71.
69. Wunthunyarat W, Seo HS, Wang YJ. Effects of germination conditions on enzyme activities and starch hydrolysis of long-grain brown rice in relation to flour properties and bread qualities. J Food Sci. 2020;85:349–57.
70. Phimolsiripol Y, Mukprasirt A, Schoenlechner R. Quality improvement of rice-based gluten-free bread using different dietary fibre fractions of rice bran. J Cereal Sci. 2012;56:389–95.
71. Fabian C, Ju YH. A review on rice bran protein: its properties and extraction methods. Crit Rev Food Sci Nutr. 2011;51:816–27.
72. Phongthai S, D'Amico S, Schoenlechner R, Rawdkuen S. Comparative study of rice bran protein concentrate and egg albumin on gluten-free bread properties. J Cereal Sci. 2016;72:38–45.
73. de la Hera E, Martinez M, Gómez M. Influence of flour particle size on quality of gluten-free rice bread. LWT Food Sci Technol 2013; 54:199-206.
74. de La Hera E, Rosell CM, Gomez M. Effect of water content and flour particle size on gluten-free bread quality and digestibility. Food Chem. 2014;151:526–31.
75. Mir SA, Shah MA, Naik HR, Zargar IA. Influence of hydrocolloids on dough handling and technological properties of gluten-free breads. Trends Food Sci Technol. 2016;51:49–57.
76. Bender D, Schönlechner R. Innovative approaches towards improved gluten-free bread properties. J Cereal Sci. 2020;91:102904.
77. Marco C, Rosell C. Breadmaking performance of protein enriched, gluten-free breads. Eur Food Res Technol. 2008;227:1205–13.
78. Cornejo F, Rosell CM. Physicochemical properties of long rice grain varieties in relation to gluten free bread quality. LWT Food Sci Technol. 2015;62:1203–10.
79. Roman L, Reguilon MP, Gomez M, Martinez MM. Intermediate length amylose increases the crumb hardness of rice flour gluten-free breads. Food Hydrocoll. 2020;100:105451.
80. Kang TY, Sohn KH, Yoon MR, Lee JS, Ko S. Effect of the shape of rice starch granules on flour characteristics and gluten-free bread quality. Int J Food Sci Technol. 2015;50:1743–9.
81. Tavares BO, Silva E, Silva VS, Junior M, Ida E, Damiani C. Stability of gluten free sweet biscuit elaborated with rice bran, broken rice and okara. Food Sci Technol. 2016;36:296–303.
82. Schober TJ, O'Brien CM, McCarthy D, Darnedde A, Arendt EK. Influence of gluten-free flour mixes and fat powders on the quality of gluten-free biscuits. Eur Food Res Technol. 2003;216:369–76.

83. Giuberti G, Rocchetti G, Sigolo S, Fortunati P, Lucini L, Gallo A. Exploitation of alfalfa seed (*Medicago sativa* L.) flour into gluten-free rice cookies: nutritional, antioxidant and quality characteristics. Food Chem. 2018:239679–87.
84. Rocchetti G, Senizza A, Gallo A, Lucini L, Giuberti G, Patrone V. In vitro large intestine fermentation of gluten-free rice cookies containing alfalfa seed (*Medicago sativa* L.) flour: a combined metagenomic/metabolomic approach. Food Res Int. 2019;120:312–21.
85. Haralampu SG. Resistant starch—a review of the physical properties and biological impact of RS3. Carbohydr Polym. 2000;41:285–92.
86. Graybosch RA. Waxy wheats: origin, properties, and prospects. Trends Food Sci Technol. 1998;9:135–42.
87. Shi MM, Gao QY. Physicochemical properties, structure and in vitro digestion of resistant starch from waxy rice starch. Carbohydr Polym. 2011;84:1151–7.
88. Van Hung P, Chau HT, Phi NTL. In vitro digestibility and in vivo glucose response of native and physically modified rice starches varying amylose contents. Food Chem. 2016;191:74–80.
89. Van Hung P, Vien NL, Phi NTL. Resistant starch improvement of rice starches under a combination of acid and heat-moisture treatments. Food Chem. 2016;191:67–73.
90. Giuberti G, Marti A, Fortunati P, Gallo A. Gluten free rice cookies with resistant starch ingredients from modified waxy rice starches: nutritional aspects and textural characteristics. J Cereal Sci. 2017;76:157–64.
91. Sharma R, Srivastava T, Saxena DC. Valorization of deoiled rice bran by development and process optimization of extrudates. Eng Agric Envron Food. 2019;12:173–80.
92. Pasqualone A, Costantini M, Coldea TE, Summo C. Use of legumes in extrusion cooking: a review. Foods. 2020;9(7):958.
93. Riaz MN. Extruded snacks. In: Hui YH, Sherkat F, editors. Handbook of food science, technology, and engineering. Boca Raton: Taylor & Francis; 2006. p. 168-1–8.
94. Alam MS, Kaur J, Khaira H, Gupta K. Extrusion and extruded products: changes in quality attributes as affected by extrusion process parameters: a review. Crit Rev Food Sci Nutr. 2016;56:445–73.
95. Hagenimana A, Ding X, Fang T. Evaluation of rice flour modified by extrusion cooking. J Cereal Sci. 2006;43:38–46.
96. Dalbhagat CG, Mahato DK, Mishra HN. Effect of extrusion processing on physicochemical, functional and nutritional characteristics of rice and rice-based products: a review. Trends Food Sci Technol. 2019;85:226–40.
97. Brennan C, Brennan M, Derbyshire E, Tiwari BK. Effects of extrusion on the polyphenols, vitamins and antioxidant activity of foods. Trends Food Sci Technol. 2011;22:570–5.
98. Guha M, Ali SZ. Changes in rheological properties of rice flour during extrusion cooking. J Texture Stud. 2011;42:451–8.
99. Chanvrier H, Pillin CN, Vandeputte G, Haiduc A, Leloup V, Gumy JC. Impact of extrusion parameters on the properties of rice products: a physicochemical and X-ray tomography study. Food Struct. 2015;6:29–40.
100. Tumuluru JS, Sokhansanj S, Bandyopadhyay S, Bawa AS. Changes in moisture, protein, and fat content of fish and rice flour coextrudates during single-screw extrusion cooking. Food Bioprocess Technol. 2013;6:403–15.
101. Zhu LJ, Shukri R, de Mesa-Stonestreet NJ, Alavi S, Dogan H, Shi YC. Mechanical and microstructural properties of soy protein–high amylose corn starch extrudates in relation to physiochemical changes of starch during extrusion. J Food Eng. 2010;100:232–8.
102. Hu Z, Tang X, Zhang M, Hu X, Yu C, Zhu Z, Shao Y. Effects of different extrusion temperatures on extrusion behavior, phenolic acids, antioxidant activity, anthocyanins and phytosterols of black rice. RSC Adv. 2018;8:7123–32.
103. Zhang R, Khan SA, Chi J, Wei Z, Zhang Y, Deng Y, et al. Different effects of extrusion on the phenolic profiles and antioxidant activity in milled fractions of brown rice. LWT Food Sci Technol. 2018;88:64–70.

104. Ramchiary M, Das AB. Vacuum-assisted extrusion of red rice (bao-dhan) flour: physical and phytochemical comparison with conventional extrusion: J Food Process Preserv. 2020. 44; e14570.
105. Lu SH, Lin TC. Rice-based snack foods. In: Lusas RW, Rooney LW, editors. Snack foods processing. Boca Raton: CRC Press; 2001. p. 439–55.
106. Hsieh F, Huff HE, Peng IC, Marek SW. Puffing of rice cakes as influenced by tempering and heating conditions. J Food Sci. 1989;54:1310–2.
107. Huff HE, Hsieh F, Peng IC. Rice cake production using long-grain and medium-grain brown rice. J Food Sci. 1992;57:1164–7.
108. Kim JD, Lee JC, Hsieh FH, Eun JB. Rice cake production using black rice and medium-grain brown rice. Food Sci Biotechnol. 2001;10:315–22.
109. Lee JC, Kim JD, Hsieh FH, Eun JB. Production of black rice cake using ground black rice and medium-grain brown rice. Int J Food Sci Technol. 2008;43:1078–82.
110. Fast RB, Perdon AA, Schonauer SL. Breakfast—forms, ingredients, and process flow. In: Perdon AA, Schonauer SL, Poutanen KS, editors. Breakfast cereals and how they are made. St Paul: American Association of Cereal Chemists International; 2020. p. 5–35.
111. Chandrasekhar PR, Chattopadhyay PK. Studies on microstructural changes of parboiled and puffed rice. J Food Process Preserv. 1990;14:27–37.
112. Mariotti M, Alamprese C, Pagani MA, Lucisano M. Effect of puffing on ultrastructure and physical characteristics of cereal grains and flours. J Cereal Sci. 2006;43:47–56.
113. Dutta A, Mukherjee R, Gupta A, Ledda A, Chakraborty R. Ultrastructural and physicochemical characteristics of rice under various conditions of puffing. J Food Sci Technol. 2015;52:7037–47.
114. Mariotti M, Pagani MA, Lucisano M. The role of buckwheat and HPMC on the breadmaking properties of some commercial gluten-free bread mixtures. Food Hydrocoll. 2013;30:393–400.
115. Mir SA, Bosco SJD, Shah MA, Mir MM, Sunooj KV. Process optimization and characterization of popped brown rice. Int J Food Prop. 2016;19:2102–12.
116. Jiamjariyatam R, Kongpensook V, Pradipasena P. Effects of amylose content, cooling rate and aging time on properties and characteristics of rice starch gels and puffed products. J Cereal Sci. 2015;61:16–25.
117. Joshi ND, Mohapatra D, Joshi DC. Varietal selection of some *indica* rice for production of puffed rice. Food Bioprocess Technol. 2014;7:299–305.
118. Pal S, Bagchi TB, Dhali K, Kar A, Sanghamitra P, Sarkar S, et al. Evaluation of sensory, physicochemical properties and Consumer preference of black rice and their products. J Food Sci Technol. 2019;56:1484–94.
119. Mir SA, Bosco SJD, Shah MA, Mir MM. Effect of puffing on physical and antioxidant properties of brown rice. Food Chem. 2016;191:139–46.
120. Abecassis J, Cuq B, Boggini G, Namoune H. Other traditional durum derived products. In: Sissons M, Abecassis J, Marchylo B, Carcea M, editors. Durum wheat: chemistry and technology. St Paul: American Association of Cereal Chemists International; 2012. p. 177–200.
121. Benatallah L, Agli A, Zidoune MN. Gluten-free couscous preparation: traditional procedure description and technological feasibility for three rice-leguminous supplemented formulae. J Food Agric Environ. 2008;6:105.
122. Chemache L, Kehal F, Namoune H, Chaalal M, Gagaoua M. Couscous: ethnic making and consumption patterns in the Northeast of Algeria. J Ethnic Foods. 2018;5:211–9.
123. Bellocq B, Ruiz T, Cuq B. Contribution of cooking and drying to the structure of couscous grains made from durum wheat semolina. Cereal Chem. 2018;95:646–59.
124. FAO/WHO. Standard for processed cereal-based foods for infants and young children. 2017; CODEX STAN 74-1981, International.
125. Jeelani P, Ghai A, Saikia N, Kathed M, Mitra A, Krishnan A, et al. Baby foods based on cereals. In: Gutiérrez TJ, editor. Food science, technology and nutrition for babies and children. Cham: Springer; 2020. p. 59–97.

126. Hojsak I, Braegger C, Bronsky J, Campoy C, Colomb V, Decsi T, et al. Arsenic in rice: a cause for concern. J Pediatr Gastroenterol Nutr. 2015;60:142–5.
127. European Commission (EU) 2015/1006 of 25 June 2015 amending Regulation (EC) No 1881/2006 as regards maximum levels of inorganic arsenic in foodstuffs. Off J L. 2015;161; 26.6.2015.
128. European Food Safety Authority Panel on Contaminants in the Food Chain (CONTAM). Scientific opinion on arsenic in food. EFSA J. 2009;7:1351.
129. González N, Calderón J, Rúbies A, Bosch J, Timoner I, Castell V, et al. Dietary exposure to total and inorganic arsenic via rice and rice-based products consumption. Food Chem Toxicol. 2020;141:111420.
130. Vanga SK, Raghavan V. How well do plant based alternatives fare nutritionally compared to cow's milk? J Food Sci Technol. 2018;55:10–20.
131. Munekata PE, Domínguez R, Budaraju S, Roselló-Soto E, Barba FJ, Mallikarjunan K, et al. Effect of innovative food processing technologies on the physicochemical and nutritional properties and quality of non-dairy plant-based beverages. Foods. 2020;9:288.
132. Silva AR, Silva MM, Ribeiro BD. Health issues and technological aspects of plant-based alternative milk. Food Res Int. 2020;131:108972.
133. Mäkinen OE, Wanhalinna V, Zannini E, Arendt EK. Foods for special dietary needs: non-dairy plant-based milk substitutes and fermented dairy-type products. Crit Rev Food Sci Nutr. 2016;56:339–49.
134. Lichtenstein AH, Ausman LM, Carrasco W, Gualtieri LJ, Jenner JL, Ordovas JM, et al. Rice bran oil consumption and plasma lipid levels in moderately hypercholesterolemic humans. Arterioscler Thromb. 1994;14:549–56.
135. Sugano M, Tsuji E. Rice bran oil and cholesterol metabolism. J Nutr. 1997;127:521S–4S.
136. Patel M, Naik SN. Gamma-oryzanol from rice bran oil–a review. J Sci Ind Res. 2004;63:569–78.
137. Danielski L, Zetzl C, Hense H, Brunner G. A process line for the production of raffinated rice oil from rice bran. J Supercrit Fluids. 2005;34:133–41.
138. Godber JS. Rice bran oil. In: Moreau RA, Kamal-Eldin A, editors. Gourmet and health-promoting specialty oils. Urbana: AOCS Press; 2009. p. 377–408.

Chapter 9
The Bright and Dark Sides of Wheat

Fatma Boukid

Abstract Wheat is a worldwide staple food for centuries. However, several debates and questions about the effect of wheat intake versus human health started to emerge in the last decades. Wheat is one of the main sources of carbohydrates and bioactive compounds, indicating the great importance of wheat nutrients and micronutrients as a crucial part of human daily diet. This is the "bright side" of the wheat. On the other hand, the raising claims toward the association between wheat consumption and several health issues urged to investigate if there is a "dark side" of wheat that could be considered a real threat for human wellbeing. Evidence sustained that wheat is involved in protein allergenicity in the case of genetically predisposed subjects, whereas wheat and overweight are still under investigation. Wheat intake impact on human health is a multivariable situation that should be studied case by case.

Keywords Wheat · Nutritional aspects · Disorders · Human health

1 Introduction

Wheat (Triticum spp.) is one of the Neolithic founder crops [1]. The genus *Triticum* consists of six species including *Triticum monococcum* L. (AA genome), *Triticum urartu* Tumanian ex Gandilyan (AA genome), *Triticumt urgidum* L. (AABB genome), *Triticum timopheevii* Zhuk. (AAGG genome), *Triticum aestivum* L. (AABBDD genome) and *Triticum zhukovskyi* Menabde & Ericz. (AAAAGG genome) which are grouped into three sections: Monococcon consisting of diploid species, Dicoccoidea consisting of tetraploid species and Triticum consisting of hexaploid species [2, 3]. Wheat is counted among the 'big three' cereal crops. The major wheat species grown throughout the world, accounting for about 700 million tonnes is *Triticum aestivum* and about 35–40 million tonnes is *Triticum durum* [4].

F. Boukid (✉)
Food Safety and Functionality Programme, Food Industry Area, Institute of Agriculture and Food Research and Technology (IRTA), Catalonia, Spain
e-mail: fatma.boukid@irta.cat

F. Boukid (ed.), *Cereal-Based Foodstuffs: The Backbone of Mediterranean Cuisine*, https://doi.org/10.1007/978-3-030-69228-5_9

In recent years, consumers keep looking for health-beneficial foods due consumers awareness about the association between food, nutrition and health. Claims and evidences regarding wheat intake and public health start to emerge among professionals. This issue has divided the scientific community on whether wheat is beneficial for healthy wellbeing or not. Controversial statements could be divided in two opposite sides; "Bright side of wheat" referring to wheat health benefits and "Dark side of wheat" referring to health issues.

The consumption of wheat is associated with a number of health benefits which may be related to the presence of a range of dietary fibers (DF), phytochemical, vitamin and mineral component [5]. Cereal fiber is now recognized to be beneficial for bowel health. The major part of DF is not digested in the small intestine; instead it will be fermented by gut microbiota in the colon producing several crucial products for the human body. It was reported an inverse association between dietary fiber and whole grain intake and risk of colorectal cancer [6]. However, limited evidence is available to support a health claim of a relationship between intake of whole grains and a reduced risk of diabetes type 2 [7]. Wheat intake contribution in lowering the risk of chronic diseases, such as coronary heart disease, cancer, body weight management and gastrointestinal health is also discussed [7–10].

On the other side, wheat is an important source of carbohydrates for the human body, yet currently overweight and obesity are often attributed to wheat intake. Furthermore, wheat has a wide range of products with different forms such as white wheat flour, whole wheat flour, and their derived products which raise more questions regarding the implications of processing on nutritional value. Moreover, the complex structure of carbohydrates is fully implicated in digestion, absorption and therefore the calories provided and burned by the body. It is also well established that wheat is classified among the "big eight". Proteins related allergies and intolerances, including celiac disease, wheat allergies and more currently the non-celiac gluten sensitivity, have gradually emerged as an epidemiologically relevant phenomenon with an estimated global prevalence around 5% [11].

Therefore, wheat intake emerges separately as subject for health benefits or health concerns, the focus of this review is on the multifaceted influence of wheat on human health. Based on scientifically sound evidences, the multivariable situation in favour or against wheat consumption was thoroughly discussed through assessing wheat intake effects on human health.

2 The Bright Side: Wheat Nutritional Aspects

The mechanisms underlying wheat nutritional aspects involve mainly the wheat essential bioactive compounds and their interaction with gut microbiota (GM). Fundamentally, wheat nutritional added value is derived from the nutrient digested, thereby absorbed going through the digestive system. Here, we will try to put together the main functional facts about wheat and their contribution in the human wellbeing (Fig. 9.1).

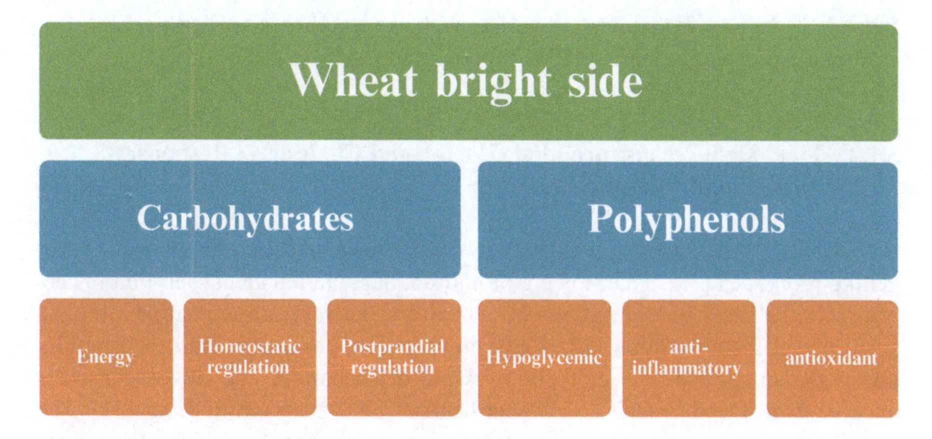

Fig. 9.1 Wheat nutritional aspects

2.1 *Dietary Fibers: Classification and HEALTH Benefits*

2.1.1 Fibers of Interest

Carbohydrates differ in their chemical properties, accordingly in the pattern and extent of fermentation [12]. Non-digestible carbohydrates, mainly known as DF, are the most appreciated for their input in health benefits. Massive volume of research focused on fibers, yet in this review, the spotlight concerns only specific benefits fully approved by the scientific community, while evidence putting forward wheat DF as a miracle ingredient are avoided. The whole wheat grain contains between 11.5 and 15.5% dry weight total DF [4]. The DF fraction consists of non-starch polysaccharides (NSPs), resistant starch (RS), oligosaccharides (mostly fructans), and the non-carbohydrate polyphenolic either lignin. The main NSPs are arabinoxylans, mixed linkage b(1,3; 1,4)-D-glucan (*b*-glucan), and cellulose [13]. Arabinoxylans (AX) are cell wall components that constitute an important part of the DF [14]. Resistant starch is defined as a portion of starch that cannot be digested in the small intestine and passes to the colon to be fermented by microbiota [15].

DF are extremely complex and diverse, often called "black box" due to their structural diversity allowing the generation of large numbers of classes and subclasses. Commonly, the classification of fibers is based on their solubility, soluble (e.g., gums, pectins), and insoluble (e.g., cellulose) [16]. Fibers can be grouped also in compounds digested in the small intestine and compounds digested in the large intestine. Based on their chains size, fiber can be subdivided in two groups, short chain carbohydrates and long chain carbohydrates. Short chain carbohydrates include the oligosaccharides (fructo-oligosaccharides and galacto-oligosaccharides) which are highly fermentable, while the long-chain carbohydrates can be classified in soluble fibers which are highly fermentable and insoluble fibers with slow fermentability [17]. The functionality of fiber is strongly associated with their functional and physiochemical properties such as solubility, viscosity, particle size, adsorption, and water-holding [18].

2.1.2 Fiber Reduces Postprandial Glucose and Cholesterol Responses in the Small Intestine

Soluble fibers (SF) health benefits are generally attributed to their viscous and/or gelling properties [19]. This slowing of nutrient degradation and absorption lowers peak serum glucose concentration after a meal and delivers nutrients further into the small bowel for absorption [20]. Only monosaccharides can be absorbed rapidly across the small intestinal epithelium, while disaccharides and oligosaccharides must be hydrolyzed to their constituent hexoses for absorption to occur [21]. The postprandial glucose level lowering effect is more of short-term rather than long-term effect [22]. Furthermore, SF is not fermented remaining gelled throughout the large bowel, providing a dichotomous [20]. Indeed, increased and regular consumption of SF might lead to significant improvements in blood glucose levels, the delay of gastric emptying, insulin resistance and metabolic profiles of patients with diabetes type II [23]. SF is believed to be able to interrupt the enterohepatic circulation of bile salts resulting in the reduction of serum cholesterol, thereby increasing hepatic conversion of cholesterol into newly synthesized bile acids and decreasing serum LDL, without affecting high density lipoprotein (HDL) concentrations [17].

2.1.3 Insights into DF Mechanism and Health Benefits

The fundamental molecular mechanism behind carbohydrates intake and metabolic homeostasis is complicated, involving several factors including food quantity and quality, host metabolism, energy needs, etc. In the case of a healthy individual without prior surgical resection of the small bowel, about 85% of carbohydrates, 66–95% of proteins, and all fats are absorbed before entering the large intestine, while the indigestible carbohydrates and proteins that the colon receives represent from 10 to 30% of the total ingested energy [24, 25]. The assessment of DF outcome on health is strongly linked with the products of their fermentation by GM. In the colon, GM gain energy mainly through the fermentation of non-digested dietary components and of host secretions [24]. The fermentation of DF and resistant starch enabled the synthesis of short chain fatty acids (SCFA); the most abundant (95%) are acetate, propionate, and butyrate [26]. SCFA production represents an adaptive process to conserve calories, fluid, and electrolytes [27]. Acetate (60% of SCFA) is a substrate for hepatic *de novo* lipogenesis and cholesterol biosynthesis [28, 29]. Propionate (20% of SCFA) is mainly a substrate for hepatic gluconeogenesis [30]. Butyrate is utilized also by microbiota and serves as the primary energy source of colonocytes [26]. Butyrate is also an effective regulator of modulation of mucin protein production at the transcriptional and translational levels [31].

Besides serving as energy sources, acetate, propionate and butyrate have been recognized as ligands of the sensing center G-protein-coupled receptors (GPCRs) expressed in enteroendocrine L cells of ileum and colon as well as in adipocytes and immune cells [32]. The sensing center in the taste buds sends signals to gastrointestinal (GI) tracts to initiate absorption and digestion; nutrients will be sensed by the GPCRs [33, 34]. There is now mounting evidences that SCFA might contribute in glucose homeostasis and insulin sensitivity improvements [35]. SCFA are supposed to activate specific GPCRs receptors which are free fatty acid receptor 2 and 3 (FFA2 and FFA3) inducing the secretion of gut appetite-restricting hormones mainly, glucagon-like peptide-1 (GLP-1) and glucose-dependent insulinotropic polypeptide (GIP) or "Incretin" [36]. The modulation of gut hormone release will influence neuronal signaling in appetite centers in the brain to mediate the appropriate feeding behavior [37]. GLP-1 stimulates insulin secretions and inhibits glucagon release, promotes satiety, reduces food intake and slows gastric emptying [38]. Butyrate and propionate were reported to induce gut hormones and reduce food intake of mince suggesting a possible novel mechanism by which GM regulates host metabolism [39]. It was also found that propionate increase significantly the secretion of PYY and GLP-1 and reduce energy intake and weight gain in overweight adults [40, 41]. On the other hand, acetate has a direct role in central appetite regulation [42]. Resistant starch was found to contribute in lowing plasma cholesterol level, probably by delaying gastric emptying, thereby limiting hepatic lipogenesis owing to less glucose as substrate and less insulin [43]. The consumption of 30 g/day of resistant starch improved insulin sensitivity in women with insulin resistance [44]. Belobrajdic et al. reported that the intake of two doses of resistant starch (low: 4%; high: 16%) were effective in reducing adiposity and weight gain in obese-prone and obese-resistant rats [45]. The positive correlation between caecal digesta total SCFA pools and plasma gut hormones (GLP-1 and PYY) were inversely associated with total body fat and visceral fat mass [45]. SCFA might be more relevant in the cases of high-fat diet-induced metabolic alterations by promoting changes in the host activity of peptides/systems from lipid synthesis to lipids oxidation [39]. The correlation between DF and weight change was reported to be dependent of many potential factors including age, baseline fiber and fat intakes, activity level, and baseline energy intake [46]. Consistently, observational studies sustained the potential of fermentable DF for weight loss and improving metabolic health in obesity [47, 48]. For instance, pectin was suggested as specific type of soluble fermentable DF promoting satiety and decreasing adiposity in diet-induced obese rats. Furthermore, SCFAs are considered strong anti-inflammatory agents. Tests conducted on human monocyte-derived dendritic cells showed that butyrate and propionate strongly reduced the release of several pro-inflammatory chemokines by the inhibition of the expression of lipopolysaccharide [49]. In addition, high-fiber diet increased circulating levels of SCFAs, which enhanced mince immunity against lung allergic inflammation; whereas a low-fiber diet decreased levels of SCFAs and increased allergic airway disease [50]. Production of SCFA enabled also the reduction of the luminal pH which by itself inhibits pathogenic microorganisms and increases the absorption of some nutrients [51].

2.2 *Dietary Polyphenols*

Wheat phenolics are present in three forms free, conjugated and bound [52]. Phenolics, lignans, phytate, tocopherols, and tocotrienols are among the compounds acting as antioxidants in whole grain [53]. Ferulic acid is the most abundant phenolic acid present in wheat bran with other phenolic acid such as vanillic acid, p-coumaric acid, syringic acid, caffeic acid, and p-hydroxybenzoic acid [54]. Ferulic acid is suggested to be an important contributor in the beneficial effects associated with wheat consumption [55]. It was reported that ferulic and p-coumaric acids to have antiproliferative effects on human Caco-2 colon cancer cells [4]. Bound polyphenols cannot be digested by human enzymes, could survive stomach and small intestine digestion, and therefore may possibly reach the colon [56]. Polyphenols which reach the colon can be metabolized by the resident GM, and then be conjugated in the intestinal cells and later in the liver by methylation, sulfation or glucuronidation, but the understanding of the microbial bioconversion processes is limited [57].

In general, around 0.5–5% of the ferulic acid is absorbed within the small intestine, mainly the soluble free fraction, and about 90% are suggested to exert a protective action of the colon from cancer [58]. The hypoglycemic effects of polyphenols are mainly attributed to the improvement of pancreatic β-cells function and insulin action, stimulation of insulin secretion [59]. Consistently, polyphenols are supposed to suppress glucose release from the liver which improves its uptake in peripheral tissues by modulating intracellular signals [56]. Overall, the benefits of dietary polyphenols as anti-inflammatory and antioxidant effects, inhibition of α-amylases or α- glucosidases and thus decrease of starch digestion, and inhibition of advanced glycation end products formation [60].

3 The Black Side: Wheat Related Disorders

Cereal-based food products have been the basis of the human diet since ancient times [2]. Despites the importance of wheat as a daily source of energy, it is not surprising the plaguing of wheat causality association with health issues; one of the main reasons is the mounting prevalence of several chronic and metabolic diseases. Wheat related disorders have become a growing area of clinical and scientific interest. In this context, this section interest is to investigate if wheat intake might trigger health issues and how (Fig. 9.2).

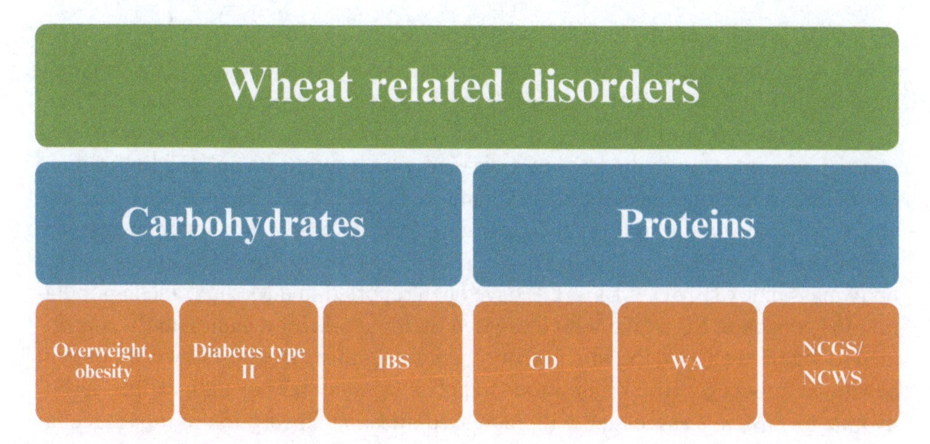

Fig. 9.2 Wheat related disorders

3.1 *Carbohydrates Related Disorders*

Obesity is a worldwide epidemic associated with significant morbidity and mortality [61]. Obesity is a disease of energy balance, characterized by a chronic disequilibrium between energy intake and expenditure [62]. As glucose is the prime fuel normally used by brain cells and many organs, carbohydrates present a daily metabolic challenge for the body [35]. Carbohydrates are considered the major source of energy for the metabolism [63]. Simple sugars and the rapidly digestible starch are the carbohydrates digested in the duodenum and proximal regions of the small intestine, their absorption may lead to a rapid elevation of blood glucose and usually a subsequent episode of hypoglycaemia [64]. Wheat carbohydrates have an important impact on postprandial blood glucose, are those that are absorbed relatively quickly from the small intestine. The effect of carbohydrates on the glycemic response is commonly assessed using glycemic index (GLI) or glycemic load (GL). Carbohydrates with a low GI are slowly digested and absorbed which have been reported to be involved in controlling postprandial plasma glucose excursions [65]. However, high carbohydrates with high GLI release glucose rapidly which increased glucose postprandial level, thereby triggers the development of hyperinsulinemia and insulin resistance [40]. Based on GLI, there are three categories, high (> 70), intermediate (>55–<70) and low (<55) [66]. Generally, GLI was developed to indicate the glucose level in blood, whereas it was noted that the GLI of food with similar carbohydrate contents did not usually have the same impact on blood glucose level [66]. Furthermore, although the quantity of the ingested carbohydrates and the time needed for the digestion and the absorption in the GI tracts are fundamental, GLI do not include this information. In the case of whole wheat flour, GLI ranges from 70 to 99, consequently, the high range of variability might be attributed to the multitude of factors (e.g. variety, cooking, starch content and fibers). On the other hand, insufficient evidence was found regarding the opposite association between insoluble DF and the risk of diabetes or the no association between soluble DF and the risk of diabetes [46]. The glycemic load (GL) is calculated as the product of GLI and the amount of carbohydrate in a serving which might be subdivided in three levels low (<10), intermediate (11–19) and high (>20) [29]. Despite knowing that wheat flour has a quite important GL ranging from 17 to 20, it is not enough to put wheat in the danger zone. Furthermore, related to obesity, no associations among body mass index (BMI), GLI or GL have been reported in a Mediterranean population [67]. Further, SF intake with increased SCFA production significantly contributed to digested energy, thereby potentially outweighing the well-known short-term beneficial effects of SF consumption [51]. The fermentation of indigestible carbohydrate by GM was associated with increased intestinal absorption of monosaccharides and SCFA which was proposed to be associated with an increase of hepatic lipogenesis [32]. Fernandes et al. sustained that colonic fermentation patterns alteration can lead to different fecal SCFA concentrations in overweight or obese compared with healthy lean humans. The high faecal concentration of SCFA was reported to be associated with metabolic risk factors and thus may influence metabolic homeostasis [68]. In the case obese patients, SCFA contribution in host metabolic disorders is still uncertain due to the implications of several parameters such as diet components, genetic and environmental factors [35, 51].

Malabsorption and intolerance to carbohydrates are also problems, frequently encountered in the GI system [69]. Notably, wheat contains quite low quantities of fructan (wheat bread, 0.61–1.94 g/100 g), but its widespread consumption contributes in the daily fructan intake [70]. Individuals with malabsorption of fructose are unable to absorb free fructose in the small intestine because the enzymes were unable to completely hydrolyze glyosidic linkages in the complex polysaccharide [71]. Consequently, the unabsorbed fructose is delivered to the colon, undergoes bacterial fermentation and induces abdominal symptoms, such as pain, bloating, and altered bowel habit [69]. Enterocytes, impaired enteric microbiome, might be also the responsible of the reduction of fructose absorption inducing the decrease of the intestinal permeability [72]. Irritable bowel syndrome (IBS) is another GI disorder characterized by symptoms including, abdominal pain and diarrhea, with no abnormal pathology affecting 15% of the population [73]. FODMAPs, when malabsorbed, are highly osmotic substances that can cause an influx of water into the colon [71]. Moreover, these short-chain carbohydrates are poorly absorbed in the small intestine and rapidly fermented by bacteria in the gut, which can lead to increased gas, distention, bloating, cramping, and diarrhoea, which are the usual symptoms of IBS [71]. Further, a large gap in quality evidence was underlined between the efficacy of the low FODMAP diet compared with a diet of normal FODMAPs content in unselected patients with IBS [71]. Due to the difficulties in the performance of high-quality cohort trials, the current understanding of the causality relationship between the intake of FODMAPs and GI symptom pattern and quality of life is still poor [74].

Overall, the implication degree carbohydrates in metabolic disorders is closely related to its type, structure and daily intake. Although several indexes are used to control and check its outcome on human health, more scientific evidences are needed. For GI disorders, carbohydrates impact is still unclear because digestion and absorption are a dynamic system involving a wide range of variables; both acute and long-term effect should be relevant. FODMAP are short-chain poorly absorbed carbohydrates and have been grouped together under this umbrella term because they all are rapidly fermented and are osmotically active, with additive effects [74].

3.2 Protein Related Disorders

Wheat as a trigger to health risks is linked commonly to a spectrum of diseases induced by proteins including celiac disease (CD), wheat allergy (WA), non-celiac gluten sensitivity (NCGS), and non-celiac wheat sensitivity (NCWS) (Fig. 9.3).

CD is a genetically determined chronic inflammatory intestinal disease induced by gluten, affecting approximately 1% of people in the world [75]. GI symptoms are mucosal inflammation, small intestine villous atrophy, increased intestinal permeability and malabsorption of macro- and micronutrients [76]. Gluten proteins are gliadins and glutenins. Gliadins are considered the primary triggers of the abnormal immune response in CD. Gluten elicits an adaptive Th1-mediated immune response in individuals carrying HLA-DQ2 or HLA-DQ8 as major genetic predisposition [77, 78]. CD is one of the most amply reviewed diseases, interplaying environmental and genetic factors [79, 80]. Despite the great knowledge and understanding of CD, here we bring some interesting points which are not fully deciphered. Gluten is CD major external factor. To date, scientists are unable to fully explain the reasons behind the currently rapid increase in the incidence of CD. Several hypotheses are putted forward. Breeding progress was the first to blame suggesting that D genome is the major location of gluten encoding genes [75, 81]. However, in *vitro* and in

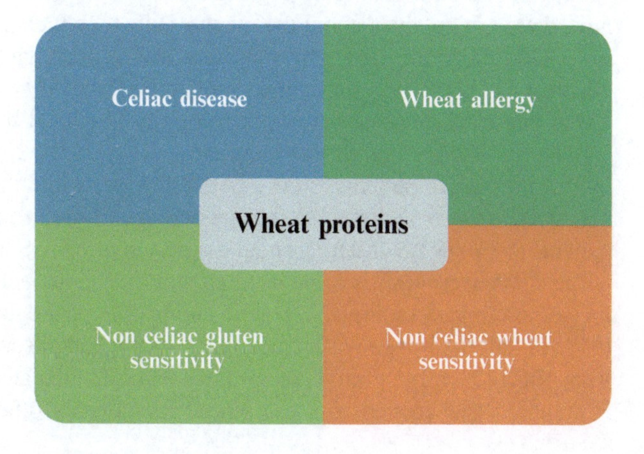

Fig. 9.3 Spectrum of diseases induced by wheat proteins

vivo trials confirmed that wheat is a trigger of CD, whether modern or ancient [80, 82, 83]. It was claimed also that introducing gluten during the first year of life may play a role in development of CD [84]. Consistently, individuals with positive HLA are genetically predisposed to develop CD symptoms when on high gluten consumption [85]. Beside genetics, other factors are assumed to be involved in triggering CD such as intestinal permeability, innate immune response to gliadin and GM activities [86]. Notably, 30% of the populations are genetically predisposed to develop CD, while only 1% of this population is CD patients [85]. Another crucial factor is the assessment tools; indeed, immune and toxic peptides triggering CD are identified and quantified using wide range of methods, but it remains ELISA the official method according to the codex alimentarius. In vitro GI digestion was usually used to simulate what happens in the human body. Although this approach was helpful to draw the full picture allowing an overall vision about CD, identified allergens sequences were often questioned because the human digestive system is much more complicated. Further, reliable prevalence data are lucking in the major countries for several reasons including low availability of diagnostic facilities and poor disease awareness [87].

Wheat allergies (WA) are also among the spectrum of wheat proteins related disorders. Depending on the route of allergen exposure and the underlying immunological mechanisms, wheat allergy is classified as a classic food allergy affecting wheat-dependent exercise-induced anaphylaxis (WDEIA), baker's asthma and rhinitis, and contact with urticaria [88]. WA is a complex disease due to the many allergenic components' wheat [74]. Amylase trypsin inhibitors could fuel inflammation and immune reactions in several intestinal and non-intestinal immune disorders [89, 90]. The inflammatory response included also several allergenic proteins, non-specific lipid transfer protein (nsLTP), gliadins, HMW glutenins germ agglutinin and peroxidase [91]. WDEIA is caused by a specific type of grain protein, ω_5-gliadins [91].

Non-celiac gluten sensitivity (NCGS) has recently been recognized by the scientific community as one of gluten-related disorders [92]. However, scientists are divided in two categories, researchers which are questioning the existence of NCGS, for the second half, distinguishing between CD, WA and NCGS is confusing. Until now, there are no clues about the mechanisms by which gluten induces symptoms in NCGS, but its difference from CD and WA appears be the available method for diagnostics [11]. It was underlined some differences between CD and NCGS such as the intestinal mucosal barrier, histology of duodenal biopsy, and mucosal gene expression. Further, genes encoding HLA DQ2 or DQ8 molecules are present in half of the NCGS patients, similarly to CD case, while these genes are present in healthy people as well (30%), but less frequently than in the case of the NCGS patients (50%) [93, 94]. Differential diagnostic of NCGS based on the exclusion of CD and WA symptoms is considered an important approach covering lots of aspects including clinical, biological, genetic and histological data, but requires validation [91]. An Italian survey revealed that the most frequent disorders associated to NCGS were IBS (47%), food intolerance (35%) and IgE-mediated allergy (22%). NCWS is another issue related to some wheat components other than gluten including amylase-trypsin inhibitors, FODMAPs, and wheat-germ agglutinin can be responsible culprits in NCWC [74]. NCWS is also commonly reported in association with IBS in patients, where around 24% of IBS patients may be sensitive to fructans [74].

4 Concluding Remarks

You are what you eat, which raise the challenge to find the balance between wheat intake and the best health outcome. Screening wheat components of interest, independently of their journey in the human body is not enough. It is indeed crucial to fully evaluate wheat nutrients and antinutrients interaction with the host in the first place, and then with the other components of the diet. Hence, because wheat intake impact on health is a multivariable equation, consumers predisposed or already subjects to wheat related disorders should be aware of the intake quantity and quality to create their own balance.

In recent years, numerous research works were committed to improve wheat. In terms of nutritional quality, to breed wheat with high amylose content can offer wheat with high health benefits specially the raise of its resistance to digestion [95]. As previously mentioned, the adverse effects of wheat are mainly associated with protein allergies or intolerance. Advanced breeding technologies can offer "healthier" wheat. Genetic editing has lot of successes showing a very low off-target. However, the issue is if genetically edited wheat is considered genetically modified or not. As a matter of fact, the changes induced by gene editing techniques do not leave fingerprint in the plant genome due to the absence of DNA insertion in the final product. The same modifications could even occur naturally, while genetic editing is more of a target approach with the potential to induce specific modification. Thus, going through the Directive 2001/18, considering these genotypes as genetically modified is not fully clear. To go beyond this uncertainty, risk assessment of the technique and the final product are required.

References

1. Peleg Z, Fahima T, Korol AB, Abbo S, Saranga Y. Genetic analysis of wheat domestication and evolution under domestication. J Exp Bot [Internet]. 2011;62:5051–61. Oxford University Press [cited 2018 Dec 20]. Available from: https://academic.oup.com/jxb/article-lookup/doi/10.1093/jxb/err206.
2. Boukid F, Folloni S, Sforza S, Vittadini E, Prandi B. Current trends in ancient grains-based foodstuffs: insights into nutritional aspects and technological applications. Compr Rev Food Sci Food Saf [Internet]. 2018;17:123–36. Wiley/Blackwell (10.1111) [cited 2018 Nov 7]. Available from: http://doi.wiley.com/10.1111/1541-4337.12315.
3. Charmet G, Storlie E, Oury FX, Laurent V, Beghin D, Chevarin L, et al. Genome-wide prediction of three important traits in bread wheat. Mol Breed [Internet]. 2014;34:1843–52. Springer; [cited 2018 Dec 20]. Available from: http://www.ncbi.nlm.nih.gov/pubmed/26316839.

4. Shewry PR, Hey S. Do "ancient" wheat species differ from modern bread wheat in their contents of bioactive components? J Cereal Sci. 2015;65:236–43.
5. Shewry PR, Ward JL. Exploiting genetic variation to improve wheat composition for the prevention of chronic diseases. Food Energy Secur [Internet]. 2012;1:47–60 [cited 2018 Nov 7]. Available from: http://doi.wiley.com/10.1002/fes3.2.
6. Aune D, Keum N, Giovannucci E, Fadnes LT, Boffetta P, Greenwood DC, et al. Whole grain consumption and risk of cardiovascular disease, cancer, and all cause and cause specific mortality: systematic review and dose-response meta-analysis of prospective studies. BMJ [Internet]. 2016;i2716 [cited 2018 Nov 7]. Available from: http://www.bmj.com/lookup/doi/10.1136/bmj.i2716.
7. Augustin LSA, Kendall CWC, Jenkins DJA, Willett WC, Astrup A, Barclay AW, et al. Glycemic index, glycemic load and glycemic response: an international scientific consensus summit from the international carbohydrate quality consortium (ICQC). Nutr Metab Cardiovasc Dis. 2015;25:795–815.
8. Judson PL, Al Sawah E, Marchion DC, Xiong Y, Bicaku E, Zgheib NB, et al. Characterizing the efficacy of fermented wheat germ extract against ovarian cancer and defining the genomic basis of its activity. Int J Gynecol Cancer [Internet]. 2012;22:960–7 [cited 2018 Mar 4]. Available from: http://www.ncbi.nlm.nih.gov/pubmed/22740002.
9. Mueller T, Jordan K, Voigt W. Promising cytotoxic activity profile of fermented wheat germ extract (Avemar®) in human cancer cell lines. J Exp Clin Cancer Res [Internet]. 2011 [cited 2018 Mar 4];30:42. Available from: http://www.ncbi.nlm.nih.gov/pubmed/21496306.
10. Ikuomola DS, Otutu OL, Oluniran DD. Quality assessment of cookies produced from wheat flour and malted barley (Hordeum vulgare) bran blends. Cogent Food Agric. 2017;3.
11. Catassi C, Bai J, Bonaz B, Bouma G, Calabrò A, Carroccio A, et al. Non-celiac gluten sensitivity: the new frontier of gluten related disorders. Nutrients [Internet]. 2013;5:3839–53 [cited 2018 Nov 9]. Available from: http://www.mdpi.com/2072-6643/5/10/3839.
12. Goff HD, Repin N, Fabek H, El Khoury D, Gidley MJ. Dietary fibre for glycaemia control: towards a mechanistic understanding. Bioact Carbohydr Diet Fibre. 2018;14:39–53.
13. Mussatto SI, Dragone G, Roberto IC. Brewers' spent grain: generation, characteristics and potential applications. J Cereal Sci. 2006;43:1–14.
14. Qiu S, Yadav MP, Yin L. Characterization and functionalities study of hemicellulose and cellulose components isolated from sorghum bran, bagasse and biomass. Food Chem. 2017;230:225–33.
15. Birt DF, Boylston T, Hendrich S, Jane J-L, Hollis J, Li L, et al. Resistant starch: promise for improving human health. Adv Nutr [Internet]. 2013;4:587–601 [cited 2018 Nov 5]. Available from: https://academic.oup.com/advances/article/4/6/587/4595564.
16. Zamora-Gasga VM, Bello-Pérez LA, Ortíz-Basurto RI, Tovar J, Sáyago-Ayerdi SG. Granola bars prepared with Agave tequilana ingredients: chemical composition and in vitro starch hydrolysis. LWT-Food Sci Technol [Internet]. 2014;56:309–14 [cited 2019 Jan 4]. Available from: https://linkinghub.elsevier.com/retrieve/pii/S0023643813004878.
17. Slavin J. Fiber and prebiotics: mechanisms and health benefits. Nutrients [Internet]. 2013;5:1417–35 [cited 2019 Jan 4]. Available from: http://www.mdpi.com/2072-6643/5/4/1417.
18. Hemdane S, Jacobs PJ, Bosmans GM, Verspreet J, Delcour JA, Courtin CM. Study on the effects of wheat bran incorporation on water mobility and biopolymer behavior during bread making and storage using time-domain 1H NMR relaxometry. Food Chem. 2017;236:76–86.
19. Papathanasopoulos A, Camilleri M. Dietary fiber supplements: effects in obesity and metabolic syndrome and relationship to gastrointestinal functions. Gastroenterology. 2010;138:65–72.e2.
20. McRorie JW. Evidence-based approach to fiber supplements and clinically meaningful health benefits, part 1. Nutr Today [Internet]. 2015;50:82–9. Lippincott Williams and Wilkins [cited 2020 Apr 10]. Available from: http://www.ncbi.nlm.nih.gov/pubmed/25972618.
21. Shepherd AJ, Mohapatra DP. Tissue preparation and immunostaining of mouse sensory nerve fibers innervating skin and limb bones. J Vis Exp [Internet]. 2012;1–6 [cited 2020 Apr 10]. Available from: http://www.ncbi.nlm.nih.gov/pubmed/22314687.

22. Meng H, Matthan NR, Ausman LM, Lichtenstein AH. Effect of macronutrients and fiber on postprandial glycemic responses and meal glycemic index and glycemic load value determinations. Am J Clin Nutr [Internet]. 2017;105:842–53. American Society for Nutrition [cited 2020 Apr 10]. Available from: https://academic.oup.com/ajcn/article/105/4/842-853/4569720.
23. Chen H, Zhao C, Li J, Hussain S, Yan S, Wang Q. Effects of extrusion on structural and physicochemical properties of soluble dietary fiber from nodes of lotus root. LWT. 2018;93:204–11.
24. Krajmalnik-Brown R, Ilhan ZE, Kang DW, DiBaise JK. Effects of gut microbes on nutrient absorption and energy regulation. Nutr Clin Pract. 2012;27:201–14.
25. Jefferson A, Adolphus K. The effects of intact cereal grain fibers, including wheat bran on the gut microbiota composition of healthy adults: a systematic review. Front Nutr. 2019;6:33.
26. Barroso E, Cueva C, Peláez C, Martínez-Cuesta MC, Requena T. Development of human colonic microbiota in the computer-controlled dynamic SIMulator of the GastroIntestinal tract SIMGI. LWT-Food Sci Technol. 2015;61:283–9.
27. Gullón B, Gullón P, Tavaria F, Pintado M, Gomes AM, Alonso JL, et al. Structural features and assessment of prebiotic activity of refined arabinoxylooligosaccharides from wheat bran. J Funct Foods. 2014;6:438–49.
28. Moore J, Gunn P, Fielding B. The role of dietary sugars and de novo lipogenesis in non-alcoholic fatty liver disease. Nutrients [Internet]. 2014;6:5679–703 [cited 2018 Nov 13]. Available from: http://www.mdpi.com/2072-6643/6/12/5679.
29. Hu FB, Malik VS. Sugar-sweetened beverages and risk of obesity and type 2 diabetes: epidemiologic evidence. Physiol Behav. 2010;100:47–54.
30. Ríos-Covián D, Ruas-Madiedo P, Margolles A, Gueimonde M, De los Reyes-Gavilán CG, Salazar N. Intestinal short chain fatty acids and their link with diet and human health. Front. Microbiol. 2016;7:185.
31. Hu J, Lin S, Zheng B, Cheung PCK. Short-chain fatty acids in control of energy metabolism. Crit Rev Food Sci Nutr [Internet]. 2018;58:1243–9 . Taylor and Francis Inc. [cited 2020 Apr 10]. Available from: http://www.ncbi.nlm.nih.gov/pubmed/27786539.
32. Braune A, Bunzel M, Yonekura R, Blaut M. Conversion of dehydrodiferulic acids by human intestinal microbiota. J Agric Food Chem [Internet]. 2009;57:3356–62 [cited 2018 Dec 27]. Available from: http://pubs.acs.org/doi/abs/10.1021/jf900159h
33. Raka F, Farr S, Kelly J, Stoianov A, Adeli K. Metabolic control via nutrient-sensing mechanisms: role of taste receptors and the gut-brain neuroendocrine axis [Internet]. Am J Physiol Endocrinol Metab NLM (Medline). 2019:E559–72 [cited 2020 Apr 10]. Available from: http://www.ncbi.nlm.nih.gov/pubmed/31310579.
34. Roper SD, Chaudhari N. Taste buds: cells, signals and synapses. Nat Rev. 2017;18:485–97.
35. Canfora EE, Jocken JW, Blaak EE. Short-chain fatty acids in control of body weight and insulin sensitivity. Nat Rev Endocrinol. 2015;11:577–91.
36. Priyadarshini M, Kotlo KU, Dudeja PK, Layden BT. Role of short chain fatty acid receptors in intestinal physiology and pathophysiology. Compr Physiol. 2018;8:1065–90.
37. Miyamoto J, Hasegawa S, Kasubuchi M, Ichimura A, Nakajima A, Kimura I. Nutritional signaling via free fatty acid receptors. Int J Mol Sci. 2016;17:450.
38. Falcinelli S, Rodiles A, Unniappan S, Picchietti S, Gioacchini G, Merrifield DL, et al. Probiotic treatment reduces appetite and glucose level in the zebrafish model. Sci Rep. 2016;6:18061.
39. Lin H V., Frassetto A, Kowalik EJ, Nawrocki AR, Lu MM, Kosinski JR, et al. Butyrate and propionate protect against diet-induced obesity and regulate gut hormones via free fatty acid receptor 3-independent mechanisms. PLoS One [Internet]. 2012;7:e35240 [cited 2020 Apr 10]. Available from: http://www.ncbi.nlm.nih.gov/pubmed/22506074.
40. Rahat-Rozenbloom S, Fernandes J, Cheng J, Wolever TMS. Acute increases in serum colonic short-chain fatty acids elicited by inulin do not increase GLP-1 or PYY responses but may reduce ghrelin in lean and overweight humans. Eur J Clin Nutr. 2017;71:953–8.
41. Chambers ES, Viardot A, Psichas A, Morrison DJ, Murphy KG, Zac-Varghese SEK, et al. Effects of targeted delivery of propionate to the human colon on appetite regulation, body weight maintenance and adiposity in overweight adults. Gut BMJ. 2015;64:1744–54.

42. Frost G, Sleeth ML, Sahuri-Arisoylu M, Lizarbe B, Cerdan S, Brody L, et al. The short-chain fatty acid acetate reduces appetite via a central homeostatic mechanism. Nat Commun. 2014;5:3611.
43. Lockyer S, Nugent AP. Health effects of resistant starch. Nutr Bull [Internet]. 2017;42:10–41. Blackwell Publishing Ltd [cited 2020 Apr 11]. Available from: http://doi.wiley.com/10.1111/nbu.12244.
44. Gower BA, Bergman R, Stefanovski D, Darnell B, Ovalle F, Fisher G, et al. Baseline insulin sensitivity affects response to high-amylose maize resistant starch in women: a randomized, controlled trial.[Erratum appears in Nutr Metab (Lond). 2016;13:6; PMID: 26839576]. Nutr Metab (Lond) [Internet]. 2016;13:2 [cited 2020 Apr 11]. Available from: http://proxycheck.lib.umanitoba.ca/libraries/online/proxy.php?http://ovidsp.ovid.com/ovidweb.cgi?T=JS&CSC=Y&NEWS=N&PAGE=fulltext&D=prem&AN=26766961.
45. Belobrajdic DP, King RA, Christophersen CT, Bird AR. Dietary resistant starch dose-dependently reduces adiposity in obesity-prone and obesity-resistant male rats. Nutr Metab. 2012;9:93.
46. Lattimer JM, Haub MD. Effects of dietary fiber and its components on metabolic health [Internet]. Nutrients. 2010:1266–89. MDPI AG [cited 2020 Apr 11]. Available from: http://www.pubmedcentral.nih.gov/articlerender.fcgi?artid=3257631&tool=pmcentrez&rendertype=abstract.
47. Adam CL, Williams PA, Garden KE, Thomson LM, Ross AW. Dose-dependent effects of a soluble dietary fibre (pectin) on food intake, adiposity, gut hypertrophy and gut satiety hormone secretion in rats. Blachier F, editor. PLoS One [Internet]. 2015;10:e0115438. Public Library of Science [cited 2020 Apr 11]. Available from: https://dx.plos.org/10.1371/journal.pone.0115438.
48. Adam CL, Williams PA, Dalby MJ, Garden K, Thomson LM, Richardson AJ, et al. Different types of soluble fermentable dietary fibre decrease food intake, body weight gain and adiposity in young adult male rats. Nutr Metab. 2014;11:36.
49. Nastasi C, Candela M, Bonefeld CM, Geisler C, Hansen M, Krejsgaard T, et al. The effect of short-chain fatty acids on human monocyte-derived dendritic cells. Sci Rep. 2015;5:1–10.
50. Harsch I, Konturek P. The role of gut microbiota in obesity and type 2 and type 1 diabetes mellitus: new insights into "old" diseases. Med Sci. 2018;6:32.
51. Ríos-Covián D, Ruas-Madiedo P, Margolles A, Gueimonde M, De los Reyes-Gavilán CG, Salazar N. Intestinal short chain fatty acids and their link with diet and human health [Internet]. Front. Microbiol. 2016:185. Frontiers Media S.A. [cited 2020 Apr 10]. Available from: http://www.ncbi.nlm.nih.gov/pubmed/26925050.
52. Leoncini E, Prata C, Malaguti M, Marotti I, Segura-Carretero A, Catizone P, et al. Phytochemical profile and Nutraceutical value of old and modern common wheat cultivars. PLoS One. 2012;7:e45997.
53. Ramírez-Maganda J, Blancas-Benítez FJ, Zamora-Gasga VM, García-Magaña M d L, Bello-Pérez LA, Tovar J, et al. Nutritional properties and phenolic content of a bakery product substituted with a mango (Mangifera indica) 'Ataulfo' processing by-product. Food Res Int. 2015;73:117–23.
54. Boukid F, Dall'Asta M, Bresciani L, Mena P, Del Rio D, Calani L, et al. Phenolic profile and antioxidant capacity of landraces, old and modern Tunisian durum wheat. Eur Food Res Technol. 2019;245:73–82.
55. Abdel-Aal E-SM, Rabalski I. J Cereal Sci. [Internet]. 2013. Academic Press [cited 2018 Dec 20]. Available from: http://agris.fao.org/agris-search/search.do?recordID=US201500065466.
56. Okarter N, Liu R. Health benefits of whole grain phytochemicals. Crit Rev Food Sci Nutr. 2010;50:193–208.
57. Conlon MA, Bird AR. The impact of diet and lifestyle on gut microbiota and human health. Nutrients. 2015;7:17–44.
58. Fardet A. How can both the health potential and sustainability of cereal products be improved? A French perspective. J Cereal Sci. 2014;60:540–8.

59. Golzarand M, Bahadoran Z, Mirmiran P, Sadeghian-Sharif S, Azizi F. Dietary phytochemical index is inversely associated with the occurrence of hypertension in adults: a 3-year follow-up (the Tehran Lipid and Glucose Study). Eur J Clin Nutr. 2015;69:392–8.
60. Fardet A, Rock E, Rémésy C. Is the in vitro antioxidant potential of whole-grain cereals and cereal products well reflected in vivo? J Cereal Sci. 2008;48:258–76.
61. Springmann M, Wiebe K, Mason-D'Croz D, Sulser TB, Rayner M, Scarborough P. Health and nutritional aspects of sustainable diet strategies and their association with environmental impacts: a global modelling analysis with country-level detail. Lancet Planet Heal. 2018;2:e451–61.
62. Deutz NEP, Bauer JM, Barazzoni R, Biolo G, Boirie Y, Bosy-Westphal A, et al. Protein intake and exercise for optimal muscle function with aging: recommendations from the ESPEN expert group. Clin Nutr Churchill Livingstone. 2014;33:929–36.
63. Lafiandra D, Riccardi G, Shewry PR. Improving cereal grain carbohydrates for diet and health [Internet]. J Cereal Sci. 2014:312–26 Academic Press [cited 2020 Apr 11]. Available from: http://www.ncbi.nlm.nih.gov/pubmed/24966450.
64. Lecerf JM, Clerc E, Jaruga A, Wagner A, Respondek F. Postprandial glycaemic and insulinaemic responses in adults after consumption of dairy desserts and pound cakes containing short-chain fructo-oligosaccharides used to replace sugars. J Nutr Sci [Internet]. 2015;4:e34 [cited 2018 Dec 23]. Available from: http://www.journals.cambridge.org/abstract_S2048679015000221.
65. Scazzina F, Siebenhandl-Ehn S, Pellegrini N. The effect of dietary fibre on reducing the glycaemic index of bread. Br J Nutr. 2013;109:1163–74.
66. Eleazu CO. The concept of low glycemic index and glycemic load foods as panacea for type 2 diabetes mellitus; prospects, challenges and solutions. Afr Health Sci [Internet]. 2016;16:468–79. Makerere University, Medical School [cited 2020 Apr 11]. Available from: http://www.ncbi.nlm.nih.gov/pubmed/27605962.
67. Di Angelantonio E, Bhupathiraju SN, Wormser D, Gao P, Kaptoge S, de Gonzalez AB, et al. Body-mass index and all-cause mortality: individual-participant-data meta-analysis of 239 prospective studies in four continents. Lancet [Internet]. 2016;388:776–86. Lancet Publishing Group [cited 2020 Apr 11]. Available from: http://www.ncbi.nlm.nih.gov/pubmed/27423262.
68. Maslowski KM, Vieira AT, Ng A, Kranich J, Sierro F, Yu D, et al. Regulation of inflammatory responses by gut microbiota and chemoattractant receptor GPR43. Nature. 2009;461:1282–6.
69. Fedewa A, Rao SSC. Dietary fructose intolerance, fructan intolerance and FODMAPs. Curr Gastroenterol Rep [Internet]. 2014;16:370. Current Medicine Group LLC 1 [cited 2020 Apr 11]. Available from: http://www.ncbi.nlm.nih.gov/pubmed/24357350.
70. Whelan K, Abrahmsohn O, David GJP, Staudacher H, Irving P, Lomer MCE, et al. Fructan content of commonly consumed wheat, rye and gluten-free breads. Int J Food Sci Nutr [Internet]. 2011;62:498–503 [cited 2020 Apr 11]. Available from: http://www.ncbi.nlm.nih.gov/pubmed/21428719.
71. Rao SSC, Yu S, Fedewa A. Systematic review: dietary fibre and FODMAP-restricted diet in the management of constipation and irritable bowel syndrome. Aliment Pharmacol Ther. 2015;41:1256–70.
72. Tengjaroenkul B, Smith BJ, Caceci T, Smith SA. Distribution of intestinal enzyme activities along the intestinal tract of cultured Nile tilapia, Oreochromis niloticus L. Aquaculture. 2000;182:317–27.
73. Salari-Moghaddam A, Keshteli AH, Esmaillzadeh A, Adibi P. Empirically derived food-based inflammatory potential of the diet, irritable bowel syndrome, and its severity. Nutrition. 2019;63–64:141–7.
74. Fasano A, Sapone A, Zevallos V, Schuppan D. Nonceliac gluten sensitivity. Gastroenterology. 2015;148:1195–204.
75. Boukid F, Mejri M, Pellegrini N, Sforza S, Prandi B. How looking for celiac-safe wheat can influence its technological properties. Compr Rev Food Sci Food Saf [Internet]. 2017;16:797–807 [cited 2018 Nov 9]. Available from: http://doi.wiley.com/10.1111/1541-4337.12288
76. de Punder K, Pruimboom L. The dietary intake of wheat and other cereal grains and their role in inflammation. Nutrients [Internet]. 2013;5:771–87 [cited 2018 Mar 4]. Available from: http://www.ncbi.nlm.nih.gov/pubmed/23482055.

77. Almeida LM, Gandolfi L, Pratesi R, Uenishi RH, de Almeida FC, Selleski N, et al. Presence of DQ2.2 associated with DQ2.5 increases the risk for celiac disease. Autoimmune Dis [Internet]. 2016;2016:5409653. Hindawi Limited [cited 2018 Nov 9]. Available from: http://www.ncbi.nlm.nih.gov/pubmed/28042478.
78. Pisapia L, Camarca A, Picascia S, Bassi V, Barba P, Del Pozzo G, et al. HLA-DQ2.5 genes associated with celiac disease risk are preferentially expressed with respect to non-predisposing HLA genes: implication for anti-gluten T cell response. J Autoimmun [Internet]. 2016;70:63–72 [cited 2018 Nov 9]. Available from: http://www.ncbi.nlm.nih.gov/pubmed/27083396.
79. Boukid F, Prandi B, Buhler S, Sforza S. Effectiveness of germination on protein hydrolysis as a way to reduce adverse reactions to wheat. J Agric Food Chem. 2017;65:9854–60.
80. Boukid F, Prandi B, Sforza S, Sayar R, Seo YW, Mejri M, et al. Understanding the effects of genotype, growing year, and breeding on Tunisian durum wheat Allergenicity. 2. The celiac disease case. J Agric Food Chem. 2017;65:5837.
81. Shewry PR, Tatham AS. Improving wheat to remove coeliac epitopes but retain functionality. J Cereal Sci. 2016;67:12–21.
82. Prandi B, Tedeschi T, Folloni S, Galaverna G, Sforza S. Peptides from gluten digestion: a comparison between old and modern wheat varieties. Food Res Int [Internet]. 2017;91:92–102. Elsevier [cited 2018 Nov 9]. Available from: https://www.sciencedirect.com/science/article/pii/S0963996916305816?via%3Dihub
83. Kaur A, Bains NS, Sood A, Yadav B, Sharma P, Kaur S, et al. Molecular characterization of α-gliadin gene sequences in Indian wheat cultivars Vis-à-Vis celiac disease eliciting epitopes. J Plant Biochem Biotechnol. 2017;26:106–12.
84. Balakireva AV, Zamyatnin AA. Properties of gluten intolerance: gluten structure, evolution, pathogenicity and detoxification capabilities. Nutrients. 2016;8:644.
85. Gujral N, Freeman HJ, Thomson ABR. Celiac disease: Prevalence, diagnosis, pathogenesis and treatment. World J Gastroenterol. 2012;18:6036–59.
86. Mohan Kumar BV, Prasada Rao UJS, Prabhasankar P. Immunogenicity characterization of hexaploid and tetraploid wheat varieties related to celiac disease and wheat allergy. Food Agric Immunol. 2017;28:888–903.
87. Catassi C, Kryszak D, Bhatti B, Sturgeon C, Helzlsouer K, Clipp SL, et al. Natural history of celiac disease autoimmunity in a USA cohort followed since 1974. Ann Med. 2010;42:530–8.
88. Colgrave ML, Byrne K, Blundell M, Howitt CA. Identification of barley-specific peptide markers that persist in processed foods and are capable of detecting barley contamination by LC-MS/MS. J Proteome. 2016;147:169–76.
89. Uvackova L, Skultety L, Bekesova S, McClain S, Hajduch M. MSE based multiplex protein analysis quantified important allergenic proteins and detected relevant peptides carrying known epitopes in wheat grain extracts. J Proteome Res. 2013;12:4862–9.
90. Boukid F, Prandi B, Sforza S, Sayar R, Seo YW, Mejri M, et al. Understanding the effects of genotype, growing year, and breeding on Tunisian durum wheat Allergenicity. 1. The Baker's asthma case. J Agric Food Chem. 2017;65:5831–6.
91. Sapone A, Bai JC, Ciacci C, Dolinsek J, Green PHR, Hadjivassiliou M, et al. Spectrum of gluten-related disorders: consensus on new nomenclature and classification. BMC Med. 2012;10:13.
92. Scherf KA. Immunoreactive cereal proteins in wheat allergy, non-celiac gluten/wheat sensitivity (NCGS) and celiac disease. Curr Opin Food Sci. 2019;25:35–41.
93. Valerii MC, Ricci C, Spisni E, Di Silvestro R, De Fazio L, Cavazza E, et al. Responses of peripheral blood mononucleated cells from non-celiac gluten sensitive patients to various cereal sources. Food Chem. 2015;176:167–74.
94. Fallahbaghery A, Zou W, Byrne K, Howitt CA, Colgrave ML. Comparison of gluten extraction protocols assessed by LC-MS/MS analysis. J Agric Food Chem [Internet]. 2017;65:2857–66. [cited 2018 Nov 9]. Available from: http://www.ncbi.nlm.nih.gov/pubmed/28285530.
95. Carroccio A, Di Prima L, Noto D, Fayer F, Ambrosiano G, Villanacci V, et al. Searching for wheat plants with low toxicity in celiac disease: between direct toxicity and immunologic activation. Dig Liver Dis. 2011;43:34–9.

Chapter 10
Gluten-Free Breadmaking: Facts, Issues, and Future

Serap Vatansever and Clifford Hall

Abstract Gluten-free foods have attracted increased attention in the food industry due to health issues (i.e., celiac disease and other gluten-related disorders) associated with gluten and the changes in eating habits, such as following gluten-free diet (GFD). Among this food segment, gluten-free (GF) bread plays a crucial role because of the stable consumption of bread in many parts of the world. The increasing demands for GF breads require new approaches to mimic wheat-based bread in terms of the quality, sensory attributes, shelf-life stability, and nutritional value. Therefore, various ingredients have been used for GF bread formulation to mimic gluten-network and to obtain a favorable bread quality along with a good dough characteristic. Alternative ingredients (i.e., flours from pulses and pseudocereals, and proteins sources) have been engineered to improve nutritional profile and health benefits of GF bread as well as sensory and texture attributes of bread. Sourdough biotechnology and novel technologies are potential approaches to address the need for GF bread processing system. This chapter summarizes the issues associated with health, GF ingredients for breadmaking process, and technologies employed for GF breadmaking process.

Keywords Gluten-free · Celiac disease · Bread · Pulses · Pseudocereals · Hydrocolloids · Baking technologies

1 Introduction

Bread, a cereal-based product, has served as an essential food product in the human diet and is consumed as a staple food in many parts of the world. A major ingredient of bread is flour, which is mostly produced from wheat (*Triticum aestivum*). The high level of gluten-forming proteins (GFPs) [1] is the reason wheat flour is used for bread manufacturing. In addition to common bread wheat, other cereals, including

---

S. Vatansever, Ph.D. (✉) · C. Hall, Ph.D.
South Dakota State University, Department of Dairy and Food Science, Brookings, SD, USA
e-mail: serap.vatansever@sdstate.edu

F. Boukid (ed.), *Cereal-Based Foodstuffs: The Backbone of Mediterranean Cuisine*, https://doi.org/10.1007/978-3-030-69228-5_10

durum wheat (*T. turgidum*), ancient wheats, such as spelt (*T. aestivum* var. *spelta*), emmer and einkorn (*T. monococcum*), barley (*Hordeum vulgare*), and rye (*Secale cereale*) that contain GFPs. Among cereals, GFPs are classified as prolamins (i.e., gliadins, hordeins, secalins, and avenins in wheat, barley, rye, and oat, respectively) and glutenins [2, 3]. These two proteins form gluten and provide a viscoelastic feature. The ingestion of GFP, in particular prolamins, causes celiac disease (CD) and also other gluten-related disorders [2–4]. This chapter provides a summary of issues related to gluten-free bread baking.

2 Celiac Disease (CD) and Other Gluten-Related Disorders

Celiac disease (CD), also termed as gluten sensitive enteropathy or celiac sprue, is an autoimmune disease with the life-long intolerance of genetically susceptible individuals to the ingestion of GFPs [5]. GFPs are present in cereals that are wheat, durum wheat, any ancient *Triticum* species (e.g., spelt, kamut, einkorn), rye, barley, and oat [5–8]. Among these cereal grains, oat is considered a tolerated gluten-containing cereal by most people. Thus, this grain is allowable for use in gluten-free food formulation only when precautions have been made to prevent its contamination by wheat species, rye, and barley [9], which contain prolamins (e.g., gliadin, hordein) associated with CD [5, 6]. Prolamins are high in proline and glutamine [6, 10] and these amino acids are resistant to gastrointestinal digestion and trigger the deamination through tissue transglutaminase (tTG) [10].

CD is a T-cell mediated autoimmune disorder that occurs through an immune response to ingested cereal gluten proteins. This immune response mechanism occurs through the interactions of Major Histocompatibility Complex II (MHC II) molecules, in the presence of peptides formed through the ingestion of gluten proteins, with $CD4^{+}$ T-cells in the epithelium [7]. The CD pathogenesis involves in interactions among genetic susceptibility (MHC class II Human Leukocyte Antigen (HLA)-DQ2 and HLA-DQ8 alleles on the antigen-presenting cells), environmental exposure (i.e., gluten intake), and immunological response (cell surface receptors of T-cells and other immune system cells) [7, 11]. Hence, this autoimmune enteropathy causes the epithelial damage, typically small intestinal mucosa damage with the loss of absorptive villi and hyperplasia of the crypts that appears with common symptoms, such as chronic diarrhea, vomiting, abdominal pain, and fatigue [5, 7, 11]. The damage in the small intestine through CD results in maldigestion and malabsorption of most nutrients that leads to vitamin and mineral deficiencies, anemia, night blindness, and weakness of bones [5, 6]. Furthermore, slower children and adolescent growth, and women reproductive health issues related to CD have been reported. Diagnosis of CD may be delayed due to its symptoms that mimic other diseases. In particular, CD's symptoms are similar to irritable bowel syndrome, gastric ulcers, Crohn's disease, parasite infections, and anemia [3]. The number of studies on CD has increased significantly since it has been classified as an emerging, common public health problem. Recent developments in the medicine have contributed to better diagnosis and clarifications of the CD cases.

Based on recent epidemiological studies, 1% of people globally suffers from CD [12, 13]. Recently, the pooled global prevalence of CD has been recorded as 1.4% in 275,818 individuals while 0.7% in 138,792 in individuals based on serologic tests and biopsy results has been determined [13]. Singh et al. [13] indicated that differences in the prevalence of CD based on sex, age, and location. Female had greater prevalence rate (0.6%) than male (0.4%). The prevalence values in children (0.9%) were significantly ($p < 0.001$) higher than in adults (0.5%). CD was prevalent in 0.4% in South America, 0.5% in Africa and North America, 0.6% in Asia, and 0.8% in Europe and Oceania [13].

Besides CD, there are other diseases caused by the ingestion of gluten and can be categorized under the umbrella term "gluten-related disorders". Gluten-related disorders include non-celiac gluten sensitivity, wheat allergy, dermatitis herpetiformis, and gluten ataxia [4, 12]. Additionally, inflammatory bowel diseases (IBD) have similar genetic, immunological, and environmental conditions as CD [4]. Currently, fundamental and effective treatment for these five forms of gluten-based diseases is a life-long exclusion of gluten-containing foods from the diet. Nevertheless, the patients with CD and other gluten-related disorders might need generalizable treatment; thus, these diseases need to be categorized based on their conditions [12].

Non-celiac gluten sensitivity (NCGS) is a gluten-mediated disorder with a world prevalence of 3–6%. Among gluten-related disorders, NCGS has common symptoms with both CD and also wheat allergy, such as gastrointestinal complaints, diarrhea, bloating, weight loss, muscular disturbances, bone pain, and tiredness. Recent developments in the medicine have provided better abilities for diagnosing NCGS and differentiate it from CD [12]. Wheat allergy, also known as baker's asthma, is an Immunoglobulin E (IgE)-mediated food allergy with the world prevalence of 0.5–9%. This allergy is the response of antibodies to cereal proteins, specifically wheat proteins, impacting the gastrointestinal and respiratory tract, and the skin [12, 14]. Dermatitis herpetiformis (or Duhring-Brocy disease) is a skin disease with the world prevalence of 0.0001–0.05% [12] and existing in approximately 10% of celiac patients [3]. This disease shows distinctive symptoms, including urticarial plaques and blisters on the elbows, buttocks, and knees [15]. Gluten ataxia or sporadic cerebellar ataxia is another most common gluten-related disorder with the world prevalence of 14% [12]. Gluten ataxia is an immune-mediated disease occurring through the ingestion of gluten-containing foods in genetically susceptible individuals. This disease is associated with antigliadin antibodies (AGAs) of the IgG and/or IgG classes and identified via a serological screening for AGAs, tissue transglutaminase antibody (TG2), and anti-TG6 [16].

3 Gluten Free Diet and Products: A Therapy and Growing Food Segment

Currently, adhering to a lifelong gluten-free diet (GFD) through strictly exclusion of gluten containing foods is an essential therapy for the patients, suffering from CD and other gluten-related disorders [4, 8, 12, 17]. Therefore, product requirements for GFD have been rapidly growing with increased demand for the gluten-free (GF) food segment. The foods ready for consumptions to follow the GFD is dependent on the following food standards [8]:

1. Gluten-free foods, consisting of or made only from one or more ingredients which do not contain wheat, all *Triticum* species (e.g., durum wheat, spelt, and kamut), rye, barley, oats or their crossbred varieties with the gluten level of ≤20 mg/kg (or parts per million, ppm) in total [9]; or
2. Gluten-free foods, consisting of one or more ingredients from wheat, all *Triticum* species (e.g., durum wheat, spelt, and kamut), rye, barley, oats or their crossbred varieties, specifically processed to remove gluten, with the gluten level of ≤20 mg/kg in total [9]; or
3. Foods, consisting of one or more ingredients from wheat, all *Triticum* species (e.g., durum wheat, spelt, and kamut), rye, barley, oats or their crossbred varieties, specially processed to reduce gluten content up to 100 mg/kg in total [9, 18].

Thus, the Codex Alimentarius Commission [9] and the European Union (EU) gluten-free legislation established the safe amount of gluten content for GF products at ≤20 mg/kg [10, 17]. Nevertheless, the cutoff of GF foods varies. For example, North America, Spain, Italy, and UK follow 20 mg/kg; however, it is 10 mg/kg in Argentina, and 3 mg/kg in Australia, New Zealand, and Chile. Additionally, the EU regulates low gluten products at 21–100 mg/kg [17].

Over recent decades, the global GF products market has exhibited a steady global growth [12], valued at USD 4.3 billion in 2019, and is predicted to reach USD 7.5 billion by 2027 [19]. The rising population with CD and NCGS that require avoiding gluten in the diet is likely due to improved diagnosis and is the major driver of the GF food market. In addition, people diagnosed with other health conditions (i.e., autism, multiple sclerosis) and some consumers, who do not have health issues associated with gluten but consume GF products as a lifestyle, are other significant consumer groups of this market. Therefore, increasing growth in this food segment has led to intense development and research efforts in creating GF foods for CD and NCGS patients and other groups. Ultimately, these efforts have enabled a broad spectrum of GF categories (e.g., baked goods, pasta, noodle) in this food segment. Among GF food categories, GF bread plays a key role due to common consumption of bread in the world [20]. Thus, more research efforts for GF food innovation might be essential to meet the demand of the GF food market in terms of quality and quantity [3, 5, 12].

4 Gluten-Free (GF) Breadmaking: A Brief History to Recent Developments

4.1 *The Role of Gluten in Breadmaking*

Gluten is a cereal protein made up of two specific proteins, glutelin (i.e., glutenin) and prolamin (e.g., gliadin in wheat). Gluten is a highly techno-functional protein for the breadmaking process through (1) forming a continuous three-dimensional network between protein and starch during dough development, (2) giving visco-elasticity and cohesiveness for the dough system, and (3) promoting carbon dioxide retention during fermentation [3, 6, 21]. In particular, wheat gluten proteins distinctively provide a strong, cohesive, viscoelastic dough compared to other cereal grains. Glutenin contributes to the dough resistance to extension, and elasticity; thus, it provides structure during proofing as the dough undergoes stretching. On the flip side, gliadin provides the dough viscosity. Combination of these two proteins generates a unique viscoelastic structure [3, 6]. Thus, gluten becomes a fundamental component for dough rheology that shapes the overall bread quality (e.g., loaf volume, crumb texture) [21]. In addition, the solid matrix of the crumbs for the wheat breads occurs as a result of continuous network between gluten, enclosing the starch granules and fibers, and gelatinized during the baking process [12].

4.2 *Technological Challenges for Mimicking Gluten Functionality in GF Breadmaking and Nutritional Quality Concerns*

Gluten is an essential part of bread dough through formation of a protein network, generating a viscoelastic dough with gas retaining capacity, mixing tolerance, resistance to extension, when mixed with water [12]. Formulation of GF bread lacks this protein network; thus, it is a less cohesive and elastic system and ultimately becomes a liquid batter vs a dough.

In general, GF breads are produced using refined flours or pure starches, such as rice and corn starch. Using these ingredients in the GF bread formulations to replace wheat flour, in particular gluten, impacts the breadmaking process (i.e., fully gelatinization of starch-rich ingredients during baking, addition of high amount of water to increase the viscosity and improve gas retaining capacity). These modifications in the production stages cause differences in the consistency of the batter, resulting in poor and inconsistent final bread quality (i.e., crumbling texture, poor color, faster staling, and many other defects). GF batters are difficult to handle due to the lack of elasticity and cohesiveness along with poor gas retaining capacity that results in poor specific loaf volumes for GF breads [12, 22].

Among GF products, the production of good quality GF breads compared to conventional wheat breads has been one of the most technological challenges due to the lack of alternative ingredients that mimic wheat gluten techno-functionalities [23] and consequently not contributing a three-dimensional protein-starch network [22]. Various functional ingredients have been increasingly used in the GF bread formulation to replace wheat gluten functionalities along with representing the nutritional profile [3, 5, 8, 23, 24]. These alternative ingredients have been categorized as: (1) starches and flours, (2) hydrocolloids or gums, (3) fibers, (4) proteins (i.e., egg proteins, enzymes), and (5) lipids [23]. In general, these ingredients have been used to aid in the GF bread making processing and an approach to improve GF bread properties. Therefore, Bender and Schoenlechner [22] suggested that these ingredients can be classified into following groups based on their functionalities in the GF bread formulation:

1. Water-binding and film-forming ingredients to build an internal network (e.g., hydrocolloids or thickening agents, including various gums, starches);
2. Structure-forming, volume-filling, taste-enhancing ingredients (i.e., proteins, lipids, and low molecular weight carbohydrates);
3. Surface-active substances (i.e., emulsifiers) and;
4. Ingredient modification and internal network formation (i.e., enzymes).

GF bread formulations have been created with the combination of starch or flour sources with gluten substitutes, such as hydrocolloids, protein incorporation, enzymes, (Table 10.1) to yield desired GF breads that address the expectations of customers, such as texture and appearance [25]. However, differences in the GF base formula have been recorded in the literature due to variations of the base formula to adopt the batter for better handling, such as moisture content adjustment [23]. In addition to the improvement of technological quality of GF breads, their nutritional value has been investigated and many concerns have been reported corresponding to the unsatisfactory nutritional profile [8].

5 GF Breadmaking and Improvement of GF Bread Quality

5.1 *Alternative Flours and Starches*

Alternate flours with certain attributes have been used to replace the wheat flour in breadmaking. Rice (*Oryza sativa*) and corn (*Zea mays* L. ssp. *mays*) are the most commonly used cereal flours for GF bread formulations, depending on their high availability, hypoallergenic properties, and good flavor profile [3, 12, 25]. Sorghum (*Sorghum bicolor*) flour is another potential GF cereal flour that exhibits good performance in GF bread formulation [25]. In addition to cereal flours, pseudocereal flours (e.g., buckwheat, amaranth, quinoa), root and tuber flours (e.g., yam, cassava), legume flours (e.g., soy, beans, peas, lentil, chickpea, vinal, carob), and other

Table 10.1 Effects of different ingredients on the GF bread quality

Ingredient	Role	Outcomes	References
Rice flour, corn flour, soy flour, xanthan gum, carrageenan, alginate, CMC, and gelatin	Investigating the effect of different hydrocolloids on the dough rheological characteristics and enhance bread quality (i.e., crumb structure and loaf volume)	Among all hydrocolloids, xanthan gum had the best improvement for the dough and bread quality followed by CMC	[31]
Rice flour, potato starch, guar gum, HPMC, and CMC	Investigating the effect of different hydrocolloids on the dough rheological characteristics and enhance bread quality (i.e., crumb structure and loaf volume)	Among selected gums, HMPC was the most effective to improve the bread quality. Also, combination of HPMC and CMC enhanced the dough quality with a greater viscosity, increasing gas retaining capacity, and resulted in a rigid bread, exhibiting porous cell structure with better loaf volume	[33]
Rice flour, corn starch, sodium caseinate, xanthan, CMC, pectin, agarose, and oat beta-glucan	Investigating the effect of different hydrocolloids on the dough rheological characteristics and enhance bread quality (i.e., crumb structure and loaf volume)	Xanthan gum presented the most favorable effect on the dough rheology by giving a good viscoelastic feature but reduced loaf volume with greater crumb firmness. Beta- glucan increased loaf volume, and crumb firmness and porosity. Agarose and pectin improved the loaf volume. Among all hydrocolloids, CMC and pectin had pronounced improvement for GF bread	[34]
Corn starch, potato starch, guar gum, pectin, and protein sources (i.e., albumin, lupin, soy protein concentrate, pea protein isolate, and collagen)	The effect of protein source addition (10%) on the dough rheology and bread quality (i.e., crumb structure and loaf volume)	All protein sources affected dough rheology. Soy and pea protein added bread had smaller loaf volume than other, while albumin addition resulted in the greatest loaf volume. Protein addition changed crumb structure with higher porosity and lower cell density and led to firmer bread and darker crust color. Addition of pea and lupine protein improved sensory quality with higher acceptability for color and smell, while soy addition decreased the consumer's acceptability of bread	[38]

(continued)

Table 10.1 (continued)

Ingredient	Role	Outcomes	References
Corn starch, HPMC, pea protein isolate and egg white protein powder	The effect of different protein source on the corn-based GF bread quality and in vitro digestibility	Addition of pea protein exhibited higher dough viscosity compared to egg white protein added dough. Egg white protein led higher crumb hardness with a compact structure compared to GF bread enriched with pea protein. Pea protein enriched bread had greater protein digestibility than that of egg white protein	[43]
Red rice flour, cassava starch, inulin, microbial TG	The effect of TG, a network forming agent, on the bread quality	TG increased loaf volume, bread firmness and chewiness. The use of red rice flour along with inulin and TG (1%) was found alternative ingredients for GF bread formulations	[44]
Buckwheat, brown rice, corn, oat, sorghum and teff flours, and TG	The effect of TG, a network forming agent, on the bread quality produced with different GF flours	TG enhanced the baking characteristics of buckwheat and brown rice flour-based breads but did not affect sorghum, oat and teff flours-based breads. Corn-based bread had improvements with addition of TG that caused negative effects on the batter quality	[46]
Chickpea flour, cassava starch, psyllium, CGT, and TG	The effect of different enzyme for chickpea flour-based bread enriched with psyllium	Increased ratio of chickpea flour to psyllium had better dough consistency with increasing loaf volume and crumb softness. CGT had a good interaction with chickpea protein and resulted in decreased crumb firmness during storage, while the addition of TG had no effects	[50]

CMC Carboxymethyl cellulose, *HPMC* Hydroxypropyl methylcellulose, *TG* Transglutaminase, *CGT* Cyclodextrin glycosyltransferase

plants (e.g., acorn, flaxseed, chia seed, chestnut, unripe banana) can be used in GF flour blends [26]. These non-cereal gluten-free flour ingredients can provide a good nutrient profile, including high protein and dietary fibers, for GF breads [3]. In particular, legume flours are of interest in GF products to enhance nutrient composition by providing rich sources of proteins, starch, dietary fibers, vitamin and minerals [27, 28]. However, these flours do not meet functionality behavior of gluten that is important for good dough and bread quality (e.g., flavor, texture, loaf volume). Different starch characteristics (e.g., functional properties based on the type of polymorph due to the botanical origin) of GF flours can impact bread processing [21, 26].

The use of network building ingredients to mimic gluten functionality is required once alternative flours are employed in the GF bread formulations. Boukid et al. [27] blended a variety of cannellini bean (CB) flours (i.e., flour, flour fraction rich in fractured cells, and flour fraction rich in intact cells) with white rice flour at 20%. HPMC was used in this formula as a hydrocolloid to improve dough handling properties. The addition of CB flours improved the bread quality by decreasing breadcrumb hardness and increasing crust yellowness. Also, CB flour fraction rich in intact cells exhibited a softer bread compared to other GF breads [27].

Starches, separated through a wet-milling process from other components of grains or crops, display a different ingredient profile compared to flours with a high purity. The purity of starches provides a neutral flavor and white or off-white color profile, which meet the sensory criteria of GF products. But, the low nutrient density and fiber content of starches decrease the nutritional profile and health benefits of GF breads, which may negatively impact the long-term well-being of people following the GFD [3].

Among starch categories, maize and potato starches have been commonly used for GF bread processing along with other starches (i.e., rice, tapioca, wheat) [29]. The primary function of starch in breadmaking process is the formation of the crumb structure [3]. But, using starches as a base in GF bread formulation may cause a minimal structure formation along with exhibiting a poor nutritional composition (e.g., low protein, minerals, and vitamins contents) compared to wheat bread. Therefore, starches are combined with other functional ingredients, such as hydrocolloids (water binding) and proteins (nutritional enhancement, and physical and textural improvement of final products) to improve dough and bread quality [5, 29]. Furthermore, the blends of varied starches and flours are utilized in the GF bread formula with the gas-retaining ingredients to provide proper gelatinization, pasting, and retrogradation properties [3].

The effects of different starch (i.e., wheat, potato, and maize) and flour (i.e., rice and maize) sources on the GF bread quality were investigated [21]. Martinez and Gomez [21] showed that GF bread produced with starch had higher specific volume than those produced with flours due to the greater particle size and the presence of a protein layer in flour source. Among starch sources, wheat starch-based GF bread had the highest specific volume and good textural profile (i.e., low hardness and high elasticity, cohesiveness and resilience) compared to other starch-based breads owing to the formation of better uniform continuous starch-hydrocolloid matrix, resulting in better viscoelastic dough property.

5.2 *Hydrocolloids/Gums*

Hydrocolloids, which are hydrophilic long-chain polysaccharides, obtained from seaweed, plants, microorganisms, and modified cellulose and starches by enzymatic or chemical modification are used as functional ingredients in the GF bread formulations [2]. There are numerous researchers who have reported the incorporation of hydrocolloids in the GF bread formulation to improve dough handling properties, enhance GF bread quality, and extend the shelf-life of bread. Based on the literature, these polymers have been considered as the most promising ingredients to replace gluten functionalities in the GF bread system [12, 22, 25, 30, 31]. The primarily usage of hydrocolloids in GF breads is to mimic gluten viscoelastic properties through forming gel and structure-building properties once hydrated. This gel network increases batter viscosity and strengthens the boundaries of expanding cells that improves gas retention during proofing and baking. Eventually, these functional polymers enhance the volume, structure, texture, and appearance of GF breads [3, 5, 12, 22, 25, 30]. Hydrocolloids play major roles in the moisture retention and inhibition of starch retrogradation through their ability to bind great amounts of water (i.e., up to 100 times its mass) [2, 3]. Furthermore, these polymers possess "a water release" property for starch gelatinization during baking [30].

Incorporation of hydrocolloids in the GF breadmaking depends on their abilities to increase the water-holding capacity, viscosity, hydration rate, as well as processing parameters, such as temperature (i.e., the decrease in viscosity with the increase of temperature), pH, and shearing, and their interactions with other food components (e.g., protein, starch) [2, 22]. In GF formulations, these polymers have been used commonly alone or combined in the amounts of 0.3–5% of the formula weight (f_w) but the general use of hydrocolloids is up to 2% $f_{w,}$ [5]. There have been many research publications that illustrate the effects of different hydrocolloids, their amounts, their combinations, and processing conditions on dough rheology and fresh bread quality. Among hydrocolloids, xanthan gum and HMPC appear to be the most promising polymers in the complex GF bread formulation based on the research findings [12, 25, 31–33].

Lazaridou et al. [34] presented the effects of different hydrocolloids (i.e., xanthan, carboxymethyl cellulose (CMC), pectin, agarose, and oat beta-glucan) in composite GF blend, containing rice flour, corn starch, and sodium caseinate, on dough and bread properties. In this study, xanthan gum had the most pronounced effect on the dough rheology by providing significant viscoelastic properties. However, xanthan gum resulted in lower loaf volume and higher crumb firmness compared to another hydrocolloids supplementation. Beta-glucan increased the loaf volume, crump firmness, and crust porosity and lightness values. Addition of CMC and pectin in GF breads improved fresh GF bread quality (i.e., loaf volume, crumb porosity and elasticity) compared to other hydrocolloids [34]. Similarly, Sciarini et al. [31] evaluated the addition of different hydrocolloids (i.e., xanthan gum, carrageenan, alginate, CMC, and gelatin) on the GF batter and bread quality produced with a composite blend of 40% rice flour, 40% corn flour, and 20% soy flour. These researchers reported that the best batter quality was obtained with the addition of xanthan gum, followed by CMC and xanthan gum improved the fresh bread quality better than other hydrocolloids [31]. In another study, a composite GF blend, including rice flour (45%), corn starch (35%), and cassava starch (20%) was used to produce GF bread. This GF bread had a high acceptability with the addition of xanthan gum (0.5%). Using xanthan gum improved the crumb structure and texture of bread by providing uniform distributed cells, pleasant flavor, and appearance [32].

Likewise, Cato et al. [33] investigated the effect of HPMC, guar gum, and CMC on the GF bread quality produced with a blend of rice flour and potato starch. HPMC produced bread with the most desirable quality (i.e., loaf volume, texture, and crust and crumb color).

Furthermore, a combination of two or more hydrocolloids (e.g., xanthan gum with guar gum) had better performance in the production of GF breads due to their synergistic effect, leading to the formation of stronger gel with increased viscosity. Cato et al. [33] indicated that the combination of HPMC and CMC exhibited the best performance for dough rheology properties by providing a good viscosity, increasing gas retaining capacity, and bread quality by developing a rigid but porous cell structure with a good loaf volume.

5.3 Proteins

Proteins from a variety of sources (e.g., egg, dairy, legume) have been incorporated into the GF bread formulation as multifunctional ingredients. The use of proteins in the GF bread formulation enhances the nutritional value (e.g., increasing total protein content), physical and sensory properties of GF breads (e.g., crumb softening, texture, crust color, taste, and shelf-life extension) [2, 3, 24, 29, 35].

Proteins and polysaccharides are the major macromolecules of the food systems, including wheat breads. Therefore, these macromolecules together play an essential role to form the structure, texture, and stability of foods through their functional properties, such as emulsifying, thickening, and gelling properties [29, 36]. In particular, the common use of the starches in the GF bread formulations requires another main component, proteins, both to create structure along with the addition of hydrocolloids and also to enhance the nutritional composition of GF breads [3, 29]. Furthermore, proteins involved in Maillard reactions, through interacting with reducing sugars, contribute to color development of the crust and aroma of the bread [23]. Greater water holding capacity of some proteins (i.e., legume proteins) [27], which requires more water addition into the batter, improves texture and shelf-life of the breads [23].

Animal-based proteins (i.e., egg and dairy proteins) have been employed in the GF bread formulations to improve the dough handling properties and final bread characteristics as well as nutritional value of GF breads [3, 22]. Egg and dairy proteins have a great importance due to their complete amino acid profile and high degree of digestibility [3]. The incorporation of egg proteins to the GF bread formulation facilitates the cohesive film formation needed to generate a stable foam and improve the gas retention during baking. Thus, these improvements help to create a structure in the GF bread [24]. In addition, egg proteins help to provide breads with a finer and uniform cell structure [37] as well as producing breads with low crumb cohesiveness [23].

The use of dairy proteins in the GF bread formulation has been reported to improve the GF bread parameters (e.g., texture, loaf volume, taste, and crust color) and to extend shelf-life of breads due to their high-water binding capacity and good gelling property that improve dough handling [2, 25, 38, 39]. In particular, milk components improve the crust color as a result of Maillard reactions between dairy proteins and lactose [2]. Among dairy proteins, casein favorably enhances the nutritional profile as a rich source of essential amino acids (i.e., lysine, methionine, tryptophan) and calcium. Furthermore, the emulsifying property of casein helps to stabilize other components of GF dough [38]. Whey proteins and skim milk powder are other dairy protein sources for the GF bread formulations with an ability to create gel-like structure and possessing a high-water binding capacity, respectively. However, the use of dairy protein sources in GF product formulation has been limited since they must be labeled as an allergen on food labels. In addition, CD is accompanied by lactose intolerance [2, 38].

Legume proteins, cereal proteins (non-gluten containing proteins, such as corn and rice proteins), and tuber proteins, are plant-based proteins that can be applicable in GF products. Among plant-based protein categories, legume proteins (specifically soybean and pea proteins), have become recognizable protein supplements in GF products [23] due to their high bioavailability along with great functional properties (e.g., water binding capacity, emulsion, gelation) [40, 41]. Marco and Rosell [42] reported that soybean protein enrichment (14%) in the rice-based GF breads increased water absorption and modified textural characteristics (i.e., decrease in the specific volume) of breads. These authors suggested that the combination of soybean protein, HPMC, and transglutaminase (TG) can be used for protein enriched GF breads [42]. In a recent study, the addition of pea protein and egg white protein in the corn starch-based GF formulation was assessed on the specific volume and weight loss of the bread [43]. The researchers found that enriching GF bread with pea protein resulted in higher dough viscosity due to greater water binding capacity of pea protein than egg white protein. Furthermore, egg white protein increased crumb hardness and had a compact structure compared to pea protein. Also, breads with pea protein had greater protein digestibility than those of egg white protein [43].

5.4 *Enzymes*

Enzyme technology as a food technology intervention has been employed to address dough rheological properties, and the quality and shelf-life of the GF breads through modifying the main components (e.g., starch, protein) of GF ingredients. Extensively studied enzymes in the GF bread formulations have been categorized into three groups: protein-connecting enzymes (TG and glucose oxidase [GO]), starch-hydrolyzing enzymes (α-amylase and cyclodextrin glycosyltransferase [CGT]), and proteases [3–5, 22, 24].

Among the enzyme categories, the most commonly employed enzymes in the GF breads are protein-networking enzymes or crosslinking enzymes, in particular microbial TG [22]. These enzymes stabilize the GF batter, improve the batter handling and rheological properties (e.g., viscoelasticity, gas retention capacity), and result in the increased GF bread quality. TG catalyzes the acyl-transfer reactions between lysine and glutamine residues, leading to the conversion of soluble proteins into insoluble high molecular weight protein polymer/networks [5, 24]. This protein polymer can mimic gluten functionality (e.g., gas-retention capacity, viscoelasticity) and thus can enhance batter and bread properties (e.g., viscoelastic batter, loaf volume, and crumb softness), depending on the proper use of protein substrates and enzyme concentration to form an optimal protein network [22, 44].

The use of microbial TG increased loaf volume and reduced the crumb hardness and chewiness of jasmine flour-based GF bread [45]. Another research showed that microbial TG increased loaf volume and also crumb hardness and chewiness of GF bread formulated with red rice flour and inulin [44]. Additionally, based on the protein source of GF bread formula, the addition of TG had positive effects on the buckwheat and brown rice flour-based, no effects on sorghum, oat and teff flours-based and negative effects on the maize-based GF breads (Fig. 10.1) [46].

Likewise, GO catalyzes crosslinking between protein residues through disulfide binding, causing protein polymerization that enhances viscoelastic properties of GF batter and resulting in improved loaf volume and crumb properties [3, 22]. The

Fig. 10.1 GF bread slices from buckwheat flour (BW), brown rice flour (BR), and corn flour (CR) with the addition of TG at 0, 1, and 10 U [46]

efficiency of GO depends on the base formula of GF bread. The use of GO improved the quality (e.g., loaf volume and crumb softness) of rice [47], sorghum and corn GF breads, but did not impact buckwheat and teff breads [48].

Starch-hydrolyzing enzymes are less commonly used in GF breadmaking compared to protein-connecting enzymes [22]. CGT and α-amylase are employed to hydrolyze starch, retard starch retrogradation, and improve shelf-life of GF breads [49]. Hydrolysis of starch by CGT forms a cyclodextrin substrate, acting as an emulsifier that promotes formation of a protein-lipid complex. Gujral et al. [49] reported that the use of CGT and α-amylase improved the loaf volume of rice bread. Nevertheless, the addition of CGT yielded a product with a good crumb softness and more pronounced anti-firming effect on the crumb than that of amylase treatment, resulting in a sticky crumb texture. Similarly, CGT reduced crumb firmness of chickpea GF bread [50].

Proteases and lipases have been used in the GF breadmaking. Proteases hydrolyze proteins and cause the modifications in protein-starch interactions, resulting in suitable porosity, greater loaf volume and softer crumb texture, and also retaining bread staling [51]. However, the addition of proteases has been reported to have both positive and negative effects in GF breads based on the GF base formula. Renzetti and Arendt [48] stated that the use of protease impacted adversely the structure and texture of buckwheat, maize, and sorghum breads due to causing more liquid-like batter with a poor viscoelasticity as a result of protein degradation. On the other hand, another study showed that protease addition improved volume, crumb appearance, and texture of rice bread [52]. Azizi et al. [51] tested protease and lipase effects on GF blends that included rice and quinoa flour and corn starch and obtained breads with increased loaf volume.

6 Novel Technologies in the GF Breadmaking

6.1 Sourdough

Historically, the use of sourdough for food production has existed since ancient times as a leaving agent for breadmaking process [53]. Sourdough is a bread-leaving agent that is a metabolically active unique microbial ecosystem of lactobacilli and yeasts [4, 54]. The sourdough ecosystem is obtained from fermentation of flour and water by lactic acid bacteria (LAB) and yeasts [55]. In this ecosystem, yeasts actively produce CO_2 and various aroma compounds, while LAB contribute in the acidification of dough through the production of lactic and acetic acids, and other antimicrobial compounds, extending the shelf-life of GF breads [8, 22]. Among sourdough biota, *Lactobacillus fermentum, L. plantarum,* and *L. paralimentarius* are strains generally isolated from GF sourdoughs (i.e., amaranth, teff, quinoa, buckwheat, rice, and maize) [22].

Sourdough technology has been extensively studied to optimize required proportion of sourdough to produce GF breads. Since sourdough is a highly heterogenous material, its use can dramatically impact the manufacturing conditions and eventually dough and bread quality. Therefore, the use of sourdough in the GF breadmaking (Table 10.2) can be purposed to enhance the following: (1) the quality of dough (i.e., gas retention) and bread (texture and flavor), (2) nutritional profile regarding mineral bioavailability, starch digestibility, and concentration of bioactive compounds (e.g., bioconversion of phenols into biologically active compounds), and (3) shelf-life by retarding staling process and preventing mold and bacterial growth [4, 8, 12, 56]. In regard to these purposes of sourdough technology, Novotni et al.

Table 10.2 The use of novel technologies for GF breadmaking

Technology	Advantage	Disadvantage	References
Sourdough	Sourdough biotechnology has been used to reduce the amount of gluten by degrading gluten-forming proteins. This technology also improves the quality of GF bread and extends shelf-life [4]. The use of sourdough at 15 and 22.5% increased loaf volume and decreased crumb firmness and GI, and extended shelf-life of partially based frozen GF bread [57]. GF sourdough can be combined with alternative ingredients (i.e., legume flours) to enhance nutritional value and the quality of bread. Addition of chickpea and cowpea flour improved load and specific volume and acceptability of sorghum-based sourdough bread [59]	The length of the sourdough process for digestion of dietary proteins (i.e., 3–6 h) and the variations based on the different source of cereals might lead variability for fermentation time [4]. Therefore, it may create inconvenience for the production line and increase the cost. The amount of sourdough used can differ based on the GF formulation (e.g., the source of blend, time) [57]	[4, 57, 59]
High hydrostatic pressure (HHP)	HHP is an emerging green food processing. This technology has been used for physical modification of starch and protein ingredients. Modified starch induced by HHP exhibited improved swelling capacities without the loss of the granule integrity	HHP requires optimum pressure level for its efficiency. Therefore, HHP operation conditions need to be optimized based on the different source of starch	[61]
Infrared-microwave baking	A hybrid heating system, consisting of infrared-microwave baking, might be useful for GF breadmaking due to its cost- and energy-efficiency compared to conventional baking system	This new baking approach requires more research to determine the baking quality for different GF breads, produced different formulation	[65, 66]
Ohmic heating	Ohmic heating is an emerging food technology by providing rapid and uniform heat distribution compared to other heating systems	This novel baking technology also require more research efforts to provide more outcomes for the quality of GF baking produced from various sources	[67]

[57] tested the different range of sourdough supplementation in the GF bread formula, blending of rice, corn, and buckwheat flour, and corn and potato starch. They reported that addition of 15 and 22.5% sourdough improved the quality (increased bread volume and decreased crumb firmness), extended shelf-life, and reduced glycemic index (GI) of partially based frozen GF bread. Gobbetti et al. [4] stated that the reduction in GI for sourdough fortified GF bread might be related to the synthesis of lactic and acetic acid, and short chain peptides that are potentially responsible for the reduction in GI. Houben et al. [58] reported that *L. plantarum* AL30, a sourdough fermentation strain, improved the dough rheology of GF batter, containing amaranth flour, that was similar to that found in pure wheat flour.

Furthermore, Gobbetti et al. [4] stated that a combination of sourdough fermentation and the use of alternative value-added ingredients, (i.e., legume flours, rich in dietary fiber, resistant starch and protein content) might be a potential enhancement to produce nutritious, healthier, and better-quality GF bread. Legume fortification improved the specific volume and acceptability sensory attribute of sorghum-based sourdough bread (SSB). In particular, fortification of chickpea flour into the SSB had higher nutrition values (i.e., ash and protein), greater acceptability and lower crumb firmness [59].

6.2 *Emerging Food Technologies in the Gluten-Free Breadmaking*

High hydrostatic pressure (HHP), an emerging eco-friendly food technology, has been used for non-thermal preservation of foods, such as fruit juices [60]. This food technology has been considered for physical modification of starch and protein ingredients to improve their functional proporties [22]. Li et al. [61] reported high pressure induced gelatinization of starches (Table 10.2). The improvement of swelling capacities of starch by modifying microstructure (loss of crystalline structure) of starch granules without losing the granule integrity resulted from the HHP [61]. In terms of proteins, high pressure disrupts the quaternary and tertiary structures of proteins that leads an increase of the reactivity of SH groups as well as enhancing protein functional features (e.g., protein solubility) [62].

The effect of high pressure on the technological properties of GF batters (i.e., buckwheat, teff, and white rice batters) were investigated. Viscoelastic properties of GF batters were increased due to starch gelatinization and protein crosslinking induced by high pressure [63]. Recently, HHP was applied as a pre-treatment of GF raw materials (i.e., corn starch and rice flour). HHP-treated corn starch and rice flour had a greater water binding capacity [64]. The GF batters prepared with HHP-treated ingredients had good batter handling properties and exhibited better bread quality (i.e., volume, crumb softness) than those of GF with non-treated GF ingredients [64].

Novel non-conventional baking technologies have become emerging technologies in the baking industry (Table 10.2). The suitable heating methods that can improve essential quality parameters (e.g., volume, crumb and crust formation, color, and flavor) of GF bread are being investigated [22]. These new baking technologies are microwave, infrared, jet-impingement, ohmic heating or combination of these (hybrid heating) [65].

Microwave and infrared (IR) processing are of interest due to cost- and time-efficiency of these methods. However, microwave baking impacts bread quality adversely; causing a decreased volume, gummy texture, increased crumb firmness and rapid staling. Jet-impingement heating requires more energy and negatively effects the bread quality due to the formation of thick crust with a poor texture [22]. Nevertheless, Chhanwal et al. [65] reported that hybrid heating system using IR-microwave baking can be a potential approach for GF breadmaking in regard to not only providing cost- and energy-efficiency, but also enhancing the bread quality. IR-microwave heating system was employed to produce GF bread and resulted in bread with higher firmness but with greater specific loaf volume than oven baked bread [66].

Ohmic heating is another emerging food technology with an advantage of rapid and uniform heat distribution compared to other heating systems. Bender et al. [67] reported that ohmic heating was successfully applied to improve the GF bread quality. A high initial power/rapid heating (2–8 kW) was employed for early-stage baking to stabilize the crumb structure before releasing CO_2. Then, moderate (1 kW) and low power (0.3 kW) were used to fully baking the bread surface. This baking approach resulted in higher loaf volume and finer pore structure than those of conventionally baked breads [67].

7 Conclusion and Future Prospects

The GF bakery products segment has been accounted for a major segment among the global GF products market size. Major drivers of this rising food segment are millennials as a significant contributor of this food segment to follow a special diet as well as people, avoiding gluten due to health issues related to gluten [19]. Allied Market Research [19] data indicated that this segment held the largest GF market share in 2019. The Europe region accounted for 47.50% of the market share in the global GF products market [19]. Increasing global demands for GF baked goods has required sustainable and more nutritious GF baked goods.

The lack of gluten in the GF breadmaking is the primary technological challenge to overcome during the GF breadmaking process. Furthermore, the lack of gluten negatively impacts bread quality parameters, including structure, texture, appearance, and shelf-life. Low nutritional profile is also another major constraint in traditional GF breads. Therefore, innovative approaches require the improvement of GF dough rheology and handling to yield the desired quality of GF bread, which are comparable with wheat-based breads. Current literatures show that there are great efforts both to enhance nutritional value and also to improve the quality of GF bread. Combination of functional alternative ingredients (e.g., legume and pseudo-cereals flours, proteins, fibers, enzymes, and hydrocolloids) is the most commonly used approach to address the necessities of GF breadmaking. But more research efforts are necessary for the utilization of value-added ingredients (e.g., legume ingredients, specifically pulse ingredients) in GF breadmaking process.

In addition to the use of ingredients and formulation parameters, employment of novel technologies in the GF breadmaking contributes pronounced developments to produce a high-quality GF bread with longer shelf stability. Sourdough treatment plays a role to address the major constraints of GF breadmaking, such as nutrition, technological, and shelf-life problems. Furthermore, high pressure treatment might be potential technology for ingredient modifications to improve the quality of GF bread. Emerging baking technologies are also very optimistic approaches to address the industry needs (i.e., time, cost, and energy efficiencies) as well as enhancing the quality of GF bread. However, effects of different non-conventional baking technologies on the sensory attributes and shelf-life of GF bread have not determined so investigations on these parameters are also other needs.

References

1. Cunha LM, Fonseca SC, Lima RC, Loureiro J, Pinto AS, Patto MCV, et al. Consumer-driven improvement of maize bread formulations with legume fortification. Foods. 2019;8(7). https://doi.org/10.3390/foods8070235.
2. Padalino L, Conte A, Del Nobile MA. Overview on the general approaches to improve gluten-free pasta and bread. Foods. 2016;5(4). https://doi.org/10.3390/foods5040087.
3. Casper JL, Atwell WA. Gluten-free: baked products. Cambridge: Woodhead Publishing; 2014. p. 1–88.
4. Gobbetti M, Pontonio E, Filannino P, Rizzello CG, De Angelis M, Di Cagno R. How to improve the gluten-free diet: the state of the art from a food science perspective. Food Res Int. 2018;110:22–32. https://doi.org/10.1016/j.foodres.2017.04.010.
5. Capriles VD, Areas JAG. Novel approaches in gluten-free breadmaking: Interface between food science, nutrition, and health. Compr Rev Food Sci Food Saf. 2014;13(5):871–90. https://doi.org/10.1111/1541-4337.12091.
6. Malalgoda M, Simsek S. Celiac disease and cereal proteins. Food Hydrocoll. 2017;68:108–13. https://doi.org/10.1016/j.foodhyd.2016.09.024.
7. Marsh MN. Gluten, major histocompatibility complex, and the small intestine: a molecular and immunobiologic approach to the spectrum of gluten sensitivity ('celiac sprue'). Gastroenterology. 1992;102(1):330–54.
8. Melini F, Melini V, Luziatelli F, Ruzzi M. Current and forward-looking approaches to technological and nutritional improvements of gluten-free bread with legume flours: a critical review. Compr Rev Food Sci Food Saf. 2017;16(5):1101–22. https://doi.org/10.1111/1541-4337.12279.
9. Codex Alimentarius Commission. Codex Standard for foods for special dietary use for person intolerant to gluten. Codex Standard 118–1979. 2008. http://www.fao.org/fao-whocodexalimentarius/shproxy/en/?lnk=1&url=https%253A%252F%252Fworkspace.fao.org%252Fsites%252Fcodex%252FStandards%252FCXS%2B118-1979%252FCXS_118e_2015.pdf. Accessed 25 Sept 2020.

10. Cohen IS, Day AS, Shaoul R. Gluten in celiac disease-more or less? Rambam Maimonides Med J. 2019;10(1). https://doi.org/10.5041/rmmj.10360.
11. Lahdeaho ML, Vainio E, Lehtinen M, Parkkonen P, Partanen J, Koskimies S, et al. Activation of celiac disease immune system by specific ax-gliadin eptides. Cereal Chem. 1995;72(5):475–9.
12. Foschia M, Horstmann S, Arendt EK, Zannini E. Nutritional therapy - facing the gap between coeliac disease and gluten-free food. Int J Food Microbiol. 2016;239:113–24. https://doi.org/10.1016/j.ijfoodmicro.2016.06.014.
13. Singh P, Arora A, Strand TA, Leffler DA, Catassi C, Green PH, et al. Global prevalence of celiac disease: systematic review and meta-analysis. Clin Gastroenterol Hepatol. 2018;16(6):823–836.e2. https://doi.org/10.1016/j.cgh.2017.06.037.
14. Tatham AS, Shewry PR. Allergens in wheat and related cereals. Clin Exp Allergy. 2008;38(11):1712–26. https://doi.org/10.1111/j.1365-2222.2008.03101.x.
15. Caproni M, Bonciolini V, D'Errico A, Antiga E, Fabbri P. Celiac disease and dermatologic manifestations: many skin clue to unfold gluten-sensitive enteropathy. Gastroenterol Res Pract. 2012. https://doi.org/10.1155/2012/952753.
16. Hadjivassiliou M, Sanders DS, Woodroofe N, Williamson C, Grunewald RA. Gluten ataxia. Cerebellum. 2008;7(3):494–8. https://doi.org/10.1007/s12311-008-0052-x.
17. Bascunan KA, Vespa M, Araya M. Celiac disease: understanding the gluten-free diet. Eur J Nutr. 2017;56(2):449–59. https://doi.org/10.1007/s00394-016-1238-5.
18. European Commission. Commission regulation (EC) No 41/2009 of 21 January 2009 concerning the composition and labelling of foodstuffs suitable for people intolerant to gluten. Official Journal of the European Union L16/3–5. 2009. https://eur-lex.europa.eu/legal-content/EN/TXT/PDF/?uri=CELEX:32009R0041&from=EN. Accessed 25 Sept 2020.
19. Allied Market Research. Gluten free products market. https://www.alliedmarketresearch.com/gluten-free-products-market. Accessed 5 Oct 2020.
20. Mir SA, Shah MA, Naik HR, Zargar IA. Influence of hydrocolloids on dough handling and technological properties of gluten-free breads. Trends Food Sci Technol. 2016;51:49–57. https://doi.org/10.1016/j.tifs.2016.03.005.
21. Martinez MM, Gomez M. Rheological and microstructural evolution of the most common gluten-free flours and starches during bread fermentation and baking. J Food Eng. 2017;197:78–86. https://doi.org/10.1016/j.jfoodeng.2016.11.008.
22. Bender D, Schoenlechner R. Innovative approaches towards improved gluten-free bread properties. J Cereal Sci. 2020;91:102904. https://doi.org/10.1016/j.jcs.2019.102904.
23. Roman L, Belorio M, Gomez M. Gluten-free breads: the gap between research and commercial reality. Compr Rev Food Sci Food Saf. 2019;18(3):690–702. https://doi.org/10.1111/1541-4337.12437.
24. Houben A, Hoechstoetter A, Becker T. Possibilities to increase the quality in gluten-free bread production: an overview. Eur Food Res Technol. 2012;235(2):195–208. https://doi.org/10.1007/s00217-012-1720-0.
25. Rai S, Kaur A, Chopra CS. Gluten-free products for celiac susceptible people. Front Nutr. 2018;5:114. https://doi.org/10.3389/fnut.2018.00116.
26. Witczak M, Ziobro R, Juszczak L, Korus J. Starch and starch derivatives in gluten-free systems - a review. J Cereal Sci. 2016;67:46–57. https://doi.org/10.1016/j.jcs.2015.07.007.
27. Boukid F, Vittadini E, Lusuardi F, Ganino T, Carini E, Morreale F, et al. Does cell wall integrity in legumes flours modulate physiochemical quality and in vitro starch hydrolysis of gluten-free bread? J Funct Foods. 2019;59:110–8. https://doi.org/10.1016/j.jff.2019.05.034.
28. Vatansever S, Rao J, Hall C. Effects of ethanol modified supercritical carbon dioxide extraction and particle size on the physical, chemical, and functional properties of yellow pea flour. Cereal Chem. https://doi.org/10.1002/cche.10334.
29. Villanueva M, Perez-Quirce S, Collar C, Ronda F. Impact of acidification and protein fortification on rheological and thermal properties of wheat, corn, potato and tapioca starch-based gluten-free bread doughs. LWT Food Sci Technol. 2018;96:446–54. https://doi.org/10.1016/j.lwt.2018.05.069.

30. Anton AA, Artfield SD. Hydrocolloids in gluten-free breads: a review. Int J Food Sci Nutr. 2008;59(1):11–23. https://doi.org/10.1080/09637480701625630.
31. Sciarini LS, Ribotta PD, Leon AE, Perez GT. Effect of hydrocolloids on gluten-free batter properties and bread quality. Int J Food Sci Technol. 2010;45(11):2306–12. https://doi.org/10.1111/j.1365-2621.2010.02407.x.
32. Lopez ACB, Pereira AJG, Junqueira RG. Flour mixture of rice flour, corn and cassava starch in the production of gluten-free white bread. Braz Arch Biol Technol. 2004;47(1):63–70. https://doi.org/10.1590/s1516-89132004000100009.
33. Cato L, Gan JJ, Rafael LGB, Small DM. Gluten free breads using rice flour and hydrocolloid gums. Food Australia. 2004;56(3):75–8.
34. Lazaridou A, Duta D, Papageorgiou M, Belc N, Biliaderis CG. Effects of hydrocolloids on dough rheology and bread quality parameters in gluten-free formulations. J Food Eng. 2007;79(3):1033–47. https://doi.org/10.1016/j.jfoodeng.2006.03.032.
35. Capriles VD, dos Santos FG, Areas JAG. Gluten-free breadmaking: improving nutritional and bioactive compounds. J Cereal Sci. 2016;67:83–91. https://doi.org/10.1016/j.jcs.2015.08.005.
36. Doublier JL, Garnier C, Renard D, Sanchez C. Protein-polysaccharide interactions. Curr Opin Colloid Interface Sci. 2000;5(3–4):202–14. https://doi.org/10.1016/s1359-0294(00)00054-6.
37. Sahagun M, Gomez M. Assessing influence of protein source on characteristics of gluten-free breads optimising their hydration level. Food Bioprocess Technol. 2018;11(9):1686–94. https://doi.org/10.1007/s11947-018-2135-0.
38. Ziobro R, Juszczak L, Witczak M, Korus J. Non-gluten proteins as structure forming agents in gluten free bread. J Food Sci Technol Mysore. 2016;53(1):571–80. https://doi.org/10.1007/s13197-015-2043-5.
39. Gallagher E, Kunkel A, Gormley TR, Arendt EK. The effect of dairy and rice powder addition on loaf and crumb characteristics, and on shelf life (intermediate and long-term) of gluten-free breads stored in a modified atmosphere. Eur Food Res Technol. 2003;218(1):44–8. https://doi.org/10.1007/s00217-003-0818-9.
40. Vatansever S, Hall C. Flavor modification of yellow pea flour using supercritical carbon dioxide plus ethanol extraction and response surface methodology. J Supercrit Fluids. 2020;156 https://doi.org/10.1016/j.supflu.2019.104659.
41. Vatansever S, Tulbek MC, Riaz MN. Low-and high-moisture extrusion of pulse proteins as plant-based meat ingredients: a review. Cereal Foods World. 2020;10:12–4. https://doi.org/10.1094/CFW-65-4-0038.
42. Marco C, Rosell CM. Breadmaking performance of protein enriched, gluten-free breads. Eur Food Res Technol. 2008;227(4):1205–13. https://doi.org/10.1007/s00217-008-0838-6.
43. Sahagun M, Benavent-Gil Y, Rosell CM, Gomez M. Modulation of in vitro digestibility and physical characteristics of protein enriched gluten free breads by defining hydration. LWT Food Sci Technol. 2020;117:108642. https://doi.org/10.1016/j.lwt.2019.108642.
44. Souza Gusmao TA, de Gusmao RP, Moura HV, Silva HA, Rangel Moreira Cavalcanti-Mata ME, Martins Duarte ME. Production of prebiotic gluten-free bread with red rice flour and different microbial transglutaminase concentrations: modeling, sensory and multivariate data analysis. J Food Sci Technol Mysore. 2019;56(6):2949–58. https://doi.org/10.1007/s13197-019-03769-8.
45. Pongjaruvat W, Methacanon P, Seetapan N, Fuongfuchat A, Gamonpilas C. Influence of pregelatinised tapioca starch and transglutaminase on dough rheology and quality of gluten-free jasmine rice breads. Food Hydrocoll. 2014;36:143–50. https://doi.org/10.1016/j.foodhyd.2013.09.004.
46. Renzetti S, Dal Bello F, Arendt EK. Microstructure, fundamental rheology and baking characteristics of batters and breads from different gluten-free flours treated with a microbial transglutaminase. J Cereal Sci. 2008;48(1):33–45. https://doi.org/10.1016/j.jcs.2007.07.011.
47. Gujral HS, Rosell CM. Improvement of the breadmaking quality of rice flour by glucose oxidase. Food Res Int. 2004;37(1):75–81. https://doi.org/10.1016/j.foodres.2003.08.001.

48. Renzetti S, Arendt EK. Effects of oxidase and protease treatments on the breadmaking functionality of a range of gluten-free flours. Eur Food Res Technol. 2009;229(2):307–17. https://doi.org/10.1007/s00217-009-1048-6.
49. Gujral HS, Haros M, Rosell CM. Starch hydrolyzing enzymes for retarding the staling of rice bread. Cereal Chem. 2003;80(6):750–4. https://doi.org/10.1094/cchem.2003.80.6.750.
50. Santos FG, Aguiar EV, Centeno ACL, Rosell CM, Capriles VD. Effect of added psyllium and food enzymes on quality attributes and shelf life of chickpea-based gluten-free bread. LWT Food Sci Technol. 2020;134:110025. https://doi.org/10.1016/j.lwt.2020.110025.
51. Azizi S, Azizi MH, Moogouei R, Rajaei P. The effect of quinoa flour and enzymes on the quality of gluten-free bread. Food Sci Nutr. 2020;8(5):2373–82. https://doi.org/10.1002/fsn3.1527.
52. Kawamura-Konishi Y, Shoda K, Koga H, Honda Y. Improvement in gluten-free rice bread quality by protease treatment. J Cereal Sci. 2013;58(1):45–50. https://doi.org/10.1016/j.jcs.2013.02.010.
53. Gobbetti M, Minervini F, Pontonio E, Di Cagno R, De Angelis M. Drivers for the establishment and composition of the sourdough lactic acid bacteria biota. Int J Food Microbiol. 2016;239:3–18. https://doi.org/10.1016/j.ifoodmicro.2015.05.022.
54. Ozel S, Sabanoglu S, Con AH, Simsek O. Diversity and stability of yeast species during the fermentation of Tarhana. Food Biotechnol. 2015;29(1):117–29. https://doi.org/10.1080/08905436.2014.996895.
55. De Vuyst L, Van Kerrebroeck S, Harth H, Huys G, Daniel HM, Weckx S. Microbial ecology of sourdough fermentations: diverse or uniform? Food Microbiol. 2014;37:11–29. https://doi.org/10.1016/j.fm.2013.06.002.
56. Gobbetti M, De Angelis M, Di Cagno R, Polo A, Rizzello CG. The sourdough fermentation is the powerful process to exploit the potential of legumes, pseudo-cereals and milling by-products in baking industry. Crit Rev Food Sci Nutr. 2020;60(13):2158–73. https://doi.org/10.1080/10408398.2019.1631753.
57. Novotni D, Cukelj N, Smerdel B, Bituh M, Dujmic F, Curic D. Glycemic index and firming kinetics of partially baked frozen gluten-free bread with sourdough. J Cereal Sci. 2012;55(2):120–5. https://doi.org/10.1016/j.jcs.2011.10.008.
58. Houben A, Gotz H, Mitzscherling M, Becker T. Modification of the rheological behavior of amaranth (Amaranthus hypochondriacus) dough. J Cereal Sci. 2010;51(3):350–6. https://doi.org/10.1016/j.jcs.2010.02.003.
59. Olojede AO, Sanni AI, Banwo K. Effect of legume addition on the physiochemical and sensorial attributes of sorghum-based sourdough bread. LWT Food Sci Technol. 2020;118:108769. https://doi.org/10.1016/j.lwt.2019.108769.
60. Aguilar CN, Ruiz HA, Rios AR, Chavez-Gonzalez M, Sepulveda L, Rodriguez-Jasso RM, et al. Emerging strategies for the development of food industries. Bioengineered. 2019;10(1):522–37. https://doi.org/10.1080/21655979.2019.1682109.
61. Li WH, Tian XL, Liu LP, Wang P, Wu GL, Zheng JM, et al. High pressure induced gelatinization of red adzuki bean starch and its effects on starch physicochemical and structural properties. Food Hydrocoll. 2015;45:132–9. https://doi.org/10.1016/j.foodhyd.2014.11.013.
62. Puppo C, Chapleau N, Speroni F, de Lamballerie-Anton M, Michel F, Anon C, et al. Physicochemical modifications of high-pressure-treated soybean protein isolates. J Agric Food Chem. 2004;52(6):1564–71. https://doi.org/10.1021/jf034813t.
63. Vallons KJR, Ryan LAM, Arendt EK. Promoting structure formation by high pressure in gluten-free flours. LWT Food Sci Technol. 2011;44(7):1672–80. https://doi.org/10.1016/j.lwt.2010.11.024.
64. Cappa C, Barbosa-Canovas GV, Lucisano M, Mariotti M. Effect of high pressure processing on the baking aptitude of corn starch and rice flour. LWT Food Sci Technol. 2016;73:20–7. https://doi.org/10.1016/j.lwt.2016.05.028.
65. Chhanwal N, Bhushette PR, Anandharamakrishnan C. Current perspectives on non-conventional heating ovens for baking process a review. Food Bioprocess Technol. 2019;12(1):1–15. https://doi.org/10.1007/s11947-018-2198-y.

66. Demirkesen I, Sumnu G, Sahin S. Quality of gluten-free bread formulations baked in different ovens. Food Bioprocess Technol. 2013;6(3):746–53. https://doi.org/10.1007/s11947-011-0712-6.
67. Bender D, Gratz M, Vogt S, Fauster T, Wicki B, Pichler S, et al. Ohmic heating-a novel approach for gluten-free bread baking. Food Bioprocess Technol. 2019;12(9):1603–13. https://doi.org/10.1007/s11947-019-02324-9.

Chapter 11
The Holy Grail of Ancient Cereals

Sabrina Geisslitz and Katharina Scherf

Abstract The interest in ancient cereals has been revived during the last decades due to the demand for products with health benefits, better taste and favourable nutritional composition. Especially the ancient wheat species -einkorn, emmer, khorasan and spelt- play a special role within the group of ancient cereals. The ancient wheat species evolved thousands of years ago and were the dominant wheat species in former times, but today their cultivation and use are negligible compared to modern wheats. A possible higher nutritional value of ancient wheat species compared to modern ones was the subject of several studies indicating that ancient wheat species only have slightly higher contents of e.g., bioactive phytochemicals. One characteristic of einkorn, emmer and spelt is that they are gluten-containing cereals, which give them better baking quality compared to pseudocereals, but gluten ingestion can also lead to adverse reactions in susceptible individuals. In modern wheat, the gluten is responsible for the superior baking quality and changes at the molecular level of this two-component glue explain the poorer baking quality of ancient wheat species. Nevertheless, ancient wheat species are not suitable for a gluten-free diet, even if differences in their immunogenic potential were identified in the context of celiac disease and wheat sensitivity.

Keywords Ancient grains · Celiac disease · Wheat sensitivity · Gluten · Health · Nutrition

S. Geisslitz · K. Scherf (✉)
Department of Bioactive and Functional Food Chemistry, Institute of Applied Biosciences, Karlsruhe Institute of Technology (KIT), Karlsruhe, Germany
e-mail: katharina.scherf@kit.edu

F. Boukid (ed.), *Cereal-Based Foodstuffs: The Backbone of Mediterranean Cuisine*, https://doi.org/10.1007/978-3-030-69228-5_11

1 What Are Ancient Cereals?

1.1 Classification

The crops maize, wheat and rice are the most important cereals worldwide and are grown on about 60% of the agricultural land. Other cereals with minor agricultural and economic importance are barley, millet, oat, rye, sorghum and teff and the pseudocereals amaranth, buckwheat, chia and quinoa.

The term "ancient" refers to all kinds of primitive grains that were not extensively bred and selected in modern times. These grains still have positive characteristics of wild ancestors, such as large individual variety, but also negative characteristics, such as large ear height, brittle rachis and low harvest yield [1]. Purposive breeding had the aim to generate "modern", genetically homogenous lines with stable and improved traits resulting in high yields, high productivity, all kinds of stress resistance (e.g., against pests) and other positive properties (e.g., no lodging) [2]. In contrast, "landraces" are heterogeneous lines with high tolerance to different kinds of regional environmental conditions and were developed by natural and human selection, but not by extensive modern breeding [3].

Two classification systems were suggested to group ancient grains. On the one hand, they are classified according to their type into cereals, minor cereals and pseudocereals and on the other hand, they are grouped into gluten-containing and gluten-free cereals (Fig. 11.1) [4].

1.2 Ancient Cereals and the Holy Grail

Ancient cereals are often associated with positive attributes, e.g., sustainable, natural, tasty, healthy and wholesome. The cultivation of ancient cereals can be performed more sustainably, because they do not need extensive fertilization, such as

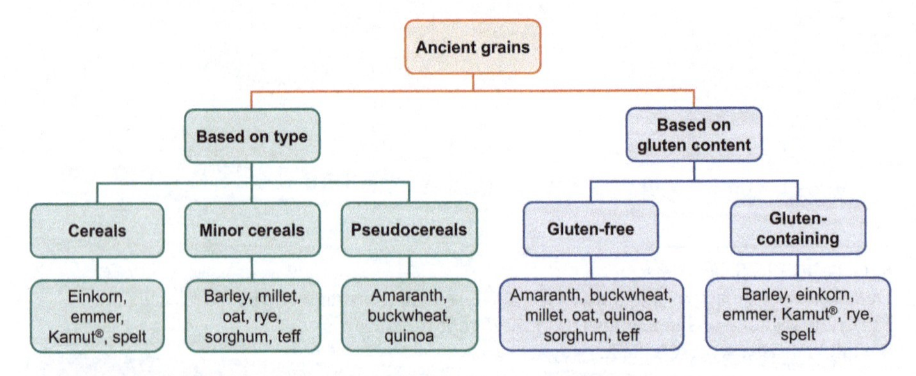

Fig. 11.1 Classification of ancient grains either based on the type or based on the gluten content. Modified from [4]

the high yielding maize and wheat. Further, ancient cereals are cultivated more often in organic farming, which prohibits the use of pesticides, synthetic fertilizer and genetically modified organisms and regulates the use of other agricultural chemicals making ancient cereals more natural. Products made of ancient cereals may be tastier than well-known products of maize and wheat, because new and innovative recipes had to be found to promote the use of ancient cereals. The traditional product spectrum of e.g., rice, polenta and white bread, was expanded by tasty and fancy ancient cereal products, such as salads, burgers, bowls and drinks made with e.g., quinoa, amaranth and chia. A potential higher nutritional value of ancient cereals may be gained by higher contents of bioactive phytochemicals, which act as antioxidants or meet the vitamin demand, or by a balanced nutrient composition, e.g., high protein and low fat. One true advantage of the ancient cereals, amaranth, buckwheat, chia and quinoa is that they are less "toxic" due to the fact that they are gluten-free cereals and are thus suitable as alternatives for diets of susceptible individuals suffering from gluten-triggered immune reactions. Especially ancient wheat species are often associated with health benefits and better wholesomeness, although they contain gluten. However, gluten is not only characterised by triggering adverse reactions, but has the unique potential to form a cohesive and viscoelastic dough resulting in breads with regular and fluffy crumb and crispy crust. This might eventually make ancient wheat suitable as healthier and better alternatives to modern wheat species with similar potential for baking. The following chapter gives background information and arguments for, but also against this assertion.

2 The Thousand Years Old Story of Ancient Wheats

The first archaeological findings of domesticated cereals were reported in the Fertile Crescent between the rivers Tigris and Euphrates in what is now Iraq, Turkey and Syria and were dated as about 10,000 years old [5]. During the "Neolithic Revolution", cereals were cultivated and selected for the first time as the population developed from hunters and gatherers moving from place to place to settlers who starting farming at the same place. The first cereals were barley (*Hordeum vulgare*) and ancient *Triticum* species, of which the modern wheat species common wheat (*Triticum aestivum* L. ssp. *aestivum*) and durum wheat (*T. turgidum* L. ssp. *durum*) evolved. The taxonomy and history of how modern wheat species developed is not clarified in total and still discussed. Some theories and suggestions are published and the following is one of the most accepted (Fig. 11.2).

Two different wild einkorn forms evolved probably from the same, unknown *Triticum* ancestor. These two diploid wheat species *T. monococcum* ssp. *boeticum* (A^mA^m) and *T. urartu* (AA) are separated by crossing barriers and have different plant morphology. The difference between the species is shown by the superscript letter "m", and the genome of *T. urartu* can also be indicated as A^uA^u. In contrast to *T. urartu*, which was never domesticated by humans, *T. monococcum* ssp. *boeticum*

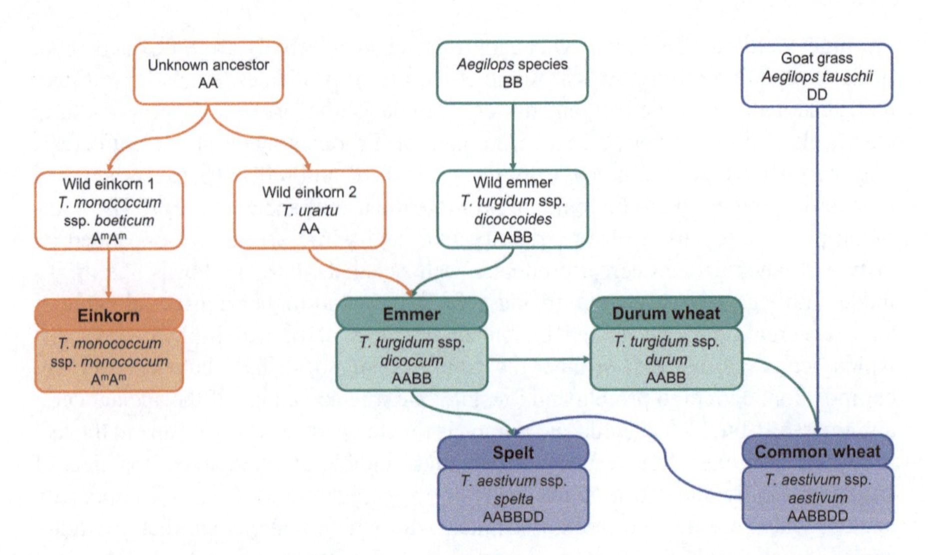

Fig. 11.2 Taxonomy of the wheat species einkorn, emmer, durum wheat, spelt and common wheat (modified from [6–8]). Einkorn, emmer and spelt are hulled wheat species and durum wheat and common wheat are free-threshing wheat species. Einkorn is a diploid wheat species (genome A^mA^m), emmer and durum wheat are tetraploid wheat species (AABB), and spelt and common wheat are hexaploid wheat species (AABBDD). Their evolutionary relationship is shown by arrows. Wild wheat species are displayed in colourless boxes

was domesticated by cultivation, selection and repeated harvesting to diploid, domesticated einkorn (*T. monococcum* ssp. *monococcum*) in the Karacadağ mountains in Southeastern Turkey [9]. The domesticated form of einkorn is characterised by larger grains compared to the wild forms and a tough rachis leading to easier harvesting, because the grains remain at the ear after maturation. However, the fragile, brittle rachis of wild einkorn forms helps to shatter and disperse the seeds on the ground for plant reproduction in nature. The term "einkorn" is originated in the German language and literally means "single grain". This shows that just one slender grain is built per spikelet. *T. urartu* was never domesticated but is involved in the development of the other polyploid *Triticum* species.

The tetraploid wild emmer with the AABB genome (*T. turgidum* ssp. *dicoccoides*) resulted from a spontaneous hybridisation of *T. urartu* (AA) and an unidentified *Aegilops* species (BB). The source of the B-allele is still not clarified at all and under discussion. In former times it was suspected that the B-allele has its origin in *Ae. speltoides*, but it was shown that this species has an S-allele and thus, cannot be the ancestor of wild emmer [10]. As alternative, it is assumed that the origin of the B-allele is an introgression of several parental *Aegilops* species including *Ae. bicornis*, *Ae. searsi*, *Ae. longissimi*, *Ae. sharonensis* and *Ae. speltoides* [11]. The domestication of emmer to *T. turgidum* ssp. *dicoccum* took place close to the Karacadağ mountains, where einkorn was domesticated as well [12]. The point in time of emmer and einkorn domestication is still not completely clarified and it is discussed, if this happened simultaneously or if one species was domesticated earlier than the other one [12]. As a result of the hybridisation to the tetraploid genome, emmer has two slender grains per spikelet.

As the next step, free-threshing tetraploid genotypes occurred spontaneously from emmer. These grains were selected by humans and about 10,000 years ago the tetraploid, free-threshing durum wheat was cultivated in the Fertile Crescent for the first time [13]. In contrast to durum wheat, einkorn and emmer are hulled wheat species with a tough glume. The husks have to be removed before milling, because husks belong to the non-eatable part of the grain and are indigestible. In times without mechanical thresher, it was labour-intensive and time-consuming to manually de-hull grains with flails. Thus, free-threshing grains were much easier to gather and clean and the use of durum wheat increased.

The botanical name of wild emmer, emmer and durum wheat (*T. turgidum* ssp.) indicates their very close relationship, and the three wheat species are interfertile with each other and as well as with other tetraploid wheat species [14]. Alternatively, emmer is called *T. dicoccum* and durum wheat *T. durum*. Both the long and the short botanical names are accepted equally, but the long names indicate the close relationship more explicitly.

Not only is the origin of the B-allele not clarified yet, but different theories are published about the origin of the D-allele and the development of hexaploid wheat species. It is generally accepted that the hexaploid wheat species common wheat and spelt (*T. aestivum* ssp. *spelta*) resulted from crossbreeding of a tetraploid and a diploid wheat species, because no wild hexaploid wheat species is known. The diploid wheat species was probably goat grass (*Ae. tauschii*) with the DD genome [7, 8, 15]. The D-allele is suggested to be a complex hybridisation of the three original A, B and D progenitors and not to be a pure genome [11]. It has been assumed that goat grass still has this hybridised D-allele. Another alternative theory is that the D-allele resulted just of a hybridisation of progenistors with the AA and BB genome [16].

Emmer has been thought to be the ancestor for hexaploid wheat species for a long time. In a long accepted theory, spelt originated from emmer and goat grass and spelt then developed into common wheat [15]. However, newer studies propose the free-threshing durum wheat as ancestor for hexaploid wheat species [8]. The appearance order of spelt and common wheat is debatable. It is not absolutely clear whether common wheat appeared earlier than spelt or spelt before common wheat. Furthermore, it is not clear, if one species is an ancestor for the other one. The last option might be that both species developed independently of each other. Newer theories suggest that common wheat is a hybridisation of durum wheat and goat grass and developed before spelt and then, spelt resulted from a hybridisation of common wheat and emmer (Fig. 11.2) [8, 17]. Beyond these questions, the time point and the location are still under discussion. It is supposed that the development of common wheat and spelt took place in Transcaucasia and in regions near the Caspian Sea about 10,000 years ago [8, 15].

There are other local, ancient hexaploid and tetraploid *Triticum* landraces known, e.g., *T. timopheevi* ssp. *timopheevi* (Zandari wheat), *T. turgidum* ssp. *paleocolchicum* (Georgian emmer) and *T. aestivum* ssp. *macha* (hulled Macha wheat), which were only minor subject of modern breeding [1]. The tetraploid khorasan wheat (*T. turgidum* ssp. *turanicum*) is of more interest and importance. One genotype of khorasan wheat has been trademarked as Kamut® in the US. Kamut® has its own website (www.kamut.com), on which the history, recipes, traders and the superior health- and agronomic-related properties are promoted. Even though the history of the brand Kamut® is very well known and emphasised, the history and taxonomy of khorasan wheat is still discussed. One theory is that khorasan is a hybridisation of *T. turgidum* ssp. *durum* and *T. polonicum* and occurred probably in a region of Eastern Mediterranean countries (e.g., Turkey) [18]. Another alternative is that the parents of Kamut® are *T. turgidum* ssp. *dicoccum* and *T. polonicum* and khorasan is not originated in Egypt, but in Western Turkey [19].

Compared to the ancient wheat species einkorn, emmer and spelt, Kamut® has a special role within ancient wheat, because it is a single cultivar of the tetraploid khorasan wheat. Kamut® has to fulfil certain criteria such as a protein content of 12–18%, a selenium (Se) content between 400 and 1000 μg/kg and it has to be grown under organic cultivation and never be hybridized or genetically modified (reviewed by [20]). Although Italy is the largest European consumer and importer (70%) of Kamut®, the nutritional value is not discussed here, because Kamut® represents only a single cultivar. An excellent and up-to-date review is given by Bordoni *et al.* [20]. Furthermore, the number of studies about gluten and its effect on baking quality is very limited for Kamut®.

There are indeed some contrary theories about the evolution of hexaploid wheat species and their exact ancestry. Depending on which theory might be true, it could be possible that common wheat is older than spelt and was therefore cultivated and used by humans earlier. Nevertheless, einkorn, emmer and spelt belong to ancient wheats due to their dominance several thousand years ago and their almost complete absence today.

3 The Myth of Higher Nutritional Value of Ancient Wheats

Human nutrition consists substantially of carbohydrates, fats, dietary fibre, proteins, water, minerals and vitamins. The macronutrients carbohydrates, fats and proteins provide energy and structural material and the micronutrients minerals and vitamins (defined due to lower intake) are required mostly for other reasons. The European Union (EU) regulates the daily reference intake of energy, macronutrients, vitamins and minerals in the food information regulation (regulation (EU) No 1169/2011). The reference energy intake (2000 kcal) should be covered by 70 g fat, 260 g carbohydrates and 50 g proteins. The high protein content of typically 8–20% and the low lipid content of typically 2–7% makes wheat grains an essential part of human nutrition worldwide. According to the "Mediterranean Diet Pyramid" the major food intake should be from vegetables, olive oil, beans, nuts, seed and grain products, lower amounts from fish, seafood, poultry, eggs, cheese and yogurt and very low amounts from meats and sweets [21]. Newer proposals differentiate between wholegrain and white flour products. Wholegrain products should be eaten with every meal and white flour products only in the same amounts as meats and sweets and thus, once a week [22].

According to the HEALTHGRAIN Consortium, wholemeal flour should be made of the intact kernels after removal of inedible parts such as the husks. The composition of this flour has to be equal to that of the intact kernel and thus, the proportions of starchy endosperm, germ and bran are present in the same amounts as in the intact kernel [23]. The wheat kernel is structured by the starchy endosperm and the germ, which are enclosed by the bran. In the case of spelt wheat species, the kernels are additionally surrounded by non-edible husks. The husks of einkorn, emmer and spelt correspond to over 30% of the kernel and this reduces significantly the kernel yield compared to free-threshing wheat species [24]. The husks can be used at least as ingredient of livestock feed. The starchy endosperm corresponds to over 80%, the germ to about 2% and the bran to about 15% of the wheat kernel. The composition of these three components varies notably. As the name starchy endosperm indicates, the endosperm is rich in starch (about 80%). Other major components of the endosperm are proteins (about 10%), lipids (1.6%) and minerals (0.5%). The germ is rich in proteins (34%), lipids (28%) and minerals (6%) and the bran mostly contains crude fibre (63%), proteins (18%) and minerals (8%). Because the bran and germ are removed for white flour, it is obvious that the composition of white flour and wholemeal flour is considerably different: The contents of proteins, lipids, crude fibre and minerals are higher in wholemeal flour than in white flour. One unpleasant property of wholemeal flour is the lower storage stability due to the higher lipid content and consequently the risk for rancidity. In contrast, the higher levels of protein, crude fibre and minerals increase the nutritional value of wholegrain products.

The ancient wheat species einkorn, emmer and spelt are mostly consumed as wholegrain products and as white flour only in lower amounts. This might be one reason why ancient wheat species are associated with health benefits. But other reasons such as the increase of the prevalence of food intolerances during the last decades and the loss of biodiversity and genetic variance due to the use of extensively bred modern wheat, might be conceivable. The questions, if ancient wheat species differ in their composition of bioactive compounds or might be even better or healthier than modern wheat species, were intensively discussed in scientific reviews by Shewry [14], by Shewry and Hey [25], by Arzani and Ashraf [26], by Hammed and Simsek [27] and by Dinu *et al.* [28]. These authors gave excellent overviews of the overall composition or were more focused on single ingredients e.g., dietary fibre, bioactive phytochemicals and minerals. However, they also concluded that there is a lack of literature data with comprehensive sample sets consisting of high numbers of cultivars per wheat species grown under the same conditions but by multiple-year and repeated location trials. The following sections will focus on studies conducted with einkorn, emmer, spelt and common wheat in the same study and will highlight bioactive compounds of various molecular size to collect clues to point out the myth of higher nutritional value of ancient wheat species.

3.1 The Giants: Dietary Fibre

According to the EU, dietary fibre includes carbohydrate polymers, which are not digestible or cannot be absorbed in the human small intestine. In addition, health claims for dietary fibre from brans of barley, rye, oat and wheat have been approved and thus, health related statements are allowed on dietary fibre enriched foods. The group of dietary fibre is divided in water-soluble and water-insoluble dietary fibre and includes different components. The major components are cell wall components such as the polysaccharides arabinoxylan, cellulose, pectin, β-glucan and Klason lignin and others such as fructans (fructooligosaccharides) and resistant starch. Resistant starch is concentrated in the endosperm, whereas the others are enriched in the bran and the outer layers of the kernel. In general, dietary fibre can positively influence the blood glucose level, the cholesterol level and the whole digestion process. The adequate intake of dietary fibre for adults should be 25 g per day in the EU [29].

The comprehensive HEALTHGRAIN project studied the changes of bioactive compounds of more than 150 common wheat cultivars of different release dates from 1900 to 2000. In this study five cultivars each of spelt, emmer and einkorn were included and all samples were cultivated in the same year at the same location. Contrary to expectations, the ancient wheat species einkorn, emmer and spelt had a lower dietary fibre content than modern common wheat [30]. Because einkorn and emmer have smaller kernels and dietary fibre is enriched in the bran, it was thought that emmer and einkorn might have a higher dietary fibre content. Further, no difference was observed for the total arabinoxylan content and the Klason lignin content between the ancient and modern wheat species. Only the content of water-extractable arabinoxylan was highest for einkorn. Last, the β-glucan content was higher in common wheat and spelt than in emmer and einkorn and this was confirmed by two other small studies [31, 32]. For 26 common wheat cultivars of the HEALTHGRAIN project, field trials were repeated at four locations throughout Europe, with one location repeated in two years [33]. For the ancient wheat species, this is still required. However, the findings indicate that ancient wheats do not have a higher nutritional value due to an increased dietary fibre content compared to modern wheat species.

3.2 The Elfs: Bioactive Phytochemicals

A comprehensive review on bioactive phytochemicals in ancient wheat species is available by Shewry [25], who concluded that data from repeated cultivation is required, but this was not fulfilled by the HEALTHGRAIN project. A German study cultivated 15 cultivars each of common wheat, durum wheat, spelt, emmer and einkorn at four different locations in 2013. Both studies reported detailed data on individual components of four different classes: Phenolic acids (e.g., ferulic acid),

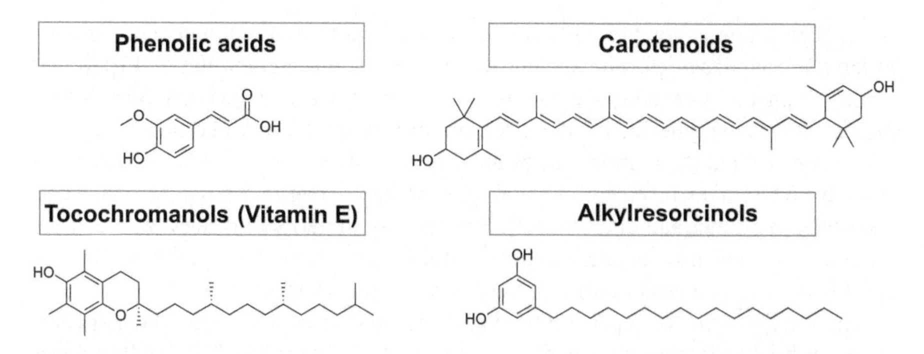

Fig. 11.3 Exemplary chemical structures of phenolic acids (ferulic acid), carotenoids (lutein), tocochromanols (α-tocopherol) and alkylresorcinols (heptadecylresorcinol), which were analysed in samples sets consisting of modern common wheat and the ancient wheat species einkorn, emmer and spelt and cultivated under the same environmental conditions

carotenoids (e.g., lutein), tocochromanols (e.g., α-tocopherol) and alkylresorcinols (phenolic lipids) (Fig. 11.3).

The most abundant phenolic acid in wheat is ferulic acid. Ferulic acid has antioxidative activity preventing from DNA damage, cancer and cell aging and is mostly bound, e.g., to arabinoxylan or sterols. During human digestion, ferulic acid is released and exerts its antioxidative effect (reviewed by [34]). The ancient wheat species had similar contents of free ferulic acid [35, 36], but einkorn had 33% higher contents of steryl ferulates on average compared to common wheat, spelt and emmer [37].

Carotenoids are synthesized only in plants and give plant-based foods an intensive yellow, orange or red colour. More than 800 carotenoids are known, of which several carotenoids act as provitamin A. The human demand for vitamin A is mainly covered by the two carotenoids retinol and β-carotene. Vitamin A plays an important role in the immune system and for growth and development, among others. Lutein is the most abundant carotenoid in wheat and has an essential role for human eye health. Einkorn had a significantly higher lutein content than common wheat, spelt and emmer [36, 38]. This is visible in the more intensive yellow colour of the flour, whereas modern common wheat was bred for its white colour. Due to a high heritability of the lutein content, the investigated einkorn cultivars are promising candidates for breeding programs resulting in high lutein contents independent of growing, climate and soil conditions [39].

Tocochromanols (vitamin E) consist of four tocopherol and tocotrienol homologues, of which α-tocopherol is the most known and best studied one. Vitamin E compounds are some of the most powerful, natural antioxidants protecting proteins and lipids due to their activity as radical scavengers. Einkorn and emmer contained lower amounts of α-tocopherols than common wheat and spelt [37, 40]. The heritability of the tocochromanols (here: sum of β-tocotrienol, β-tocopherol and α-tocopherol) was very high for four locations [37]. This is an indicator that einkorn, emmer and spelt cultivars with high levels of vitamin E can be selected leading to high levels of vitamin E independent of environmental conditions.

Alkylresorcinols are antioxidative substances and have antimutagenic and anti-inflammatory properties. The ancient wheat species einkorn, emmer and spelt had slightly higher alkylresorcinol contents than common wheat in the HEALTHGRAIN project. It was hypothesized that the reason for this are the small kernels of ancient wheat species and the enrichment of alkylresorcinols in the bran [41]. However, a study from Germany revealed higher alkylresorcinol contents in common wheat and spelt than in emmer and einkorn [42]. Further, a high heritability was observed for all wheat species also for the alkylresorcinol content. Incidentally, the alkylresorcinol pattern can be used to differentiate between ploidy levels.

The requirements of repeated cultivation of modern and ancient wheat species are not fulfilled for other bioactive phytochemicals. In the HEALTHGRAIN project only the contents of sterols and folates (vitamin B9) were analysed [43, 44]. Einkorn had higher contents of phytosterols and emmer had higher contents of folates than the respective other wheat species. In general, wheat is a valuable source of B vitamins, including thiamine (B1), riboflavin (B2), niacin (B3), folate (B9) and cobalamin (B12) (reviewed by [26]). Folate is involved in the metabolism of proteins, formation of red blood cells and plays an important role in reducing the risk of neural tube birth defects.

In summary, more studies analysing the contents of bioactive phytochemicals are still needed, which fulfil the requirements of repeated and controlled cultivation of both modern and ancient wheat species side by side as already demanded by Shewry [14] and Dinu *et al.* [28]. Nevertheless, based on the current knowledge, the ancient wheat species might be promising candidates for further breeding to increase the intake of bioactive compounds. The intake could be increased for ferulic acid and lutein with the use of einkorn, for folates with emmer and for α-tocopherol and alkylresorcinols with spelt.

3.3 The Dwarfs: Minerals

Minerals are inorganic elements and essential nutrients, which fulfil different kinds of functions in the human body, e.g., as electrolytes, as components of enzymes or as constituents of body structures such as bones and teeth. The total mineral content of food is given as the ash content and is analysed after combustion at high temperature. The individual minerals are divided into macroelements (>50 mg intake per day) and trace elements (<50 mg intake per day). For minerals, both a too low intake, e.g., iron (Fe) or iodide deficiency, but also a too high intake, e.g., carcinogenic effects of Se or toxicity of fluoride, play an essential role in nutrition.

In the HEALTHGRAIN project, the minerals Fe, zinc (Zn) and Se were analysed in the five cultivars each of einkorn, emmer and spelt [45]. The content of Se was higher in einkorn, emmer and spelt than in common wheat, but no difference was observed for Zn and Fe between the analysed wheat species. A slightly larger sample set regarding the ancient wheat species consisted of twelve einkorn, 13 emmer, five spelt and two common wheat cultivars, which were cultivated at one location

[46]. Significantly higher levels of phosphorus (P), magnesium (Mg), Zn and Fe were observed for ancient wheat species compared to modern wheat species. Due to the contrary statements in these two studies, the influence of genotype, climate, soil and the cultivation conditions is obvious.

In controlled organic cultivation, ancient wheat species had higher contents of boron (B), copper (Cu), sulphur (S), potassium (K) and Zn, but lower levels of manganese (Mn) and molybdenum (Mo) compared to modern common wheat [47]. No differences in the contents of Se, Fe, Mg, P and calcium (Ca) were present between modern and ancient wheat species. Although einkorn and emmer were not differentiated, but grouped in "primitive crops", this study was quite comprehensive and representative, including 32 primitive crops, 103 spelt cultivars and 358 common wheat cultivars in particular with repeated cultivation at different locations. The repeated cultivation indicated that the location had a significant effect on mineral concentrations for all wheat species. Only in emmer and einkorn, the effect of the genotype was higher than the location for some minerals. The intake of wholemeal flour from organic cultivation could provide more than 70% of the daily intake of Cu, Se, Fe, Mg, Zn, Mn, Mo and P. The most promising cultivars with the highest mineral contents were one cultivar each of einkorn and emmer. All in all, the mean values were comparable to studies conducted with conventional cultivation and thus, organic cultivation does not lead to lower mineral contents.

The ash content for wholemeal flours made of the ancient wheat species einkorn, emmer and spelt is significantly higher (10–20%) than for those made of the modern wheat species common wheat [48, 49]. The reason for this is the smaller kernel size of the ancient wheat species. In addition, common wheat is more often consumed in white flour products, which are characterised by even lower ash contents than wholegrain products. In summary, ancient wheat species seem to be promising candidates for increased mineral intake and thus, increased nutritional value. Furthermore, the organic cultivation commonly used to grow ancient wheats does not reduce the contents of minerals compared to conventional cultivation.

4 The Unique Potential of Wheat to Form Dough: Gluten

4.1 Gliadins and Glutenins: A Two-Component Glue

Wheat flour has the unique potential to form a cohesive and viscoelastic dough during mixing and kneading with water due to high water absorption capability. In combination with high gas holding capacity, these special techno-functional properties result in breads with high volume, regular and fluffy crumb structure and crispy crust after baking. This specialty of wheat compared to other cereals and pseudocereals is mainly induced by wheat gluten, which consists of gliadin and glutenin proteins. The contents of gliadins and glutenins and the ratio between both essentially determine the superior baking and technological properties of wheat flour. This is because of the different behaviour and role of gliadins and glutenins during

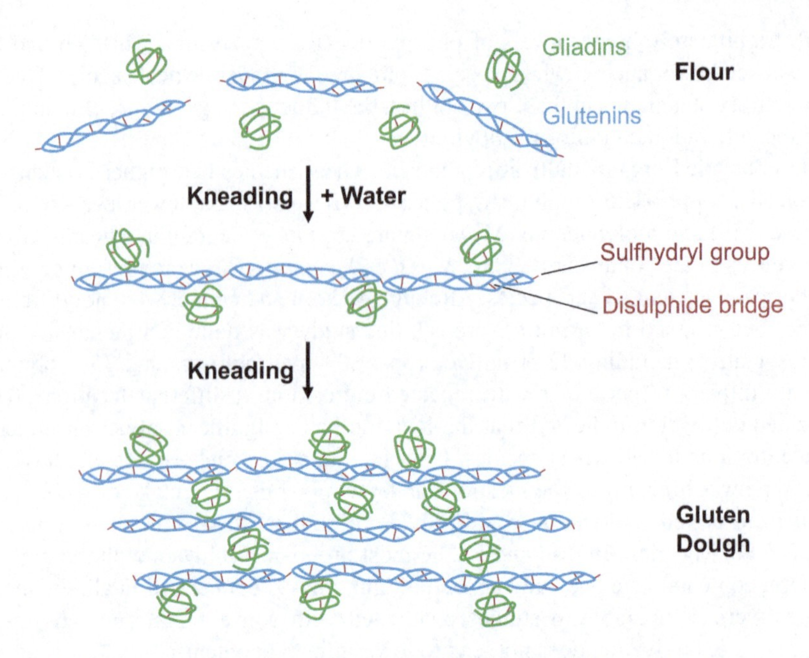

Fig. 11.4 Schematic representation of the functional effects of gliadins and glutenins in wheat gluten. During kneading new disulphide bridges between glutenins are formed. This is the reason for the elasticity and the resistance of the dough. The gliadins are the softeners of glutenins and they influence the viscosity and elasticity. The interaction of gliadins and glutenins is predominant for good baking results. Modified from [50]

kneading. The gliadins and glutenins can be seen as a two-component glue (Fig. 11.4). For one thing, the glutenins are polymers and they are responsible for the resistance to extension and elasticity of wheat dough, as they form a three-dimensional network [51, 52]. In contrast, the gliadins have a monomeric structure and are regarded as "softeners" in the dough and support viscosity and extensibility [53, 54]. Because optimal baking quality is only obtained by a balanced ratio of viscosity and elasticity, the ratio between gliadins and glutenins is one important determinant of good baking performance.

As an alternative to classical baking experiments, the baking quality may be estimated to a certain degree by indirect parameters, which are analysed by analytical methods using the essential prerequisite of a balanced ratio of gliadins and glutenins for superior baking quality. Near infrared spectroscopy has the advantage of being suitable to predict special properties (e.g., total protein content or even baking quality) from intact kernels by analysing special wavelengths in a fast and easy way, but it is only possible to predict the gliadin and glutenin contents as low, medium and high [55]. Thus, gliadins and glutenins are usually extracted from the flour according to the Osborne fractionation [56] and then, the fractions are analysed by chromatographic [57] or spectroscopic methods [58] to obtain the absolute content. In the modified Osborne fractionation, first, non-gluten proteins, the albumin/globulin fraction, are extracted with diluted salt solutions, second, the gliadins with aqueous alcohol (e.g., 60% ethanol) and last, the glutenins in aqueous alcohol (e.g., 50% 1-propanol) under reducing conditions (e.g., dithiothreitol) and heat (e.g., 60 °C) [57].

The modern common wheat has been analysed for gliadin and glutenin contents due its economic dominance and its widespread use in numerous studies (e.g., [55, 59–63]). The ratio of gliadins to glutenins was found to be relatively balanced between 1.5 and 3.1. Data about durum wheat and the ancient wheat species einkorn, emmer and spelt are very limited compared to common wheat. Gliadin and glutenin contents are available only for einkorn and spelt from a large sample set of 47 and 62 cultivars, respectively, showing a more unbalanced ratio of 4.1–14.0 and 2.2–9.0, respectively [64, 65]. One study included one cultivar of each wheat species, but these samples were cultivated at different locations in different years [66]. In fact, the samples were analysed within one laboratory with the same system guaranteeing comparability, but one cultivar is not representative for the whole wheat species. In addition, the protein composition of wheat is generally influenced by many factors. Not only the genetic background (cultivar and ploidy level), but also the cultivation and thus, the environmental and climatic conditions have a huge influence on the composition of gluten proteins (reviewed by [67]).

Two comprehensive studies were already performed on gliadin and glutenin contents of modern and ancient wheat species [61, 68], which fulfilled the requirements of simultaneous cultivation. The ancient wheat species einkorn, emmer and spelt are characterised by higher protein contents than common wheat, but the gluten composition varies exceedingly between common wheat and the ancient wheat species (Fig. 11.5). The higher gliadin contents and the lower glutenin contents of ancient wheat species compared to common wheat led to a higher ratio of gliadins to glutenins. Einkorn (3.7–12.1), emmer (3.0–11.1) and spelt (2.3–4.8) had significantly higher gliadin to glutenin ratios than common wheat (1.6–3.8) [68]. These findings have been confirmed for four different locations.

The glutenin content and the ratio of gliadins to glutenins is not only a prediction model for the baking quality of common wheat, but the prediction model was extended to ancient wheat species [48]. This means that a low, more balanced ratio of gliadins to glutenins leads to higher bread volumes of einkorn, emmer and spelt. The very high heritability observed in repeated cultivation confirmed that cultivars of einkorn, emmer and spelt can be selected with lower ratios of gliadins to glutenins and higher glutenin contents [68]. Further breeding programs might create new cultivars of ancient wheat species with improved baking quality.

4.2 Deeper Insights into Gliadin Types and Glutenin Subunits

Gluten proteins can be further classified and grouped into types and subunits. In one well-accepted nomenclature system, the gliadins are divided into ω5-, ω1,2-, α- and γ-gliadins, so-called gliadin types, according to their electrophoretic mobility. The glutenins are grouped into high-molecular-weight (HMW-GS) and

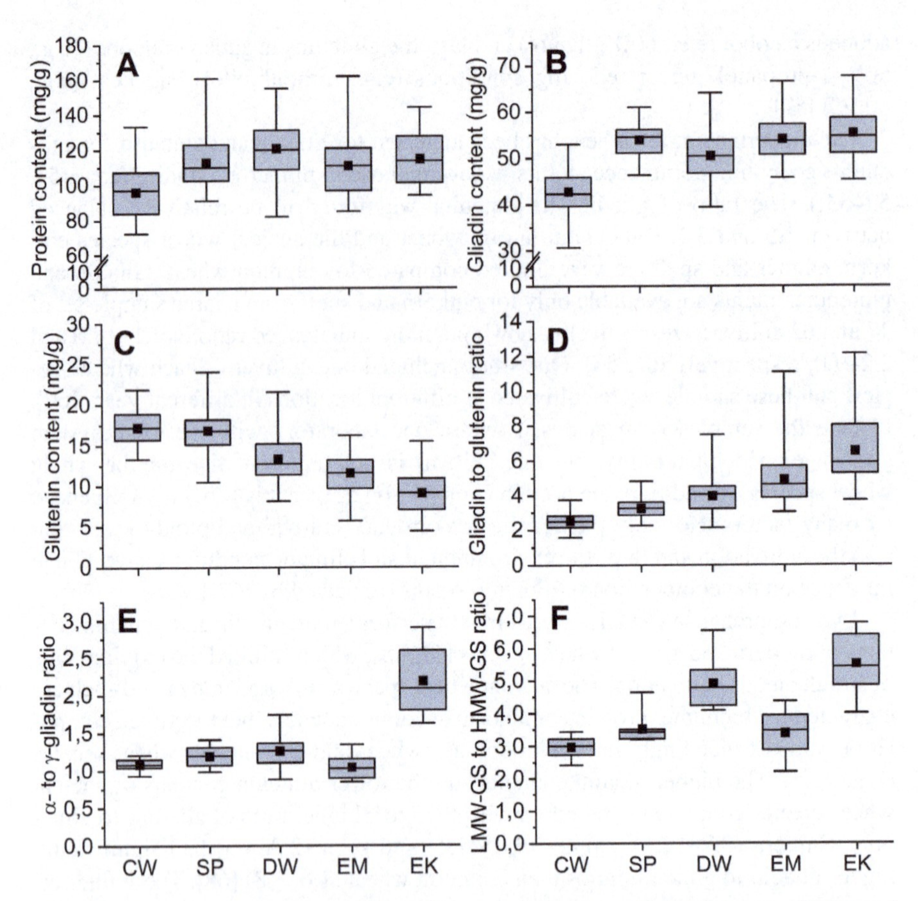

Fig. 11.5 Contents of protein (**a**), gliadins (**b**) and glutenins (**c**) and ratios between gliadins and glutenins (**d**), α- and γ-gliadins (**e**) and LMW-GS and HMW-GS (**f**) in common wheat (CW), spelt (SP), durum wheat (DW), emmer (EM) and einkorn (EK). The higher protein content of spelt, emmer and einkorn compared to common wheat increases the nutritional value of the ancient wheats. The gliadin to glutenin ratio is very important for baking quality, because the gliadins and the glutenins form the gluten network by polymerization. Data for **a–d** of 15 cultivars grown at four locations per box were taken from Geisslitz *et al.* [68] and for **e** and **f** of eight cultivars per box were taken from Geisslitz *et al.* [48]

low-molecular-weight glutenin subunits (LMW-GS) based on the molecular weight (MW) of the proteins. This classification system is inconsistent, because it uses either the electrophoretic mobility or the MW. Another classification system is based on homologous amino acid sequences and similar MW resulting in the three groups, (1) HMW group (2) medium-molecular-weight (MMW) group and (3) LMW group [66]. The HMW group includes proteins of HMW-GS (MW 67,000–88,000), the MMW group proteins of ω1,2- and ω5-gliadins (MW 40,000–55,000) and the LMW group those of LMW-GS, α- and γ-gliadins (MW 30,000–42,000). A third alternative classification system uses the term "prolamins" for all gluten proteins, even the glutenins. In this system, the gluten proteins are grouped into S-rich (α-, γ-gliadins and LMW-GS), S-poor (ω5- and ω1,2-gliadins) and HMW-prolamins (HMW-GS) [69].

4.2.1 Gliadin Types

The name "prolamins" reflects the dominance of the two amino acids *prol*ine and glut*amine* (46%) in gluten proteins [70]. The high amounts of proline and glutamine have positive physiological effects for the plants. Glutamine provides two nitrogen atoms during germination and proline causes kinks in the secondary protein structure, which allows a compact protein packing in the endosperm. Further, the high amounts of proline and as consequence, the complex secondary structure lead to prevention of degradation of the storage proteins by external enzymatic attack, e.g., against pests.

The number of cysteine residues is one important determinant for good baking quality, because of their ability to form disulphide bonds resulting in a complex gluten network. The disulphide bonds occur either inside one protein (intrachain) or between different proteins (interchain) [71]. With minor exceptions, the ω5- and ω1,2-gliadins contain no cysteine residues and are present as monomers. Most of the α- and γ-gliadins have six or eight cysteine residues, respectively, which form three or four intrachain disulphide bonds. Thus, the gliadins play only a minor role during disulphide bond and gluten network formation (reviewed by [72]). Only a small number of gliadin types has an odd number of cysteine residues due to amino acid changes and is linked to each other or to glutenin proteins and is involved in gluten network formation, but mostly as terminators and not as chain extenders [73, 74]. The presence of gliadin types with odd numbers of cysteine residues was confirmed for spelt as well [75]. No statement was included, if there is a difference between common wheat and spelt explaining the poorer baking performance of spelt.

The presence of all gliadin types was confirmed by chromatographic analysis [66] and *N*-terminal sequencing [76] in einkorn, emmer and spelt. Further, a large number of studies are available analysing the amino acid sequence of gliadin types in common wheat and ancient wheat species. Most of these studies have cysteine positions only as minor topics and focus on the elucidation of special amino acid combinations, which have an immunogenic potential (e.g., [74, 77–80]). Comprehensive genome-based studies including both ancient and modern wheat species are still required to increase the knowledge of cysteine positions in gliadins and amount of free cysteine residues available for gluten network formation.

Diploid einkorn has a different gliadin type distribution compared to the polyploid wheat species common wheat, spelt, durum wheat and emmer [48]. Especially, the ratio between α- and γ-gliadins is higher for einkorn than for the other wheat species due to the lack of an entire group of γ-gliadins [48, 61, 66] (Fig. 11.5). Compared to the high correlation between bread volume and the gliadin to glutenin ratio and the content of glutenins, the correlation between bread volume and gliadin type distribution is comparatively low over all five wheat species indicating that the distribution of gliadin types plays only a minor role for baking quality [48]. However, only little is known about the correlation between baking quality and gluten protein types in the literature and further studies with more cultivars per wheat species grown in various years and locations are required.

4.2.2 Glutenin Subunits

The glutenin fraction consists of HMW-GS and LMW-GS, which are aggregated proteins and linked by interchain disulphide bonds. After reduction of the disulphide bonds, the glutenins are soluble like the gliadins in aqueous alcohol solution. Emmer and einkorn have very low glutenin contents, as well as very low contents of LMW-GS and HMW-GS [48, 61]. In addition, the ratio of LMW-GS to HMW-GS was higher in durum wheat and einkorn than in common wheat, spelt and emmer resulting in a surplus of LMW-GS compared to HMW-GS (Fig. 11.5). The HMW-GS content is highly correlated to the bread volume of all five wheat species and also the ratio of LMW-GS to HMW-GS, even if the correlation was weaker [48].

The LMW-GS proteins have eight cysteine residues, of which six form three intrachain disulphide bonds and the remaining two are involved in interchain bonds mainly with LMW-GS, but also with HMW-GS (reviewed by [81]). Even though the LMW-GS correspond to the major part of the glutenin fraction and they are involved in gluten network formation, research and analytical studies were more focused on the characterisation of HMW-GS. So far it is well accepted that the LMW-GS are encoded on the short arms of group 1 chromosomes at the Glu-A3, Glu-B3 and Glu-D3 loci [82]. In common wheat, 7, 15 and 14 allelic variants are present on these loci, respectively. Because the genes are tightly linked, not all theoretical combinations are present in common wheat [83–85]. In two sample sets of 222 and 103 common wheat cultivars, respectively, 20 and 30 combinations were observed [84, 85]. To further increase the complexity of LMW-GS characterisation, the nomenclature of the single LMW-GS protein types is difficult and inconsistent. A good summary is given by Liu *et al.* [85], who tried to harmonize the LMW-GS nomenclature and their analysis. In brief, the nomenclature is either based on electrophoretic mobility (B, C and D subunits) or on the first amino acid of the *N*-terminal sequence (LMW-m, LMW-s and LMW-i). A uniform and simple classification system is a prerequisite for the correlation between quality parameters and LMW-GS patterns.

Nevertheless, ancient wheat species were analysed in several studies for their LMW-GS patterns. Due to the use of different gel electrophoresis systems and different nomenclature systems, these studies are hardly comparable. The variability of different LMW-GS patterns is very low for einkorn compared to the other wheat species, because most cultivars contain only two bands of LMW-GS. Only seven different LMW-GS patterns were present in 26 and 34 einkorn cultivars, respectively, but more than half of the cultivars had the same pattern [86, 87]. Emmer showed a higher number of different LMW-GS patterns with two to six LMW-GS bands. A sample set of 99 emmer cultivars represented 23 different patterns, but 60% of the cultivars had only three different patterns [88]. Spelt has the largest genetic variability of LMW-GS [89]. The two to six bands in 403 spelt cultivars showed 46 different patterns, indicating that the genetic variability is higher than in common wheat, which had only 20 different patterns in 222 cultivars [84]. However, half of the analysed spelt cultivars represented only seven patterns.

Some studies about the amino acid sequence and the presence and position of cysteine residues are available for the ancient wheat species. It was confirmed that einkorn LMW-GS contain eight cysteine residues, of which two are available for interchain disulphide bonds [90]. Nevertheless, there is the evidence that one einkorn LMW-GS has an odd number of cysteine residues, which might act as chain terminator during gluten network formation [91]. For cultivated emmer and spelt, there is a lack of data for amino acid sequences. One study identified a common wheat LMW-GS as a major wheat allergen, which has eight cysteine residues of which two are available for interchain bonds. The comparison of this common wheat LMW-GS revealed only a very low amino acid similarity of 46% compared to the spelt LMW-GS [92]. For emmer, only the wild form was analysed for the amino acid sequence of LMW-GS. All LMW-GS of wild emmer had eight cysteine residues with varying positions in the sequence, but two cysteine residues were also available for interchain disulphide bonds [93, 94]. It seems that there is a big difference between the number and position of cysteine residues in ancient wheats compared to common wheat, but research focusing on cultivated ancient wheat species is still required.

The HMW-GS are encoded on the long arms of group 1 chromosomes at the Glu-A1, Glu-B1 and Glu-D1 loci and each locus consists of two tightly linked genes encoding the x-type with higher MW (1Ax, 1Bx and 1Dx) and the y-type with lower MW (1Ay, 1By and 1Dy) [82]. In contrast to the inconsistent nomenclature of the LMW-GS, a consistent nomenclature exists for the HMW-GS, in which the individual HMW-GS are numbered according to their electrophoretic mobility [95, 96]. The genes 1Bx, 1Dx and 1Dy are constantly expressed, the genes 1Ax and 1By infrequently and 1Ay is completely silenced in hexaploid common wheat, resulting in three to five HMW-GS bands [95]. In European common wheat, three variations (NULL, Ax1, Ax2*) are known for 1Ax, five variations (Bx6, Bx7, Bx13, Bx14, Bx17) for 1Bx, five variations (By8, By9, By15, By16, By18) for 1By, three variations (Dx2, Dx3, Dx5) for 1Dx and two variations (Dy10, Dy12) for 1Dy [97]. Some variations occur only in predestined combinations, e.g., Bx14+Bx15, Bx17+Bx18, Dx2+Dy12 and Dx5+Dy10. Apart from the content, the overall presence and the combination of HMW-GS are very important for the baking quality. The deletion of some or all HMW-GS consequently led to decreased bread volumes made of common wheat [98]. Further, the combinations of Dx5+Dy10 or Bx17+By18 are associated with better baking quality of common wheat flour than the combinations of Dx2+Dy12 or Bx6+By8 [99].

Similar to common wheat, hexaploid spelt has three to five HMW-GS bands. Obviously, the D genes are not present in tetraploid emmer and 1Ay is silenced in emmer, as well, resulting in up to three HMW-GS [95]. In contrast, the 1Ay gene is not silenced in diploid einkorn, resulting in up to two HMW-GS bands [100]. The HMW-GS patterns of the ancient wheat species einkorn [100, 101], emmer [88, 102] and spelt [103, 104] were intensively studied. One comprehensive study included 1051 samples of all kinds of *Triticum* samples including 41 einkorn, 91 emmer and 162 spelt cultivars [105]. It seems that the genetic variability in spelt is lower compared to common wheat. More than 85% of 162 [105] and 403 [104] spelt cultivars had only one combination for the B genes (Bx6+By8 or Bx13+By16) and more than 90% had the combination of Dx2+Dy12. Seven and four combinations are known for the B and D genes in common wheat, respectively, which were more equally distributed over 185 cultivars [96]. In contrast, the genetic variability seems to be high in emmer. In fact, there is a lower overall number of combinations due to the lack of the DD genome, but the number of combinations for the B genes is higher with 14 (167 cultivars) [102] and 12 (99 cultivars) [88] than for common wheat. In combination with the LMW-GS pattern, there were 30 combinations for the 99 analysed emmer cultivars.

It is generally assumed that the HMW-GS form interchain disulphide bonds, because they do not occur as monomers in flour and dough. The x-type of HMW-GS has four cysteine residues and the y-type of HMW-GS has seven cysteine residues in common wheat. All cysteine residues are thought to be available for interchain disulphide bonds (reviewed by [73]). With some minor exceptions, the cysteine residues are present near the *N*- or *C*-terminal region and not within the middle region that contains mostly repetitive units.

Similar to the data availability on cysteine positions in LMW-GS, some studies about the amino acid sequence and the presence of cysteine residues are available for the ancient wheat species. It was confirmed that the einkorn 1Ax usually has the expected four cysteine residues [106]. However, the majority of einkorn 1Ay have only six cysteine residues, but there is evidence that some 1Ay of einkorn contain the expected seven [107] or even eight cysteine residues [108]. For cultivated emmer and spelt, there is a lack of data for amino acid sequences and presence of cysteine residues. For emmer only the wild form was analysed for its HMW-GS amino acid sequence. There is evidence that emmer HMW-GS types have lower numbers of cysteine residues, e.g., By16* with six cysteine residues [109]) or even higher number of deletions, e.g., Bx20 with only two cysteine residues instead of the expected four [110].

Taken together, the amino acid sequences and the positions of cysteine residues in gliadin types and glutenin subunits are already known in most cases. Nevertheless, the overall inferior baking quality of the ancient wheat species cannot be explained completely based on the current knowledge. The increasing opportunities in modern analytical methods, e.g., faster and cheaper genome sequencing and more common high resolution proteomics-based techniques, will surely help to elucidate the reasons for this on a molecular level.

4.3 *The Largest Biopolymer in the World: Glutenin Macropolymer*

The LMW-GS and HMW-GS play an important and essential role in gluten network formation. Especially for common wheat, a structure model of the glutenin proteins was postulated by Wieser [73]. In this model, the HMW-GS are bound by interchain bonds to other HMW-GS to a long, linear backbone and the LMW-GS are bound by interchain disulphide bonds to this backbone resulting in one of the greatest biopolymers in nature, the so-called glutenin macropolymer (GMP) (Fig. 11.6) [112]. The MW of GMP, is between 60,000 and 20 million and the diameter of GMP

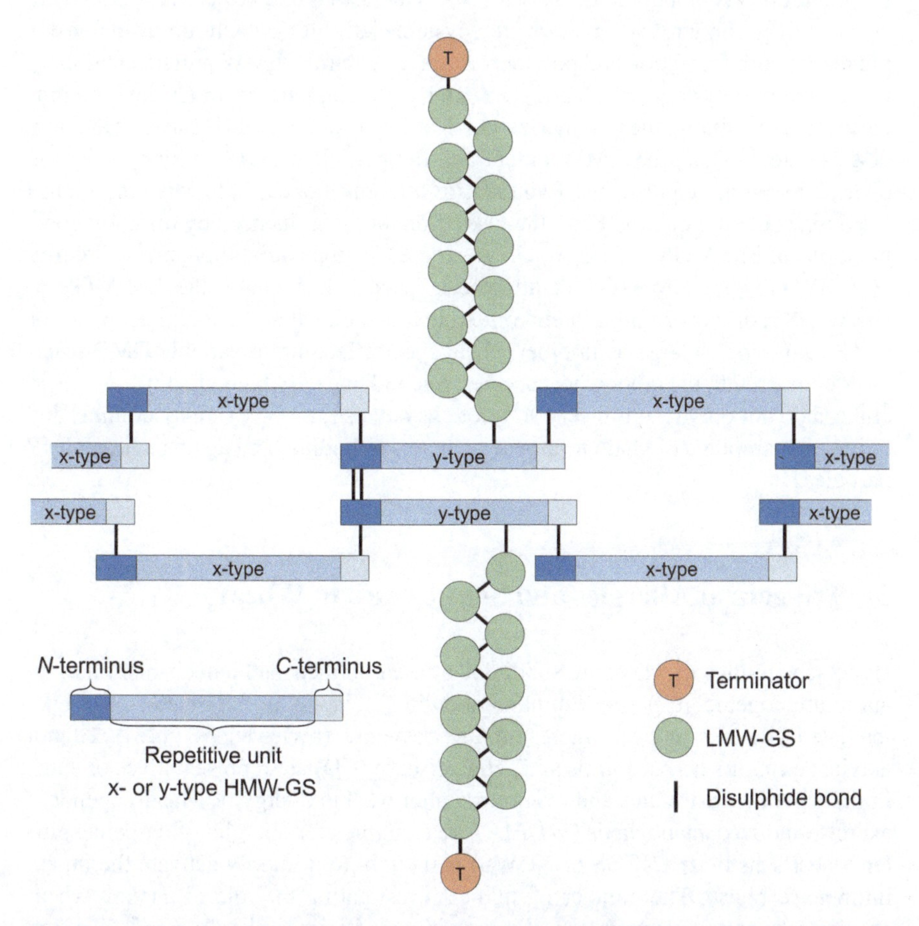

Fig. 11.6 Model system of the glutenin macropolymer as suggested by Wieser [73]. The HMW-GS (blue) are bound by interchain bonds to other HMW-GS to a long, linear backbone and the LMW-GS (green) are bound by interchain disulphide bonds to this backbone. The chain is limited by terminators (orange), which can represent proteins with an odd number of cysteine residues (e.g., gliadins), free cysteine or small molecules containing a free sulfhydryl group (e.g., glutathione). The gliadins are not displayed, but are thought to be present as shown in Fig. 11.4. Modified from [73, 111]

particles can be higher than 80 μm, although the majority of GMP particles is below 12 μm [113]. In previous times, the content of GMP was analysed gravimetrically [114], because GMP is defined as the gel layer on the surface of the starch pellet after extraction of the flour with sodium dodecyl sulphate solution (e.g., 1.5%) [115]. However, the absolute protein content is relatively low with max. 35% in this layer. The composition of the GMP layer depends on the cultivar and contains up to 70% starch [116]. For this reason, the GMP can be solubilized either with sonication or under reducing conditions and then the proteins are directly analysed by chromatographic methods [63, 117].

The amount of GMP has a notable effect on the baking quality of common wheat flours, and also that of flours of the ancient wheat species [48, 63, 73]. The background for this is thought to be that HMW-GS and LMW-GS act as chain extenders, because they contain two or more free cysteine residues, which are available for gluten network formation and polymerization. In contrast, LMW proteins with only one single free cysteine residue (i.e., α- and γ-gliadins) are chain terminators hindering and preventing the polymerization, leading to smaller GMP particles and low GMP content. Thus, the GMP content and the particle size are influenced by the distribution of gluten proteins. Two theories might be conceivable why einkorn has the poorest baking quality. First, the baking quality is influenced by different combinations of LMW-GS and HMW-GS and there is a high variability and availability of LMW-GS and HMW-GS in all wheat species. Especially, the LMW-GS to HMW-GS ratio is very high in einkorn so that it seems that the long backbone for GMP cannot be as large as in other wheat species leading to variable GMP structures compared to the other wheat species. Second, the very high gliadin contents in einkorn do not only result in very soft doughs but also in very crumbly doughs, due to the high amounts of chain terminators (here: α-gliadins) and again smaller GMP particles.

5 Triggers of Allergies and Sensitivities to Wheat

Hypersensitivities to wheat are subdivided based on their pathomechanism into an autoimmunogenic response (immunoglobulin A (IgA-) or IgG-mediated), IgE-mediated allergies and an innate immune response (reviewed by [118]). Gluten ataxia, dermatitis herpetiformis and celiac disease (CD) are representatives of autoimmunogenic sensitivities and respiratory allergy, skin allergy or wheat-dependent exercise-induced anaphylaxis (WDEIA) are examples for allergies. Non-celiac gluten/wheat sensitivity (NCGS or NCWS) is thought to primarily activate the innate immune response. The number of studies investigating the role of ancient wheat species related to IgE-mediated allergies is very limited [28, 119] and thus, the focus is on CD and NCGS in the following.

5.1 Celiac Disease

The background of CD is already well studied and whole reviews or books are necessary to show the complexity of CD [120]. CD is a lifelong inflammation of the small intestine caused by the ingestion of gluten of wheat and closely related cereals in genetically predisposed individuals. The prevalence of CD is estimated to be around 1% (0.5–1.7% confidence interval) depending on the region and the diagnosis tool used [121]. The causes for CD are a combination of environmental (consumption of gluten), genetic (human leukocyte antigens) and other factors (e.g., infections, changes of intestinal microbiota). Typical symptoms are especially intestinal manifestations, but also asymptomatic forms of CD are possible. CD is characterised by partial to total villous atrophy and crypt hyperplasia resulting in reduced absorption of nutrients. The diagnosis of CD relies on a combination of monitoring symptoms and patient history, detection of CD-specific antibodies in the blood and histologic assessment of small intestinal biopsies. So far, the only treatment is a lifelong gluten-free diet, which must not exceed 20 mg gluten per day to avoid recurrence of symptoms and ensure mucosal healing (reviewed by [118]).

The molecular reason why gluten consumption triggers CD are CD-active peptides, which are rich in glutamine and proline preventing gluten proteins from enzymatic breakdown in the human digestion. This results in large, incompletely digested peptides with more than nine amino acid residues. These intact peptides reach the small intestinal brushborder, pass through the epithelial barrier and initiate disease-specific immunogenic cascades in the lamina propria. More accurately, the immunogenic cascades are caused by specific amino acid sequences, the so-called epitopes, because of their ability to interact with human leukocyte antigens and human tissue transglutaminase resulting in activation of the adaptive immune response. These epitopes are usually units of nine amino acids and to date, 28 epitopes are known in wheat, which are present in more than 1,000 gluten derived peptides [122].

There are different ways to estimate the immunogenic potential of different wheat species: Immunoblotting, enzyme-linked immunosorbent assays (ELISA) or mass spectrometry (MS). Immunoblotting of the two α-gliadin epitopes Glia-α9 and Glia-α20 showed that there is no difference in the reactivity between spelt and common wheat [123] and between tetraploid wheat species and common wheat [124]. A study comparing two cultivars each of einkorn, emmer, spelt and common wheat revealed no difference between the wheat species for the immunogenic potential of gliadins and glutenins determined by immunoblots of positive wheat-sensitized sera [125].

One of the most important and immunodominant gluten peptides is an α-gliadin peptide with 33 amino acid (33-mer) with three overlapping epitopes (one copy of PFPQPQLPY, two copies of PYPQPQLPY and three copies of PQPQLPYPQ). The content or reactivity of the 33-mer can be estimated by the G12-ELISA. In general, common wheat and spelt had higher reactivity than emmer, Kamut® and einkorn [126–128]. In detail, emmer and Kamut® showed about 1/10 up to 1/20 lower reactivity than common wheat [126] and einkorn even had only 1/50 of that of common wheat [128]. The reason for the low reactivity is that the α-gliadins containing the 33-mer are encoded on the D-allele, which is missing in einkorn and tetraploid wheat species. The absence of the 33-mer in einkorn and tetraploid wheat species was confirmed by MS [129]. However, it has to be stated that MS does not reach the high sensitivity as provided by ELISA, but a lot more specificity.

Similar observations were made for the reactivity in R5-ELISA, which is sensitive for several epitopes of ω1,2-, γ- and α-gliadins and LMW-GS. The differences in the reactivity between common wheat and the tetraploid wheat species emmer and Kamut® were less pronounced as for the G12-ELISA and were only about 50% lower [126]. In contrast to that, the reactivity in R5-ELISA was even higher in spelt than in common wheat due to the overall higher protein and gluten content of spelt [130]. Taken together, both G12- and R5-ELISA lead to less accurate results in total gluten content in contrast to e.g., the Osborne fractionation especially for tetraploid and diploid wheat species.

One aim in CD-related research is to find cultivars with lower immunogenic potential, which are suitable for the diet of CD-patients, but have good baking properties. In addition to the absence of the 33-mer, it was shown that immunogenic peptides from γ-gliadins are present in lower amounts in einkorn compared to common wheat [131]. The reason for this is the lack of a group of γ-gliadins in einkorn, as already stated. Further, one study including 121 spelt cultivars identified spelt cultivars with lower immunogenic potential within the sample set [132]. Nevertheless, ancient wheat species contain gluten and CD-active epitopes and are thus undoubtedly not suitable for the diet of CD-patients.

5.2 *Non-Celiac Gluten/Wheat Sensitivity*

The name for the innate immune response triggered by wheat or gluten consumption is still being debated and terms such as NCGS, NCWS, wheat sensitivity or amylase/trypsin-inhibitor (ATI-) sensitivity are used [133]. Not only the name is still discussed, but other essential background knowledge such as prevalence (0.6–6%), causes (consumption of wheat or gluten-containing food), diagnosis tools (no biomarkers and tests, but exclusion of CD, wheat allergy and irritable bowel syndrome, IBS) and recommended treatment (max. 10% of the normal wheat/gluten intake or gluten-free diet) are still discussed. Once again, whole reviews are needed to display the background of this sensitivity [118, 133, 134].

Nevertheless, ATIs were identified to active the toll-like receptor 4 TLR4-MD2-CD14 complex causing secretion of proinflammatory chemokines and cytokines [135, 136]. This activation of innate immunity may lead to the symptoms typical of NCGS as well as worsening of pre-existing inflammatory reactions [136, 137]. ATIs are a group of LMW proteins (MW 12,000–16,000) soluble in aqueous salt solution. They are divided in three groups, of which two groups have mainly only inhibitory activity to α-amylase (monomeric 0.28, dimeric 0.19 and 0.53) and the third group are mainly bifunctional inhibitors to both α-amylase and trypsin (tetrameric CM1, CM2, CM3, CM16 and CM17). The monomeric and dimeric ATIs are named according to their electrophoretic mobility and the CM-types according to their solubility in a mixture of chloroform and methanol. The inhibitory activity is caused by four to five intramolecular disulphide bonds leading to a compact three-dimensional structure and resistance to digestive enzymes and heat.

The amino acid sequences of common wheat and spelt ATIs are thought to be the same due to the lack of specific entries in protein databases. In emmer, CM1 and CM17 were not found, but CM2, CM3 and CM16 had the same amino acid sequence as those of common wheat [138].The dimeric ATIs 0.19 and 0.53 of emmer were similar to common wheat with a very high amino acid sequence identity of 89–98% and the changes or deletions did not affect the regions, which are responsible for the amylase inhibitor activity [139]. The genes of einkorn ATIs are probably silenced or expressed in very low amounts, which is comprehensible due to the different set of A genes of einkorn compared to the other wheat species [6]. Interestingly, it was shown in the same study that the ATI genes are expressed in the diploid ancestor of common wheat (*T. urartu*) and the proteins were shown to have immunogenic potential.

Current research on ATIs is focusing on the inhibitory activity to α-amylase [140] and trypsin [141], bioactivity in humans [135] and contents in flour [142, 143]. Data about the correlation between these parameters are still missing, but are absolutely required to elucidate the immunogenic potential and increase the background knowledge about NCGS. However, the comparison of the mentioned studies is afflicted with difficulties and restrictions, because different samples were analysed and it is well accepted that protein content and composition are influenced both by genotype and environment. Nevertheless, the main outcomes of these studies should be summarized here. Depending on which inhibitory test was used to determine the inhibitory activity to α-amylase, no difference was observed between common wheat, spelt, emmer and Kamut® [140]. Einkorn had the highest inhibitory activity to trypsin, emmer the lowest activity and the activity of common wheat and spelt lay in between [141]. The ATI bioactivity was significantly lower for emmer and einkorn than for common wheat and the bioactivity was slightly lower for spelt and Kamut® than for common wheat [135]. In contrast to the bioactivity, spelt and emmer had higher ATI contents than common wheat and einkorn had very low ATI contents [143]. This confirmed the statement that einkorn ATI genes are silenced or expressed in very low amounts. To sum up, currently, no reliable statements are possible for the correlation between inhibitory activity, bioactivity and contents for ancient wheat species based on the available data.

Another example for a controversial observation is based on the identification of a CM3 peptide consisting of eleven amino acids (RSGNVGESGLI) as binding interface to the TLR4 [144]. The tryptic peptide (SGNVGESGLIDLPGCPR) representing the binding interface was quantitated by MS revealing higher contents in emmer and spelt than in common wheat [143]. This is in noticeable contrast to the lower observed ATI bioactivity in emmer and spelt compared to common wheat [135].

The triggers of NCGS are still under discussion and it is still unclear, whether ATIs are the only ones or if other wheat ingredients are triggers as well. Besides the lack of tests and biomarkers, one reason for this is the significant overlap between NCGS and IBS (reviewed by [145]). Another class of substances, fermentable oligo-, di-, monosaccharides and polyols (FODMAPs) were discussed as causes for NCGS, IBS and other related intestinal symptoms and manifestations. FODMAPs are poorly absorbed in the small intestine and are rapidly fermented by colon bacteria to gas, which leads to flatulence and abdominal pain in sensitive individuals. Wheat contains higher levels of FODMAPs compared to gluten-free cereals, but einkorn has similar FODMAP contents than common wheat and spelt and only emmer has lower FODMAP levels [146]. However, the use of sourdough fermentation and longer dough proofing times resulted in a clear reduction of FODMAP levels [146]. Further, it was shown that both a gluten-free diet and a low-FODMAP-diet reduced the symptoms of NCGS patients and it was suggested that NCGS is a combination of functional effects caused by FODMAPs, a mild gluten-triggered immune reaction and a microbial dysbalance [147]. These are indications why products made of ancient wheat species might be better tolerated by sensitive individuals suffering from NCGS and IBS. Products made of ancient wheat have usually longer resting times and are often manually prepared.

6 Conclusion: Opportunities to Use Ancient Cereals

The ancient wheat species and also ancient cereals in general are often sold with the marketing term "superfood" or "supergrain", because they are associated with health benefits, better taste and other positive properties. However, health related statements are strictly regulated for food by the Health Claim regulation in the EU and there are no claims available, which could be explicitly used for ancient cereals. The ancient wheat species have to be promoted with special marketing tools due to their higher price compared to modern wheat species as a result of low harvest yield, the necessity to remove the husks and challenges in cultivation. In contrast to that, ancient wheat species are more suitable for cultivation in marginal regions, which are characterised by high-stress conditions (e.g., poor soil quality, less rainfall or mountain areas). This makes ancient wheat species valuable resources for greater biodiversity and this may play an important role in nutrition security in the light of climate change.

In summary, a large number of studies is already available dealing with questions related to ancient wheat species and their ingredients, such as micronutrients (e.g., vitamins), the elucidation of the reasons for their poorer baking performance or the potential to use them in wheat-related allergies and intolerances. Further multidisciplinary research efforts are needed to screen hundreds of cultivars of ancient wheat species, evaluate promising candidates for agronomic, processing and nutritional properties, continue breeding efforts and develop products with excellent nutritious and sensory quality.

Even though ancient wheat species are only niche products, a lot of products made of ancient wheat species are available today. In Italy, the ancient wheat species are named "farro piccolo" (einkorn), "farro medio" (emmer) and "farro grande" (spelt). Typical dishes in the Italian cuisine are soups, risotto, wholegrain pasta and salads. In addition, emmer is also called "true farro" in Italy and "farro" in Spain. Other examples for the use of ancient wheats are breads, cakes and pasta made of flour and the grains can be used as alternatives for rice, when cooking the intact kernels. Flakes can be used as supplement for muesli, crackers, smoothies and drinks. Spelt milk is an alternative for cow's milk, which is not suitable in case of lactose intolerance or milk allergy. Last but not least, the famous and important beer is not only brewed with common wheat, but also with spelt, emmer and einkorn especially in German traditional breweries. Anyway, the ancient wheat species will most likely not fulfil the demand to feed the world or become suitable for the mass market, but they have great potential to enhance product diversity as specialty products or to rediscover forgotten traditional recipes.

Acknowledgments The authors thank Alexandra Axthelm (Leibniz-Institute for Food Systems Biology at the Technical University of Munich, Freising, Germany) for the excellent literature research.

References

1. Giambanelli E, Ferioli F, Koçaoglu B, Jorjadze M, Alexieva I, Darbinyan N, et al. A comparative study of bioactive compounds in primitive wheat populations from Italy, Turkey, Georgia, Bulgaria and Armenia. J Sci Food Agric. 2013;93(14):3490–501. https://doi.org/10.1002/jsfa.6326.
2. Sakuma S, Salomon B, Komatsuda T. The domestication syndrome genes responsible for the major changes in plant form in the Triticeae crops. Plant Cell Physiol. 2011;52(5):738–49. https://doi.org/10.1093/pcp/pcr025.
3. Soriano JM, Villegas D, Aranzana MJ, García del Moral LF, Royo C. Genetic structure of modern durum wheat cultivars and mediterranean landraces matches with their agronomic performance. PLoS One. 2016;11(8):e0160983. https://doi.org/10.1371/journal.pone.0160983.
4. Boukid F, Folloni S, Sforza S, Vittadini E, Prandi B. Current trends in ancient grains-based foodstuffs: insights into nutritional aspects and technological applications. Compr Rev Food Sci Food Saf. 2018;17(1):123–36. https://doi.org/10.1111/1541-4337.12315.
5. Lev-Yadun S, Gopher A, Abbo S. The cradle of agriculture. Science. 2000;288(5471):1602–3. https://doi.org/10.1126/science.288.5471.1602.
6. Zoccatelli G, Sega M, Bolla M, Cecconi D, Vaccino P, Rizzi C, et al. Expression of α-amylase inhibitors in diploid *Triticum* species. Food Chem. 2012;135(4):2643–9. https://doi.org/10.1016/j.foodchem.2012.06.123.
7. Matsuoka Y, Nasuda S. Durum wheat as a candidate for the unknown female progenitor of bread wheat: An empirical study with a highly fertile F1 hybrid with *Aegilops tauschii* Coss. Theor Appl Genet. 2004;109(8):1710–7. https://doi.org/10.1007/s00122-004-1806-6.
8. Dvorak J, Deal KR, Luo MC, You FM, Von Borstel K, Dehghani H. The origin of spelt and free-threshing hexaploid wheat. J Hered. 2012;103(3):426–41. https://doi.org/10.1093/jhered/esr152.

9. Heun M, Schaefer-Pregl R, Klawan D, Castagna R, Accerbi M, Borghi B, et al. Site of einkorn wheat domestication identified by DNA fingerprinting. Science. 1997;278(5341):1312–4. https://doi.org/10.1126/science.278.5341.1312.
10. Salse J, Chagué V, Bolot S, Magdelenat G, Huneau C, Pont C, et al. New insights into the origin of the B genome of hexaploid wheat: evolutionary relationships at the SPA genomic region with the S genome of the diploid relative *Aegilops speltoides*. BMC Genomics. 2008;9:555–67. https://doi.org/10.1186/1471-2164-9-555.
11. El Baidouri M, Murat F, Veyssiere M, Molinier M, Flores R, Burlot L, et al. Reconciling the evolutionary origin of bread wheat (*Triticum aestivum*). New Phytol. 2017;213(3):1477–86. https://doi.org/10.1111/nph.14113.
12. Luo MC, Yang ZL, You FM, Kawahara T, Waines JG, Dvorak J. The structure of wild and domesticated emmer wheat populations, gene flow between them, and the site of emmer domestication. Theor Appl Genet. 2007;114(6):947–59. https://doi.org/10.1007/s00122-006-0474-0.
13. Feldman M, Kislev ME. Domestication of emmer wheat and evolution of free-threshing tetraploid wheat. Isr J Plant Sci. 2007;55(3–4):207–21. https://doi.org/10.1560/IJPS.55.3-4.207.
14. Shewry PR. Do ancient types of wheat have health benefits compared with modern bread wheat? J Cereal Sci. 2018;79:469–76. https://doi.org/10.1016/j.jcs.2017.11.010.
15. McFadden ES, Sears ER. The origin of *triticum spelta* and its free-threshing hexaploid relatives. J Hered. 1946;37(3):81–9. https://doi.org/10.1093/oxfordjournals.jhered.a105590.
16. Marcussen T, Sandve S, Heier L, Spannagl M, Pfeifer M, Jakobsen K, et al. Ancient hybridizations among the ancestral genomes of bread wheat. Science. 2014;345:1250092. https://doi.org/10.1126/science.1250092.
17. Blatter RH, Jacomet S, Schlumbaum A. About the origin of European spelt (*Triticum spelta* L.): allelic differentiation of the HMW Glutenin B1–1 and A1–2 subunit genes. Theor Appl Genet. 2004;108(2):360–7. https://doi.org/10.1007/s00122-003-1441-7.
18. Khlestkina EK, Röder MS, Grausgruber H, Börner A. A DNA fingerprinting-based taxonomic allocation of Kamut wheat. Plant Genet Resour. 2006;4(3):172–80. https://doi.org/10.1079/PGR2006120.
19. Michalcová V, Dušinský R, Sabo M, Al Beyroutiová M, Hauptvogel P, Ivaničová Z, et al. Taxonomical classification and origin of Kamut® wheat. Plant Syst Evol. 2014;300(7):1749–57. https://doi.org/10.1007/s00606-014-1001-4.
20. Bordoni A, Danesi F, Di Nunzio M, Taccari A, Valli V. Ancient wheat and health: a legend or the reality? A review on KAMUT khorasan wheat. Int J Food Sci Nutr. 2017;68(3):278–86. https://doi.org/10.1080/09637486.2016.1247434.
21. Willett WC, Sacks F, Trichopoulou A, Drescher G, Ferro-Luzzi A, Helsing E, et al. Mediterranean diet pyramid: a cultural model for healthy eating. Am J Clin Nutr. 1995;61(6):1402S–6S. https://doi.org/10.1093/ajcn/61.6.1402S.
22. D'Alessandro A, De Pergola G. Mediterranean diet pyramid: a proposal for Italian people. Nutrients. 2014;6(10):4302–16. https://doi.org/10.3390/nu6104302.
23. van der Kamp JW, Poutanen K, Seal CJ, Richardson DP. The HEALTHGRAIN definition of 'whole grain'. Food Nutr Res. 2014;58(1):22100. https://doi.org/10.3402/fnr.v58.22100.
24. Longin CFH, Ziegler J, Schweiggert R, Koehler P, Carle R, Wuerschum T. Comparative study of hulled (einkorn, emmer, and spelt) and naked wheats (durum and bread wheat): agronomic performance and quality traits. Crop Sci. 2015;56:302–11. https://doi.org/10.2135/cropsci2015.04.0242.
25. Shewry PR, Hey S. Do "ancient" wheat species differ from modern bread wheat in their contents of bioactive components? J Cereal Sci. 2015;65:236–43. https://doi.org/10.1016/j.jcs.2015.07.014.
26. Arzani A, Ashraf M. Cultivated ancient wheats (*Triticum* spp.): a potential source of health-beneficial food products. Compr Rev Food Sci Food Saf. 2017;16(3):477–88. https://doi.org/10.1111/1541-4337.12262.

27. Hammed A, Simsek S. REVIEW: hulled wheats: a review of nutritional properties and processing methods. Cereal Chem. 2014;91:97–104. https://doi.org/10.1094/CCHEM-09-13-0179-RW.
28. Dinu M, Whittaker A, Pagliai G, Benedettelli S, Sofi F. Ancient wheat species and human health: biochemical and clinical implications. J Nutr Biochem. 2018;52:1–9. https://doi.org/10.1016/j.jnutbio.2017.09.001.
29. Authority EFS. Dietary reference values for nutrients summary report. EFSA Support Pub. 2017;14(12):e15121E. https://doi.org/10.2903/sp.efsa.2017.e15121.
30. Gebruers K, Dornez E, Boros D, Fraś A, Dynkowska W, Bedo Z, et al. Variation in the content of dietary fiber and components thereof in wheats in the HEALTHGRAIN diversity screen. J Agric Food Chem. 2008;56(21):9740–9. https://doi.org/10.1021/jf800975w.
31. Messia MC, Candigliota T, De Arcangelis E, Marconi E. Arabinoxylans and β-glucans assessment in cereals. Ital J Food Sci. 2016;29(1). https://doi.org/10.14674/1120-1770/ijfs.v573.
32. Løje H, Møller B, Laustsen AM, Hansen Å. Chemical composition, functional properties and sensory profiling of einkorn (*Triticum monococcum* L.). J Cereal Sci. 2003;37(2):231–40. https://doi.org/10.1006/jcrs.2002.0498.
33. Shewry PR, Piironen V, Lampi AM, Edelmann M, Kariluoto S, Nurmi T, et al. The HEALTHGRAIN wheat diversity screen: effects of genotype and environment on phytochemicals and dietary fiber components. J Agric Food Chem. 2010;58(17):9291–8. https://doi.org/10.1021/jf100039b.
34. Mandak E, Nyström L. Steryl ferulates, bioactive compounds in cereal grains. Lipid Technol. 2012;24(4):80–2. https://doi.org/10.1002/lite.201200179.
35. Li L, Shewry PR, Ward JL. Phenolic acids in wheat varieties in the HEALTHGRAIN diversity screen. J Agric Food Chem. 2008;56(21):9732–9. https://doi.org/10.1021/jf801069s.
36. Abdel-Aal E-SM, Rabalski I. Bioactive compounds and their antioxidant capacity in selected primitive and modern wheat species. Open Agric. 2008;2(1):7–14. https://doi.org/10.2174/1874331500802010007.
37. Ziegler JU, Schweiggert RM, Würschum T, Longin CFH, Carle R. Lipophilic antioxidants in wheat (*Triticum* spp.): a target for breeding new varieties for future functional cereal products. J Func Foods. 2016;20:594–605. https://doi.org/10.1016/j.jff.2015.11.022.
38. Hidalgo A, Brandolini A, Pompei C, Piscozzi R. Carotenoids and tocols of einkorn wheat (*Triticum monococcum* ssp. *monococcum* L.). J Cereal Sci. 2006;44(2):182–93. https://doi.org/10.1016/j.jcs.2006.06.002.
39. Ziegler JU, Wahl S, Würschum T, Longin CFH, Carle R, Schweiggert RM. Lutein and lutein esters in whole grain flours made from 75 genotypes of 5 *Triticum* species grown at multiple sites. J Agric Food Chem. 2015;63(20):5061–71. https://doi.org/10.1021/acs.jafc.5b01477.
40. Lampi A-M, Nurmi T, Ollilainen V, Piironen V. Tocopherols and Tocotrienols in wheat genotypes in the HEALTHGRAIN diversity screen. J Agric Food Chem. 2008;56(21):9716–21. https://doi.org/10.1021/jf801092a.
41. Andersson AAM, Kamal-Eldin A, Fraś A, Boros D, Åman P. Alkylresorcinols in wheat varieties in the HEALTHGRAIN diversity screen. J Agric Food Chem. 2008;56(21):9722–5. https://doi.org/10.1021/jf8011344.
42. Ziegler JU, Steingass CB, Longin CFH, Würschum T, Carle R, Schweiggert RM. Alkylresorcinol composition allows the differentiation of *Triticum* spp. having different degrees of ploidy. J Cereal Sci. 2015;65:244–51. https://doi.org/10.1016/j.jcs.2015.07.013.
43. Piironen V, Edelmann M, Kariluoto S, Bedő Z. Folate in wheat genotypes in the HEALTHGRAIN diversity screen. J Agric Food Chem. 2008;56(21):9726–31. https://doi.org/10.1021/jf801066j.
44. Nurmi T, Nyström L, Edelmann M, Lampi A-M, Piironen V. Phytosterols in wheat genotypes in the HEALTHGRAIN diversity screen. J Agric Food Chem. 2008;56(21):9710–5. https://doi.org/10.1021/jf8010678.
45. Zhao FJ, Su YH, Dunham SJ, Rakszegi M, Bedo Z, McGrath SP, et al. Variation in mineral micronutrient concentrations in grain of wheat lines of diverse origin. J Cereal Sci. 2009;49(2):290–5. https://doi.org/10.1016/j.jcs.2008.11.007.

46. Suchowilska E, Wiwart M, Kandler W, Krska R. A comparison of macro- and microelement concentrations in the whole grain of four *Triticum* species. Plant Soil Environ. 2012;58:141–7. https://doi.org/10.17221/688/2011-PSE.
47. Hussain A, Larsson H, Kuktaite R, Johansson E. Mineral composition of organically grown wheat genotypes: contribution to daily minerals intake. Int J Environ Res Public Health. 2010;7(9):3442–56. https://doi.org/10.3390/ijerph7093442.
48. Geisslitz S, Wieser H, Scherf KA, Koehler P. Gluten protein composition and aggregation properties as predictors for bread volume of common wheat, spelt, durum wheat, emmer and einkorn. J Cereal Sci. 2018;83:204–12. https://doi.org/10.1016/j.jcs.2018.08.012.
49. Brandolini A, Hidalgo A, Moscaritolo S. Chemical composition and pasting properties of einkorn (*Triticum monococcum* L. subsp. *monococcum*) whole meal flour. J Cereal Sci. 2008;47(3):599–609. https://doi.org/10.1016/j.jcs.2007.07.005.
50. Scherf KA, Koehler P. Wheat and gluten: technological and health aspects. Ernahrungs Umschau. 2016;63(08):166–75. https://doi.org/10.4455/eu.2016.035.
51. Belton PS. On the elasticity of wheat gluten. J Cereal Sci. 1999;29(2):103–7. https://doi.org/10.1006/jcrs.1998.0227.
52. Ewart JA. A modified hypothesis for the structure and rheology of glutelins. J Sci Food Agric. 1972;23(6):687–99. https://doi.org/10.1002/jsfa.2740230604.
53. Cornec M, Popineau Y, Lefebvre J. Characterization of gluten subfractions by SE-HPLC and dynamic rheological analysis in shear. J Cereal Sci. 1994;19(2):131–9. https://doi.org/10.1006/jcrs.1994.1018.
54. Khatkar BS, Bell AE, Schofield JD. The dynamic rheological properties of glutens and gluten sub-fractions from wheats of good and poor bread making quality. J Cereal Sci. 1995;22(1):29–44. https://doi.org/10.1016/S0733-5210(05)80005-0.
55. Wesley IJ, Larroque O, Osborne BG, Azudin N, Allen H, Skerritt JH. Measurement of Gliadin and Glutenin content of flour by NIR spectroscopy. J Cereal Sci. 2001;34(2):125–33. https://doi.org/10.1006/jcrs.2001.0378.
56. Osborne TB. The proteins of the wheat kernel. Washington: Carnegie Institution; 1907.
57. Wieser H, Antes S, Seilmeier W. Quantitative determination of gluten protein types in wheat flour by reversed-phase high-performance liquid chromatography. Cereal Chem. 1998;75(5):644–50. https://doi.org/10.1094/CCHEM.1998.75.5.644.
58. Thanhaeuser SM, Wieser H, Koehler P. Spectrophotometric and fluorimetric quantitation of quality-related protein fractions of wheat flour. J Cereal Sci. 2015;62:58–65. https://doi.org/10.1016/j.jcs.2014.12.010.
59. Marti A, Augst E, Cox S, Koehler P. Correlations between gluten aggregation properties defined by the GlutoPeak test and content of quality-related protein fractions of winter wheat flour. J Cereal Sci. 2015;66:89–95. https://doi.org/10.1016/j.jcs.2015.10.010.
60. Plessis A, Ravel C, Bordes J, Balfourier F, Martre P. Association study of wheat grain protein composition reveals that gliadin and glutenin composition are trans-regulated by different chromosome regions. J Exp Bot. 2013;64(12):3627–44. https://doi.org/10.1093/jxb/ert188.
61. Ozuna CV, Barro F. Characterization of gluten proteins and celiac disease-related immunogenic epitopes in the *Triticeae*: cereal domestication and breeding contributed to decrease the content of gliadins and gluten. Mol Breed. 2018;38(3). https://doi.org/10.1007/s11032-018-0779-0.
62. Hajas L, Scherf KA, Török K, Bugyi Z, Schall E, Poms RE, et al. Variation in protein composition among wheat (*Triticum aestivum* L.) cultivars to identify cultivars suitable as reference material for wheat gluten analysis. Food Chem. 2018;267:387–94. https://doi.org/10.1016/j.foodchem.2017.05.005.
63. Thanhaeuser SM, Wieser H, Koehler P. Correlation of quality parameters with the baking performance of wheat flours. Cereal Chem. 2014;91(4):333–41. https://doi.org/10.1094/CCHEM-09-13-0194-CESI.
64. Wieser H, Mueller K-J, Koehler P. Studies on the protein composition and baking quality of einkorn lines. Eur Food Res Technol. 2009;229(3):523–32. https://doi.org/10.1007/s00217-009-1081-5.

65. Koenig A, Konitzer K, Wieser H, Koehler P. Classification of spelt cultivars based on differences in storage protein compositions from wheat. Food Chem. 2015;168:176–82. https://doi.org/10.1016/j.foodchem.2014.07.040.
66. Wieser H. Comparative investigations of gluten proteins from different wheat species. I. Qualitative and quantitative composition of gluten protein types. Eur Food Res Technol. 2000;211(4):262–8. https://doi.org/10.1007/s002170000165.
67. Dupont FM, Altenbach SB. Molecular and biochemical impacts of environmental factors on wheat grain development and protein synthesis. J Cereal Sci. 2003;38(2):133–46. https://doi.org/10.1016/S0733-5210(03)00030-4.
68. Geisslitz S, Longin FHC, Scherf AK, Koehler P. Comparative study on gluten protein composition of ancient (einkorn, emmer and spelt) and modern wheat species (durum and common wheat). Foods. 2019;8(9):409. https://doi.org/10.3390/foods8090409.
69. Shewry PR, Miflin BJ, Kasarda DD. The structural and evolutionary relationships of the prolamin storage proteins of barley, rye and wheat. Philos Trans R Soc B. 1984;304(1120):297. https://doi.org/10.1098/rstb.1984.0025.
70. Rombouts I, Lamberts L, Celus I, Lagrain B, Brijs K, Delcour JA. Wheat gluten amino acid composition analysis by high-performance anion-exchange chromatography with integrated pulsed amperometric detection. J Chromatogr A. 2009;1216(29):5557–62. https://doi.org/10.1016/j.chroma.2009.05.066.
71. Grosch W, Wieser H. Redox reactions in wheat dough as affected by ascorbic acid. J Cereal Sci. 1999;29(1):1–16. https://doi.org/10.1006/jcrs.1998.0218.
72. Qi PF, Wei YM, Yue YW, Yan ZH, Zheng YL. Biochemical and molecular characterization of gliadins. Mol Biol. 2006;40(5):713–23. https://doi.org/10.1134/S0026893306050050.
73. Wieser H. Chemistry of gluten proteins. Food Microbiol. 2007;24(2):115–9. https://doi.org/10.1016/j.fm.2006.07.004.
74. Wang D-W, Li D, Wang J, Zhao Y, Wang Z, Yue G, et al. Genome-wide analysis of complex wheat gliadins, the dominant carriers of celiac disease epitopes. Sci Rep. 2017;7(1):44609. https://doi.org/10.1038/srep44609.
75. Dubois B, Bertin P, Mingeot D. Molecular diversity of α-gliadin expressed genes in genetically contrasted spelt (*Triticum aestivum* ssp. *spelta*) accessions and comparison with bread wheat (*T. aestivum* ssp. *aestivum*) and related diploid *Triticum* and *Aegilops* species. Mol Breed. 2016;36(11):152. https://doi.org/10.1007/s11032-016-0569-5.
76. Wieser H. Comparative investigations of gluten proteins from different wheat species. III. N-terminal amino acid sequences of α-gliadins potentially toxic for coeliac patients. Eur Food Res Technol. 2001;213(3):183–6. https://doi.org/10.1007/s002170100365.
77. Juhász A, Belova T, Florides CG, Maulis C, Fischer I, Gell G, et al. Genome mapping of seed-borne allergens and immunoresponsive proteins in wheat. Sci Adv. 2018;4(8):eaar8602. https://doi.org/10.1126/sciadv.aar8602.
78. Salentijn EMJ, Goryunova SV, Bas N, van der Meer IM, van den Broeck HC, Bastien T, et al. Tetraploid and hexaploid wheat varieties reveal large differences in expression of alpha-gliadins from homoeologous Gli-2 loci. BMC Genomics. 2009;10(1):48. https://doi.org/10.1186/1471-2164-10-48.
79. Huang Z, Long H, Wei Y-M, Yan Z-H, Zheng Y-L. Allelic variations of α-gliadin genes from species of *Aegilops* section Sitopsis and insights into evolution of α-gliadin multigene family among *Triticum* and *Aegilops*. Genetica. 2016;144(2):213–22. https://doi.org/10.1007/s10709-016-9891-4.
80. Ma ZC, Wei YM, Yan ZH, Zheng YL. Characterization of α-gliadin genes from diploid wheats and the comparative analysis with those from polyploid wheats. Russ J Genet. 2007;43(11):1286–93. https://doi.org/10.1134/S1022795407110117.
81. Shewry PR, Tatham AS. Disulphide bonds in wheat gluten proteins. J Cereal Sci. 1997;25(3):207–27. https://doi.org/10.1006/jcrs.1996.0100.
82. Singh NK, Shepherd KW. Linkage mapping of genes controlling endosperm storage proteins in wheat. Theor Appl Genet. 1988;75(4):628–41. https://doi.org/10.1007/BF00289132.

83. Ibba MI, Kiszonas AM, Guzmán C, Morris CF. Definition of the low molecular weight glutenin subunit gene family members in a set of standard bread wheat (*Triticum aestivum* L.) varieties. J Cereal Sci. 2017;74:263–71. https://doi.org/10.1016/j.jcs.2017.02.015.
84. Gupta RB, Shepherd KW. Two-step one-dimensional SDS-PAGE analysis of LMW subunits of glutelin. Theor Appl Genet. 1990;80(1):65–74. https://doi.org/10.1007/BF00224017.
85. Liu L, Ikeda TM, Branlard G, Peña RJ, Rogers WJ, Lerner SE, et al. Comparison of low molecular weight glutenin subunits identified by SDS-PAGE, 2-DE, MALDI-TOF-MS and PCR in common wheat. BMC Plant Biol. 2010;10(1):124. https://doi.org/10.1186/14712229-10-124.
86. Rodríguez-Quijano M, Nieto-Taladriz MT, Carrillo JM. Variation in B-LMW glutenin subunits in einkorn wheats. Genet Resour Crop Evol. 1997;44(6):539–43. https://doi.org/10.1023/A:1008691524039.
87. Lee YK, Bekes F, Gupta R, Appels R, Morell MK. The low-molecular-weight glutenin subunit proteins of primitive wheats. I. Variation in A-genome species. Theor Appl Genet. 1999;98(1):119–25. https://doi.org/10.1007/s001220051048.
88. Pflüger LA, Martín LM, Alvarez JB. Variation in the HMW and LMW glutenin subunits from Spanish accessions of emmer wheat (*Triticum turgidum* ssp. *Dicoccum* Schrank). Theor Appl Genet. 2001;102(5):767–72. https://doi.org/10.1007/s001220051708.
89. Caballero L, Martín LM, Alvarez JB. Genetic variability of the low-molecular-weight glutenin subunits in spelt wheat (*Triticum aestivum* ssp. *spelta* L. em Thell.). Theor Appl Genet. 2004;108(5):914–9. https://doi.org/10.1007/s00122-003-1501-z.
90. Cuesta S, Alvarez JB, Guzmán C. Identification and molecular characterization of novel LMW-m and -s glutenin genes, and a chimeric -m/–i glutenin gene in 1A chromosome of three diploid *Triticum* species. J Cereal Sci. 2017;74:46–55. https://doi.org/10.1016/j.jcs.2017.01.010.
91. An X, Zhang Q, Yan Y, Li Q, Zhang Y, Wang A, et al. Cloning and molecular characterization of three novel LMW-i glutenin subunit genes from cultivated einkorn (*Triticum monococcum* L.). Theor Appl Genet. 2006;113(3):383–95. https://doi.org/10.1007/s00122-006-0299-x.
92. Baar A, Pahr S, Constantin C, Scheiblhofer S, Thalhamer J, Giavi S, et al. Molecular and immunological characterization of tri a 36, a low molecular weight Glutenin, as a novel major wheat food allergen. J Immunol. 2012;189(6):3018–25. https://doi.org/10.4049/jimmunol.1200438.
93. Qin L, Liang Y, Yang D, Sun L, Xia G, Liu S. Novel LMW Glutenin subunit genes from wild emmer wheat (*Triticum turgidum* ssp. *dicoccoides*) in relation to Glu-3 evolution. Dev Genes Evol. 2015;225(1):31–7. https://doi.org/10.1007/s00427-014-0484-x.
94. Li X, Wang A, Xiao Y, Yan Y, He Z, Appels R, et al. Cloning and characterization of a novel low molecular weight glutenin subunit gene at the Glu-A3 locus from wild emmer wheat (*Triticum turgidum* L. var. *dicoccoides)*. Euphytica. 2008;159(1):181–90. https://doi.org/10.1007/s10681-007-9471-x.
95. Payne PI, Lawrence GJ. Catalogue of alleles for the complex gene loci, Glu-A1, Glu-B1, and Glu-D1 which code for high-molecular-weight subunits of glutenin in hexaploid wheat. Cereal Res Commun. 1983;11(1):29–35.
96. Payne PI, Holt LM, Law CN. Structural and genetical studies on the high-molecular-weight subunits of wheat glutenin. Theor Appl Genet. 1981;60(4):229–36. https://doi.org/10.1007/BF02342544.
97. Payne PI, Corfield KG, Holt LM, Blackman JA. Correlations between the inheritance of certain high-molecular weight subunits of glutenin and bread-making quality in progenies of six crosses of bread wheat. J Sci Food Agric. 1981;32(1):51–60. https://doi.org/10.1002/jsfa.2740320109.
98. Jiang P, Xue J, Duan L, Gu Y, Mu J, Han S, et al. Effects of high-molecular-weight glutenin subunit combination in common wheat on the quality of crumb structure. J Sci Food Agric. 2019;99(4):1501–8. https://doi.org/10.1002/jsfa.9323.
99. Shewry PR, Halford NG, Tatham AS. High molecular weight subunits of wheat glutenin. J Cereal Sci. 1992;15(2):105–20. https://doi.org/10.1016/S0733-5210(09)80062-3.

100. Waines JG, Payne PI. Electrophoretic analysis of the high-molecular-weight glutenin subunits of *Triticum monococcum*, *T. urartu*, and the a genome of bread wheat (*T. aestivum*). Theor Appl Genet. 1987;74(1):71–6. https://doi.org/10.1007/BF00290086.
101. Ciaffi M, Dominica L, Lafiandra D. High molecular weight glutenin subunit variation in wild and cultivated einkorn wheats (*Triticum* spp., Poaceae). Plant Syst Evol. 1998;209(1–2):123–37. https://doi.org/10.1007/BF00991528.
102. Vallega V, Waines JG. High molecular weight glutenin subunit variation in *Triticum turgidum* var. *dicoccum*. Theor Appl Genet. 1987;74(6):706–10. https://doi.org/10.1007/BF00247545.
103. Rodríguez-Quijano M, Vázquez JF, Carrillo JM. Variation of high molecular weight glutenin subunits in Spanish landraces of *Triticum aestivum* ssp. *vulgare* and ssp. *spelta*. J Genet Breed. 1990;44(2):121–6.
104. Caballero L, Martin LM, Alvarez JB. Allelic variation of the HMW glutenin subunits in spanish accessions of spelt wheat (*Triticum aestivum* ssp. *spelta* L. em. Thell.). Theor Appl Genet. 2001;103(1):124–8. https://doi.org/10.1007/s001220100565.
105. Xu L-L, Li W, Wei Y-M, Zheng Y-L. Genetic diversity of HMW glutenin subunits in diploid, tetraploid and hexaploid Triticum species. Genet Resour Crop Evol. 2009;56(3):377–91. https://doi.org/10.1007/s10722-008-9373-3.
106. Li HY, Li ZL, Zeng XX, Zhao LB, Chen G, Kou CL, et al. Molecular characterization of different *Triticum monococcum* ssp. *monococcum* Glu-A1mx alleles. Cereal Res Commun. 2016;44(3):444. https://doi.org/10.1556/0806.44.2016.006.
107. Guo X-H, Wu B-H, Hu X-G, Bi Z-G, Wang Z-Z, Liu D-C, et al. Molecular characterization of two y-type high molecular weight glutenin subunit alleles 1Ay12_{*} and 1Ay8_{*} from cultivated einkorn wheat (*Triticum monococcum* ssp. *monococcum*). Gene. 2013;516(1):1–7. https://doi.org/10.1016/j.gene.2012.12.037.
108. Li Z, Li H, Chen G, Kou C, Ning S, Yuan Z, et al. Characterization of a novel y-type HMW-GS with eight cysteine residues from *Triticum monococcum* ssp. *monococcum*. Gene. 2015;573(1):110–4. https://doi.org/10.1016/j.gene.2015.07.040.
109. Jin M, Xie ZZ, Ge P, Li J, Jiang SS, Subburaj S, et al. Identification and molecular characterisation of HMW glutenin subunit 1By16* in wild emmer. Theor Appl Genet. 2012;53(3):249–58. https://doi.org/10.1007/s13353-012-0101-5.
110. Margiotta B, Colaprico G, Urbano M. Polymorphism of high Mr glutenin subunits in wild emmer *Triticum turgidum* subsp. *dicoccoides*: chromatographic, electrophoretic separations and PCR analysis of their encoding genes. Genet Resour Crop Evol. 2014;61(2):331–43. https://doi.org/10.1007/s10722-013-0037-6.
111. Shewry PR, Halford NG, Belton PS, Tatham AS. The structure and properties of gluten: an elastic protein from wheat grain. Philos Trans R Soc Lond Ser B Biol Sci. 2002;357(1418):133–42. https://doi.org/10.1098/rstb.2001.1024.
112. Wrigley CW. Biopolymers: giant proteins with flour power. Nature. 1996;381(6585):738–9. https://doi.org/10.1038/381738a0.
113. Li W, Yang B, Shao Q, Xu F, Yan S. Glutenin macropolymer particles size distribution of six wheat varieties in eastern China. Wuhan Univ J Nat Sci. 2017;22(5):455–60. https://doi.org/10.1007/s11859-017-1272-z.
114. Don C, Lichtendonk W, Plijter JJ, Hamer RJ. Glutenin macropolymer: a gel formed by Glutenin particles. J Cereal Sci. 2003;37(1):1–7. https://doi.org/10.1006/jcrs.2002.0481.
115. Weegels PL, Hamer RJ, Schonfield JD. Functional properties of wheat glutenin. J Cereal Sci. 1996;23(1):1–18. https://doi.org/10.1006/jcrs.1996.0001.
116. Mueller E, Wieser H, Koehler P. Preparation and chemical characterisation of glutenin macropolymer (GMP) gel. J Cereal Sci. 2016;70:79–84. https://doi.org/10.1016/j.jcs.2016.05.021.
117. Gupta RB, Khan K, Macritchie F. Biochemical basis of flour properties in bread wheats. I. Effects of variation in the quantity and size distribution of polymeric protein. J Cereal Sci. 1993;18(1):23–41. https://doi.org/10.1006/jcrs.1993.1031.
118. Scherf KA. Immunoreactive cereal proteins in wheat allergy, non-celiac gluten/wheat sensitivity (NCGS) and celiac disease. Curr Opin Food Sci. 2019;25:35–41. https://doi.org/10.1016/j.cofs.2019.02.003.

119. Spisni E, Imbesi V, Giovanardi E, Petrocelli G, Alvisi P, Valerii MC. Differential physiological responses elicited by ancient and heritage wheat cultivars compared to modern ones. Nutrients. 2019;11:2879. https://doi.org/10.3390/nu11122879.
120. Koehler P, Wieser H, Konitzer K. Chapter 1 - celiac disease—a complex disorder. In: Koehler P, Wieser H, Konitzer K, editors. Celiac disease and gluten. Boston: Academic; 2014. p. 1–96. https://doi.org/10.1016/b978-0-12-420220-7.00001-8.
121. Singh P, Arora A, Strand TA, Leffler DA, Catassi C, Green PH, et al. Global prevalence of celiac disease: systematic review and meta-analysis. Clin Gastroenterol Hepatol. 2018;16(6):823–36.e2. https://doi.org/10.1016/j.cgh.2017.06.037.
122. Sollid LM, Tye-Din JA, Qiao S-W, Anderson RP, Gianfrani C, Koning F. Update 2020: nomenclature and listing of celiac disease–relevant gluten epitopes recognized by CD4+ T cells. Immunogenetics. 2020;72(1):85–8. https://doi.org/10.1007/s00251-019-01141-w.
123. van den Broeck H, de Jong H, Salentijn EJ, Dekking L, Bosch D, Hamer R, et al. Presence of celiac disease epitopes in modern and old hexaploid wheat varieties: wheat breeding may have contributed to increased prevalence of celiac disease. Theor Appl Genet. 2010;121(8):1527–39. https://doi.org/10.1007/s00122-010-1408-4.
124. Van den Broeck H, Chen H, Lacaze X, Dusautoir J-C, Gilissen L, Smulders MJM, et al. In search of tetraploid wheat accessions reduced in celiac disease-related gluten epitopes. Mol BioSyst. 2010;6:2206–13. https://doi.org/10.1039/c0mb00046a.
125. Sievers S, Rohrbach A, Beyer K. Wheat-induced food allergy in childhood: ancient grains seem no way out. Eur J Nutr. 2019. https://doi.org/10.1007/s00394-019-02116-z.
126. Gélinas P, McKinnon C. Gluten weight in ancient and modern wheat and the reactivity of epitopes towards R5 and G12 monoclonal antibodies. Int J Food Sci Technol. 2016;51(8):1801–10. https://doi.org/10.1111/ijfs.13151.
127. Schopf M, Scherf KA. Wheat cultivar and species influence variability of gluten ELISA analyses based on polyclonal and monoclonal antibodies R5 and G12. J Cereal Sci. 2018;83:32–41. https://doi.org/10.1016/j.jcs.2018.07.005.
128. Grausgruber H, Štěrbová L, Thrackl K, Bradová J, Baumgartner S, Dvořáček V, editors. Content of the immunodominant 33-mer peptide from a2-gliadin in common and ancient wheat flours determined by the G12 sandwich ELISA. 13th International Gluten Workshop, Mexiko; 2018.
129. Schalk K, Lang C, Wieser H, Koehler P, Scherf KA. Quantitation of the immunodominant 33-mer peptide from α-gliadin in wheat flours by liquid chromatography tandem mass spectrometry. Sci Rep. 2017;7:45092. https://doi.org/10.1038/srep45092.
130. Ribeiro M, Rodriguez-Quijano M, Nunes FM, Carrillo JM, Branlard G, Igrejas G. New insights into wheat toxicity: breeding did not seem to contribute to a prevalence of potential celiac disease's immunostimulatory epitopes. Food Chem. 2016;213:8–18. https://doi.org/10.1016/j.foodchem.2016.06.043.
131. Prandi B, Tedeschi T, Folloni S, Galaverna G, Sforza S. Peptides from gluten digestion: a comparison between old and modern wheat varieties. Food Res Int. 2017;91:92–102. https://doi.org/10.1016/j.foodres.2016.11.034.
132. Dubois B, Bertin P, Hautier L, Muhovski Y, Escarnot E, Mingeot D. Genetic and environmental factors affecting the expression of α-gliadin canonical epitopes involved in celiac disease in a wide collection of spelt (*Triticum aestivum* ssp. *spelta*) cultivars and landraces. BMC Plant Biol. 2018;18(1):262. https://doi.org/10.1186/s12870-018-1487-y.
133. Schuppan D, Pickert G, Ashfaq-Khan M, Zevallos V. Non-celiac wheat sensitivity: differential diagnosis, triggers and implications. Best Pract Res Clin Gastroenterol. 2015;29(3):469–76. https://doi.org/10.1016/j.bpg.2015.04.002.
134. Tanveer M, Ahmed A. Non-celiac gluten sensitivity: a systematic review. J Coll Physicians Surg Pak. 2019;29(1):51–7. https://doi.org/10.29271/jcpsp.2019.01.51.
135. Zevallos VF, Raker V, Tenzer S, Jimenez-Calvente C, Ashfaq-Khan M, Ruessel N, et al. Nutritional wheat amylase-trypsin inhibitors promote intestinal inflammation via activation of myeloid cells. Gastroenterology. 2017;152(5):1100–13.e12. https://doi.org/10.1053/j.gastro.2016.12.006.

136. Junker Y, Zeissig S, Kim SJ, Barisani D, Wieser H, Leffler DA, et al. Wheat amylase trypsin inhibitors drive intestinal inflammation via activation of toll-like receptor 4. J Exp Med. 2012;209(13):2395–408. https://doi.org/10.1084/jem.20102660.
137. Zevallos VF, Raker VK, Maxeiner J, Scholtes P, Steinbrink K, Schuppan D. Dietary wheat amylase trypsin inhibitors exacerbate murine allergic airway inflammation. Eur J Nutr. 2018. https://doi.org/10.1007/s00394-018-1681-6.
138. Capocchi A, Muccilli V, Cunsolo V, Saletti R, Foti S, Fontanini D. A heterotetrameric alpha-amylase inhibitor from emmer (*Triticum dicoccon* Schrank) seeds. Phytochemistry. 2013;88:6–14. https://doi.org/10.1016/j.phytochem.2012.12.010.
139. Fontanini D, Capocchi A, Muccilli V, Saviozzi F, Cunsolo V, Saletti R, et al. Dimeric inhibitors of human salivary α-amylase from emmer (*Triticum dicoccon* Schrank) seeds. J Agric Food Chem. 2007;55(25):10452–60. https://doi.org/10.1021/jf071739w.
140. Gélinas P, Gagnon F. Inhibitory activity towards human α-amylase in wheat flour and gluten. Int J Food Sci Technol. 2018;53(2):467–74. https://doi.org/10.1111/ijfs.13605.
141. Call L, Kapeller M, Grausgruber H, Reiter E, Schoenlechner R, D'Amico S. Effects of species and breeding on wheat protein composition. J Cereal Sci. 2020. https://doi.org/10.1016/j.jcs.2020.102974.
142. Bose U, Juhász A, Broadbent JA, Byrne K, Howitt CA, Colgrave ML. Identification and quantitation of amylase trypsin inhibitors across cultivars representing the diversity of bread wheat. J Proteome Res. 2020. https://doi.org/10.1021/acs.jproteome.0c00059.
143. Geisslitz S, Ludwig C, Scherf KA, Koehler P. Targeted LC–MS/MS reveals similar contents of α-amylase/trypsin-inhibitors as putative triggers of nonceliac gluten sensitivity in all wheat species except einkorn. J Agric Food Chem. 2018;66(46):12395–403. https://doi.org/10.1021/acs.jafc.8b04411.
144. Cuccioloni M, Mozzicafreddo M, Bonfili L, Cecarini V, Giangrossi M, Falconi M, et al. Interfering with the high-affinity interaction between wheat amylase trypsin inhibitor CM3 and toll-like receptor 4: in silico and biosensor-based studies. Sci Rep. 2017;7(1):13169. https://doi.org/10.1038/s41598-017-13709-1.
145. Dieterich W, Zopf Y. Gluten and FODMAPS—sense of a restriction/when is restriction necessary? Nutrients. 2019;11:1957. https://doi.org/10.3390/nu11081957.
146. Ziegler JU, Steiner D, Longin CFH, Würschum T, Schweiggert RM, Carle R. Wheat and the irritable bowel syndrome – FODMAP levels of modern and ancient species and their retention during bread making. J Funct Foods. 2016;25:257–66. https://doi.org/10.1016/j.jff.2016.05.019.
147. Dieterich W, Schuppan D, Schink M, Schwappacher R, Wirtz S, Agaimy A, et al. Influence of low FODMAP and gluten-free diets on disease activity and intestinal microbiota in patients with non-celiac gluten sensitivity. Clin Nutr. 2019;38(2):697–707. https://doi.org/10.1016/j.clnu.2018.03.017.

Chapter 12
Safety of Cereals in the Mediterranean: An Update on EU Legislation

Federica Cheli, Francesca Fumagalli, Matteo Ottoboni, and Luciano Pinotti

Abstract Cereals are still by far the world's most important source of food, both for direct human consumption and indirectly, as inputs to livestock production. Cereals, vegetables, and citrus fruits account for over 85% of the Mediterranean's total agricultural production. Cereals undergo several and different primary and secondary processing procedures. As a result, highly processed final products from cereals are used in cuisine for human nutrition. Cereal contamination has an important impact on human health. The European Union has established the most comprehensive regulations for the safety of cereal and cereal derived products to protect consumer health and facilitate world trade. This paper reviews the existing European Union legislation associated with cereal safety, with a focus on mycotoxin, heavy metal and pesticide contamination. A synoptic presentation of the main legal acts related to cereal safety is given, and the main points of each law are reported. Moreover, data regarding the occurrence, in the Mediterranean area, of mycotoxins, heavy metals and pesticide residues in cereals are reported.

Keywords Cereal safety · Mycotoxin · Heavy metal · Pesticides · Legistlation

F. Cheli (✉) · L. Pinotti
Department of Health, Animal Science and Food Safety, Università degli Studi di Milano, Milan, Italy

CRC I-WE (Coordinating Research Centre: Innovation for Well-Being and Environment), Università degli Studi di Milano, Milan, Italy
e-mail: federica.cheli@unimi.it; luciano.pinotti@unimi.it

F. Fumagalli · M. Ottoboni
Department of Health, Animal Science and Food Safety, Università degli Studi di Milano, Milan, Italy
e-mail: francesca.fumagalli1@unimi.it; matteo.ottoboni@unimi.it

F. Boukid (ed.), *Cereal-Based Foodstuffs: The Backbone of Mediterranean Cuisine*, https://doi.org/10.1007/978-3-030-69228-5_12

1 Introduction

Cereals are the world's most important source of food, both for direct human consumption and, indirectly, as inputs to livestock production [1]. In humans, about 80% of the protein and more than 50% percent of the daily energy intake come from cereals [2]. In 2019, FAO estimated a world cereal production at nearly 2.721 million tons (Mt), 2.4% higher than in 2018 [3]. Prospects for 2020 are difficult to calculate, as the effects of the COVID-19 pandemic on international cereal prize and trading could negatively influence crop-sowing decisions [4].

Ensuring that such high volumes of grain products are conformed to adequate quality and safety standards is a major undertaking of the European Union (EU) and world legislation. Moreover, it is necessary to ensure the grain safety all along the cereal chain at every stage, from production to processing and putting on the market. Cereal contamination can be heterogeneous including biological, chemical and physical contaminants [5]. In terms of food safety, mycotoxins, heavy metals and pesticide residues are among the most potential harmful for human and animal health [1, 6]. The knowledge and control of contaminants in cereals are a worldwide objective of producers, manufacturers, regulatory agencies and researchers due to the high economic and sanitary impacts on the food chain. Since it is impossible to fully eliminate the presence of contaminants, ensuring cereal safety means: (1) setting maximum limits to reduce the risk related to food consumption, (2) performing adequate analysis to ensure quality and safety standards for raw materials for direct consumption or industrial processes, and (3) regularly update the legislative framework to face the continuous evolution of the regulatory aspects at national and international level.

The aim of this paper is to provide an updated compilation of EU legislation associated with cereal safety, with a focus on mycotoxin, heavy metal and pesticide contamination.

2 Cereals in the Mediterranean Area

Cereals, vegetables, and citrus fruits account for over 85% of the Mediterranean's total agricultural production [7]. Last COCERAL (European Association of cereal, rice, feedstuffs, oilseeds, olive oil, oils and fats and agro supply trade) forecast for the 2020 crop sees the total grain crop in the EU-27 + UK at 299.2 Mt [8]. In the EU Mediterranean partner countries (Algeria, Egypt, Israel, Jordan, Lebanon, Morocco, Occupied Palestinian territory, Syria and Tunisia) cereal production accounted for 32 Mt, representing the 33% of the main cultivated crops [9]. From an agronomical perspective, the Mediterranean regions are mainly characterized by semiarid and sub-humid areas, and water availability represents the major constrain. In the last years, in the Mediterranean region, significant crop yield losses have been reported due to severe and repeated drought events during the growing season [10]. In terms of cereal production, sustainability, security and safety, cereal management must include efficient strategies to increase soil water accumulation, reduce runoff and soil evaporation losses. To improve crop production in the arid Mediterranean climate, the main strategic techniques are focused on agronomical and breeding strategies, including advanced irrigation systems, choice and selection of cereal species and varieties, which enable a better use of rainwater and rationalize fertilization [11]. In this scenario, the impact of climate change must be considered since the global warming will decrease the production yield of crops [12]. Besides the impact on crop production, short term and long-term patterns of climate are expected to affect plant pathogens, associated diseases and related food safety hazard [13, 14].

3 Cereal Contaminants

In the Council Regulation (EEC) No 315/93, the definition of contaminant is given: *"Contaminant means any substance not intentionally added to food which is present in such food as a result of the production (including operations carried out in crop husbandry, animal husbandry and veterinary medicine), manufacture, processing, preparation, treatment, packing, packaging, transport or holding of such food, or as a result of environmental contamination"* [15].

In terms of ensuring the highest possible level of safety of the food chain and compliance with EU food and feed legislation, The Rapid Alert System for Food and Feed (RASFF) (http://ec.europa.eu/food/food/rapidalert/index_en.htm) was launched in 1979. The Regulation (EC) 178/2002 established RASFF as a network involving the Member States, the Commission and the European Food Safety Authority (EFSA) [16]. The RASFF is a tool to exchange information between competent authorities on consignments of food and feed in cases, where a risk to human and animal health has been identified and measures have been taken. In 2018, 156 RASSF notifications have been reported for cereals and cereal products, 31% accounted for mycotoxins, pesticide residues and heavy metals [17]. An interesting factsheet on how food contaminants must be managed has been published by the European Commission: "Managing food contaminants: how the EU ensures that our food is safe" [18].

In the European Union and worldwide, contaminant maximum levels are laid down for cereal raw materials and final food products. Cereals undergo several and different processing steps, like primary processing (*e.g.* cleaning, debranning and milling) and secondary processing procedures (*e.g.* fermentation and thermal treatments). As a result, highly processed final products from cereals are used in cuisine for human nutrition. From a safety perspective, considering the levels indicated by the European legislation, results from literature indicate that sometimes the limits proposed for cereal-derived products may be not warranted by the limit for unprocessed cereals. The fate of contaminants during the primary and secondary processing of cereals are important topics to improve the framework of food safety. The milling process affects mycotoxin and heavy metal distribution, where the fractions rich in contaminants are eliminated from the food chain and commonly used as animal feed [6, 19]. However, these fractions may represent promising novel food ingredients with a high value for human nutrition, too. In the production of bread, bakery products, and pasta, secondary food processing procedures and the use of additives, different ingredients, fermentation and cooking can degrade, transform, bind or release mycotoxins [20]. For example, during the entire pasta production and cooking, deoxynivalenol (DON) can be reduced by an average value of 44% [21]. These topics are not the main focus of this paper since more details regarding mycotoxin repartitioning can be found in previously published literature [6, 20, 22–25]. However, the knowledge of the fate of contaminants during the processing of cereals and the development of improved processing technology are important topics to find effective mitigation strategies of contaminants in foods and strengthen the safety of the food chain.

3.1 Mycotoxin Occurrence

Mycotoxins are toxic secondary fungal metabolites that can cause a variety of adverse health effects in humans and animals, depending on the type of mycotoxin and the contamination levels [26, 27]. It has been estimated that up to 25% of the world's crops may be contaminated with mycotoxins [28, 29]. Cereal and cereal products produced worldwide can be found contaminated with mycotoxins produced by a variety of fungi that colonize crops in the field or post-harvest. The most important mycotoxins in cereals, because of their worldwide occurrence and health concern, are mainly Aflatoxins (AFs), *Fusarium* toxins, such as deoxynivalenol (DON), zearalenone (ZEN), nivalenol (NIV), fumonisins (FUM), T-2 and HT-2 toxins [6, 30–32]. As it is impossible to fully eliminate mycotoxin contamination at pre- and post-harvest level and cereal processing, several factors must be taken into account to manage the challenge of mycotoxins in cereals and in the food chain. Environmental factors and good agronomic/cultural practices are the main preharvest factors affecting mycotoxin contamination. Cereal and cereal product storage and industrial processing are the major critical points at post-harvest level. The strategies used for mycotoxin mitigation must not hinder grain quality and safety, and must comply with the existing regulatory requirements.

Numerous surveys on cereal mycotoxin occurrence in food and feed have been published [30–35]. From these surveys, several hot topic emerged. Extensive mycotoxin contamination in cereals has been reported in both developing and developed countries. In general, worldwide, FUM and DON are the top threat, with several samples exhibiting a co-occurrence of these mycotoxins. Differences in mycotoxin occurrence and concentration between geographical areas are uncontroversial, confirming that contamination is strongly dependent on regional climatic conditions. Within each geographical area, annual weather fluctuations, seasonal and local weather conditions during the crop-growing season are of great importance to explain the variation in mycotoxin occurrence. Environmental conditions, such as excessive moisture, temperature extremes, humidity, drought conditions, insect damage, crop systems and some agronomic practices can cause stress and predispose crops in the field to mould, and determine the severity of mycotoxin contamination [36–38]. The prevalence and the risk level of mycotoxin occurrence in cereals in the Mediterranean area is reported in Fig. 12.1.

Europe ranked as a moderate to severe risk region, with more than half of the samples testing above the risk threshold levels. In the Mediterranean area, the risk of mycotoxins in 2018 was 61%. In particular, the countries with the highest risk were Spain, Italy, Croatia, Turkey, Egypt, Tunisia and Morocco, due in part to their high temperatures and humidity. Samples of southern Europe showed a very high incidence of FUM, at 84% and an average of 1031 ppb. Deoxynivalenol was especially high in cereals, like wheat, barley, etc., with an average of 912 ppb. The risk was linked to the presence of multiple mycotoxins in the same sample. In order of presence in the samples, high concentrations of FUM, ZEN, DON, ochratoxin A (OTA) T-2 and finally AFs were found. These results confirm that multi-mycotoxin contamination is the most common type of contamination. This is a topic of great concern, as co-contaminated samples might exert adverse health effects due to additive/synergistic interactions of the single mycotoxins. In addition to health risks, fungal growth and mycotoxins have a detrimental effect on the quality and the processing performance of cereals. *Fusarium* damage may reduce wheat milling

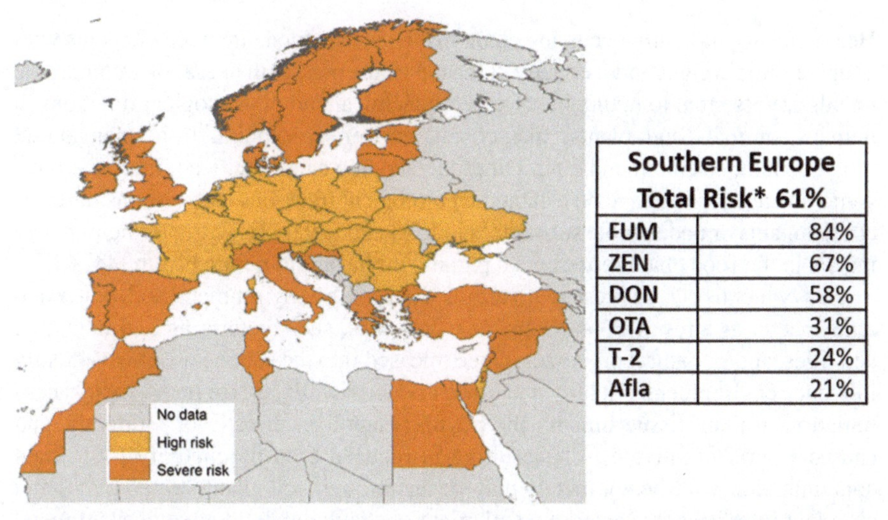

Southern Europe Total Risk* 61%	
FUM	84%
ZEN	67%
DON	58%
OTA	31%
T-2	24%
Afla	21%

*** Risk: percentage of samples testing positive for at least one mycotoxin above the threshold level (ppb).**
Severe risk: > 50% of samples may represent a risk to productivity and disease susceptibility.

Fig. 12.1 Mycotoxin risk level in the Mediterranean countries in 2018 (modified from https://www.romerlabs.com/en/knowledge-center/knowledge-library/articles/news/biomin-world-mycotoxin-survey-2018/)

performance, affect flour yield and flour ash, with a strong negative effect on flour brightness, and baking performance [39–41]. Data on the occurrence of *Fusarium* mycotoxins in durum wheat are quite limited. Visconti and Pascale (2010) found that durum wheat was generally more contaminated than common wheat, but, with very few exception, analysed durum wheat samples were as compliant to the maximum permitted level for DON and ZEN. A further scenario is represented by the climate changes. In the Mediterranean areas, great changes in temperature and rainfall patterns are forecasted. For Southern Europe, climate changes with an increase of 4–5 °C with longer drought periods, resulting in increasing desertification, and decreasing in crop yields have been reported [42]. Climate change will affect mycotoxins in food with higher food safety risks for humans and animals [13]. Particularly, it has been predicted, in Southern Europe, a great impact of climate change on *Aspergillus* and *Fusarium* infection of crops, with increased occurrence and levels of AFs and DON contamination in grains [14]. The Authors conclude that there is evidence for the emergence of new mycotoxin-commodity combinations and new fungal genotypes with higher levels of aggressiveness and altered mycotoxin production.

3.2 Heavy Metal Occurrence

Heavy metals, naturally occurring elements in various food, are trace elements with a high atomic weight and a density at least 5 times greater than that of water. Some metals are essential to maintain various biochemical and physiological functions in humans, animals and plants, like cobalt, copper, chromium, iron, manganese, molybdenum, selenium and zinc. Other metals, like cadmium, lead, mercury, inorganic tin and arsenic, have no established biological functions and are considered as contaminants or undesirable substances in food and feed [43]. The entering of heavy metals in the food chain represents a paramount hazard to public health [44, 45].

Heavy metals can be present at various levels in the environment, soil, water and atmosphere, as a result of anthropogenic activities, such as farming and industrial activities, and/or natural processes. Once released into the soil, heavy metals greatly influence environmental quality and crop productivity. Excessive heavy metal accumulation in plant tissue impairs the phyto-metabolism directly or indirectly, and causes several negative effects resulting in reduced crop productivity [46]. Plant contamination with heavy metals may occur through soil–plant, water–plant, and air–plant interfaces; however, the soil–plant one is the major source of plant metal accumulation. Plants accumulate significant concentrations of heavy metals in their edible and non-edible parts. Most of the heavy metals absorbed is stored in the roots. Many factors affect the content of these elements, like plant species, transpiration rate, soil characteristics, agricultural practices, interactions among elements and contamination from anthropogenic sources [47]. Based on their uptake, plants are classified as accumulator or hyper-accumulator, indicators and excluders [48]. A strong relationship between heavy metals in soil and food crops have been reported

[49]. It is essential to have reliable information on the concentration of heavy metal in soil. Tóth et al. [50, 51]. published detailed maps of heavy metals in the soils of the European Union discussing their implications for food safety. The Authors conclude that, although most of the European agricultural land can be considered adequately safe for food production, there are areas where precautionary measures are needed, like historical and recent industrial and mining ones. In particular, assessment and monitoring are suggested in Western Central Europe, Central Italy, Greece and South-East Ireland.

Cereals, as important components of the human daily diet, must receive considerable attention because of their potential role as vehicles of heavy metals. Specifically, in cereals, heavy metals are not exceeding tolerance levels, but they could present a problem of bio-accumulation [49]. Cereals are considered hyper-accumulators, even though the extent of accumulation depends on the type of heavy metal, as each species can show different reactions [52]. Wheat absorbs these pollutants at different rates depending on the plant organs, with decreasing rates as follows: roots>shoots>grains [53]. Heavy metals undergo a series of processes including bioaccumulation, transformation, and bio-magnification after entering the food chain; which make them very difficult to remove from living organisms [54].

Climatic changes, with reference to the elevated concentration of atmospheric CO_2, enhance the biomass production and metal accumulation in plants and help these to support greater microbial populations and/or protect the microorganisms against the impacts of heavy metals [55]. Predicting how plant–metal interaction responds to altering climatic change is critical to select suitable crop plants that would be able to produce more yields and tolerate multi-stress conditions without accumulating toxic heavy metals for future food security. The effects of global warming, drought or combined climatic stress on plant growth and metal accumulation vary substantially across physical,–chemical and–biological properties of the environment (e.g., soil pH, heavy metal type and its bio-available concentrations, microbial diversity, and interactive effects of climatic factors) and the plant used.

3.3 *Pesticide Occurrence*

The definition of 'pesticide residues' is given in Regulation (EC) No 396/2005 [56]. *"Pesticide residues include active substances and their metabolites and/or breakdown or reaction products, currently or formerly used in plant protection products…., which are present in or on the ones covered by Annex I to this Regulation, including in particular those which may arise as a result of use in plant protection, in veterinary medicine and as a biocide"*. At the website of the European Commission, Food Safety, more precise definitions are given for pesticide, plant protection products and active substances [57].

Pesticides are used in agriculture for different applications, such as controlling insect or fungal infestations or weeds-growth, either to handle immediate damages or to anticipate long-lasting problems. In the EU, the authorisation process requires the evaluation of their toxic pattern to determine acceptable levels of exposure for farm operators and of its persistency on plants for the consumers. However, their excessive use/misuse, especially in the developing countries, their volatility and long-distance transportation eventually result in widespread environmental impacts [58].

Most plant protection products are classified as dangerous to human health, even though the consequences depend on the dose taken. Two methods of pesticides distribution can be used: one consists in the spray application of the products on the plants, while the other one is covering the seeds with the pesticides. Both methods cause the presence of residues in different parts of the plants, due to the natural uptake in their metabolism. In fact it possible to find traces of pesticides in crop residues, pollen, root material, the remains related to the natural decay of the vegetal organism [59]. Parts of these treated plants can enter the food chain and the ecosystem, through their leaching, run-off and degradation. As the result of the combined action of these phenomena, residues of pesticides can be found in the soil and nearby waters, thus affecting also future crops. The concept of residues of pesticides is very important for the safety of agri-food products. The legislation allows that residues of this substances may be present, but it also establishes the maximum tolerance limit. If pesticides are used in the correct doses and manners indicated on the label, thus respecting the safety range, they degrade and, in the harvesting-period, there will be no residues in concentrations exceeding the maximum residue levels (MRLs) imposed by the legislation. An EU pesticide database, where pesticide and active substances according to the Regulations are reported, is available at: https://ec.europa.eu/food/plant/pesticides/eupesticides-db_en.

The last EU report on pesticide residues describes in detail the official control activities carried out for pesticide residues by the EU Member States, Iceland and Norway in 2017 [60]. Overall, 6000 cereal samples were analysed. The MRLs were exceeded in 2.8% of the samples. In Fig. 12.2, a comparison of residues in organic and conventional cereals is reported. In 2018, 4771 samples were collected on conventional cereals and 949 on organic cereals. Results indicate that, in organic farming, health safety could be more guaranteed.

The overall findings indicate that the dietary risk of pesticide residues is very low and unlikely to pose concern for the consumers' health. However, a number of recommendations have been proposed to increase the efficiency of European control systems (e.g. optimizing traceability), thereby continuing to ensure a high level of consumer protection.

4 EU Food Law: The Legislative Framework

The principles of EU legislation on contaminants in food are contained in Regulation 315/93/EEC laying down Community procedures for contaminants in food [61]. In the Article 2 of the Regulation, three important points are addressed: *(1) food containing a contaminant to an amount unacceptable from the public health viewpoint,*

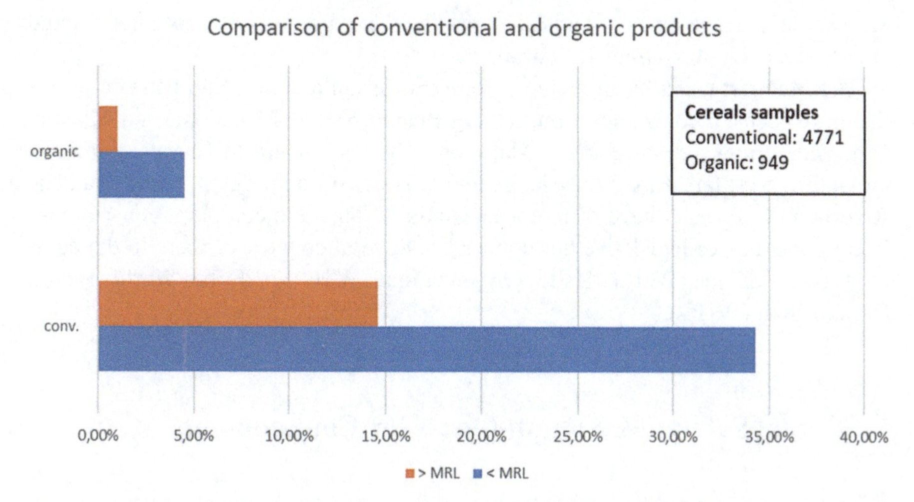

Fig. 12.2 European Union: a comparison of pesticide residues in organic and conventional cereals (data from https://efsa.onlinelibrary.wiley.com/doi/epdf/10.2903/j.efsa.2019.5743)

and in particular at a toxicological level, is not to be placed on the market; (2) contaminant levels must be kept as low as can reasonably be achieved following recommended good working practices; and (3) maximum levels must be set for certain contaminants in order to protect public health.

Since cereal contamination is unavoidable and unpredictable, contaminant regulations have been established worldwide where maximum acceptable limits have been set. However, these regulations greatly vary from country to country. The globalization of the agricultural commodity trade and the lack of legislative harmonization open a wide discussion about the awareness of contaminants entering the food supply chain. A review on EU legislation on cereal safety, with a focus on mycotoxins has been published in 2014 [62]. In the EU, the milestone of food law is Regulation (EC) No 178/2002, laying down the general principles and requirements of food law, establishing the European Food Safety Authority (EFSA) and laying down procedures in matters of food safety, the pursuit of Community policies [16]. The consolidated version is available at https://eur-lex.europa.eu/eli/reg/2002/178/oj. After this legal act, the EU harmonized regulations in an integrated strategy and approach to ensure a high level of food safety, human and animal health and welfare, and plant health from field to consumers. The integrated approach ensures food safety from primary production to placing on the market or export. This approach involves, in a general framework, the development of legislative and other actions to define maximum levels of contaminants, based on science-based risk evaluation and management, to define control systems, to evaluate compliance with EU standards in the food chain, and to manage international relations with third countries and international organizations [63]. Legislation in force, most recent legislative documents, summaries of EU legislation, overview of legislation by subject area, legislation under preparation, overview of current legislative procedures in the European

Parliament, current status of legislative packages and preparatory acts can be found at EUR-Lex site: https://eur-lex.europa.eu.

The Mediterranean basin includes some European and non-European countries that have been called for inclusion, such as Bosnia, Serbia, Macedonia and Albania. Other countries, such as Turkey, Malta and Cyprus, adhere to European laws in terms of food safety. Finally, Syria, Lebanon, Israel, Jordan, Egypt, Libya, Tunisia, Algeria and Morocco have different legislations. This is a complex scenario confirming the lack of legislative harmonisation. Several countries adhere to the common rules of the World Trade Organization (WTO) and the World Health Organization (WHO).

5 Cereal Safety: Maximum Levels for Contaminants

In order to protect public health, maximum levels of contaminants must be introduced. It is well known that it is impossible to fully eliminate crop contamination, therefore maximum levels should be set at a strict level, according to risk management, which is reasonably achievable by following good agricultural and manufacturing practices and taking into account the risks related to food the consumption.

The main reference regulations setting maximum limits, for cereal contaminants are reported in Table 12.1.

The maximum levels for mycotoxins and heavy metals in cereals and cereal-based products are reported in Table 12.2.

In the integrated approach, necessary to ensure food safety "from farm to table", EU legislation on cereal safety for animal feed has been developed. Maximum levels of undesirable substances has been defined for animal feed (Directive 2002/32/EC) [67], Commission Recommendation No 2006/576 [68] and Commission Recommendation of 27 March 2013 on the presence of T-2 and HT-2 toxin in cereals and cereal products [66].

Regarding pesticides, the topic is regulated by the EU by Regulation (EC) No 1107/2009 concerning the placing of plant protection products on the market (Table 12.1) [65]. The maximum residual levels (MRLs) are reported in the Regulation (EC) No 396/2005 (Table 12.2) [59]. The Commission approves active substances, sets maximum residue levels (MRLs) for pesticides. Cereals are included in the list of products of plant and animal origin to which MRLs apply. An EU pesticide database is available to search for pesticide residues MRLs in cereals (http://ec.europa.eu/sanco_pesticides/public/index.cfm). Starting from 2016, the Commission is carrying out a Regulatory Fitness and Performance programme (REFIT): Evaluation of the EU legislation on plant protection products and pesticides residues. The aim is to keep the entire EU pesticide legislation under review in order to assess if the regulations meet the needs of citizens, businesses and public institutions in an efficient manner.

Today, worldwide food production and food trade globalization have a huge impact on the occurrence contaminants and their entry in the food chain. The WTO

Table 12.1 Main EU Regulations related to cereal contaminants

Document	Main points	Current consolidated version (EurLex link)
Council Regulation (EEC) no 315/93 [15]	General community procedures for contaminants in food Definition of contaminant This regulation shall not apply to contaminants which are the subject of more specific community rules	https://eur-lex.europa.eu/legal-content/EN/TXT/?uri=CELEX:01993R0315-20090807
Commission Regulation (EC) No 1881/2006 [64]	Setting maximum levels for certain contaminants in foodstuffs General rules regarding prohibitions on use, mixing and detoxification Specific provisions for cereals Reference regulations for sampling and the analysis for the official control of the maximum levels Rules for monitoring and reporting According to different foodstuffs, maximum levels for Nitrate, Mycotoxins (aflatoxins, ochratoxin A, patulin, deoxynivalenol, zearalenone, fumonisins, T-2 and HT-2 toxins), Metals, Dioxins and PCBs, Polycyclic aromatic hydrocarbons, Melamine and its structural analogues are reported The foodstuffs listed in the Annex of the Regulation cannot be placed on the market where they contain a contaminant at a level exceeding the maximum level set out	https://eur-lex.europa.eu/legal-content/EN/TXT/?uri=CELEX:02006R1881-20200401
Regulation (EC) No 1107/2009 [65]	Concerning the placing of plant protection products on the market and repealing Council Directives 79/117/EEC and 91/414/EEC Setting requirements, conditions for approval and approval procedures	https://eur-lex.europa.eu/eli/reg/2009/1107/2019-12-14

(continued)

Table 12.1 (continued)

Document	Main points	Current consolidated version (EurLex link)
Regulation (EC) no 396/2005 [59]	Maximum residue levels for pesticides set in accordance with this regulation shall not apply to products covered by Annex I intended for export to third countries and treated before export, where it has been established by appropriate evidence that the third country of destination requires or agrees with that particular treatment in order to prevent the introduction of harmful organisms into its territory. Establishment of a list of groups of products for which harmonised MRLs shall apply Extensive list of commodities covered by the regulation (plant and animal origin) (Annex I) Annex II contains the list of definitive MRLs Annex III establishment of temporary MRLs Establishment of a list of active substances for which no MRLs are required (Annex IV) Procedure for applications for MRLs Rules for national and community controls	https://eur-lex.europa.eu/legal-content/EN/TXT/?uri=CELEX:02005R0396-20200131
Commission Recommendation No 2013/165/EU [66]	Recommendation on the presence of T-2 and HT-2 toxin in cereals and cereal products (for the purpose of this recommendation rice is not included in cereals and rice products are not included in cereal products) Samples should be simultaneously analysed for the presence of T-2 and HT-2 and other Fusarium-toxins such as deoxynivalenol, zearalenone and fumonisin B1 + B2 to allow the extent of co-occurrence to be assessed. Setting of indicative levels for the sum of T-2 and HT-2 from which onwards/above which investigations should be performed, certainly in case of repetitive findings	https://eur-lex.europa.eu/eli/reco/2013/165/oj

recognizes the standards, guidelines, and recommendations established by the *Codex Alimentarius* Commission as the basis for food safety harmonization at the international level. Legislation on food safety has been adopted worldwide, but food is governed by a complexity of laws and regulations and different approaches for ensuring food safety can be adopted. The result is a lack of harmonization that opens a discussion on food safety increasing health and economic problems worldwide with contaminants entering the food chain.

Table 12.2 Maximum levels for mycotoxins and heavy metals in cereals and cereal-based products for food [64, 66]

Contaminant	Cereal type	Maximum levels (μg/kg)
Mycotoxins		
Aflatoxins	All cereals and all products derived from cereals, including processed cereal products, with the exception of foodstuffs listed below	B1: 2 Sum B1, B2, G1, G2: 4
	Maize and rice to be subjected to sorting or other physical treatment before human consumption or use as an ingredient in foodstuffs	B1: 5 Sum B1, B2, G1, G2: 10
	Processed cereal-based foods and baby foods for infants and young children	B1: 0.1
	Dietary foods for special medical purposes intended specifically for infants	B1: 0.1
Ochratoxin A	Unprocessed cereals	5
	All products derived from unprocessed cereals, including processed cereal products and cereals intended for direct human consumption with the exception of foodstuffs listed below	3
	Processed cereal-based foods and baby foods for infants and young children	0.5
	Dietary foods for special medical purposes intended specifically for infants	0.5
	Wheat gluten not sold directly to the consumer	8
Deoxynivalenol	Unprocessed cereals other than durum wheat, oats and maize	1250
	Unprocessed durum wheat and oats	1750
	Unprocessed maize, with the exception of unprocessed maize intended to be processed by wet milling	1750
	Cereals intended for direct human consumption, cereal flour, bran and germ as end product marketed for direct human consumption, with the exception of foodstuffs listed below	750
	Pasta (dry)	750
	Bread (including small bakery wares), pastries, biscuits, cereal snacks and breakfast cereals	500
	Processed cereal-based foods and baby foods for infants and young children	200
	Milling fractions of maize with particle size >500 μm falling within CN code 1103 13 or 1103 20 40 and other maize milling products with particle size >500 μm not used for direct human consumption falling within CN code 1904 10 10	750

(continued)

Table 12.2 (continued)

Contaminant	Cereal type	Maximum levels (μg/kg)
	Milling fractions of maize with particle size ≤500 μm falling within CN code 1102 20 and other maize milling products with particle size ≤500 μm not used for direct human consumption falling within CN code 1904 10 10	1250
Zearalenone	Unprocessed cereals other than maize	100
	Unprocessed maize with the exception of unprocessed maize intended to be processed by wet milling	350
	Cereals intended for direct human consumption, cereal flour, bran and germ as end product marketed for direct human consumption, with the exception of foodstuffs listed below	75
	Refined maize oil	400
	Bread (including small bakery wares), pastries, biscuits, cereal snacks and breakfast cereals, excluding maize-snacks and maize-based breakfast cereals	50
	Maize intended for direct human consumption, maize-based snacks and maize-based breakfast cereals	100
	Processed cereal-based foods (excluding processed maize-based foods) and baby foods for infants and young children	20
	Processed maize-based foods for infants and young children	20
	Milling fractions of maize with particle size >500 μm falling within CN code 1103 13 or 1103 20 40 and other maize milling products with particle size >500 μm not used for direct human consumption falling within CN code 1904 10 10	200
	Milling fractions of maize with particle size ≤500 μm falling within CN code 1102 20 and other maize milling products with particle size ≤500 μm not used for direct human consumption falling within CN code 1904 10 10	300
Fumonisins (Sum B1 and B2)	Unprocessed maize, with the exception of unprocessed maize intended to be processed by wet milling	4000
	Maize intended for direct human consumption, maize-based foods for direct human consumption, with the exception of foodstuffs listed below	1000
	Maize-based breakfast cereals and maize-based snacks	800
	Processed maize-based foods and baby foods for infants and young children	200
	Milling fractions of maize with particle size >500 μm falling within CN code 1103 13 or 1103 20 40 and other maize milling products with particle size >500 μm not used for direct human consumption falling within CN code 1904 10 10	1400

(continued)

Table 12.2 (continued)

Contaminant	Cereal type	Maximum levels (μg/kg)
	Milling fractions of maize with particle size ≤500 μm falling within CN code 1102 20 and other maize milling products with particle size ≤500 μm not used for direct human consumption falling within CN code 1904 10 10	2000
		Indicative levels, μg/kg
T-2 and HT-2 toxin (Sum of T-2 and HT-2 toxin)	*Unprocessed cereals:* Barley (including malting barley) and maize Oats (with husk) Wheat, rye and other cereals	200 1000 100
	Cereal grains for direct human consumption: Oats Maize Other cereals	200 100 50
	Cereal products for human consumption: Oat bran and flaked oats Cereal bran except oat bran, oat milling products other than oat bran and flaked oats, and maize milling products Other cereal milling products Breakfast cereals including formed cereal flakes Bread (including small bakery wares), pastries, biscuits, cereal snacks, pasta Cereal-based foods for infants and young children	200 100 50 75 25 15
Heavy metals		Maximum levels, mg/kg wet weight
Lead	Processed cereal-based foods and baby foods for infants and young children	0.050
	Cereals and pulses	0.20
Cadmium	Cereal grains excluding wheat and rice	0.10
	Wheat grains, rice grains Wheat bran and wheat germ for direct consumption	0.20
	Processed cereal-based foods and baby foods for infants and young children	0.040

6 Cereal Contaminants: Prevention and Reduction

As already stated, despite efforts, cereal contamination is unavoidable and unpredictable. The overall strategy for contamination mitigation is based on prevention. This means that, in order to reduce the presence of contaminants, several measures should be adopted at all relevant stages of production, processing and distribution, particularly at the level of primary production. From a legislative perspective, this topic is very complex. People who are interested should check the EurLex site.

Cheli et al. (2014) [62] reported the main Recommendations for cereals. Examples of specific acts for cereals are represented by Commission Recommendation 2006/583/EC, stating the principles for the prevention and reduction of *Fusarium* toxin [69] and Directive 2004/107/EC [70] establishing a target value for the concentration of arsenic, cadmium, nickel and benzo(a)pyrene in ambient air so as to avoid, prevent or reduce harmful effects of arsenic, cadmium, nickel and polycyclic aromatic hydrocarbons on human health and the environment as a whole.

In any case, a conclusive statement on the quality of a cereal lot can only be made when its contaminant content is known to a sufficient degree of accuracy.

7 Cereal Safety: Sampling and Analysis for Official Controls

The Regulation (EU) 2017/625 regulates the official controls of food and feed [71]. According to this Regulation, *"official controls means activities performed by the competent authorities, or by the delegated bodies or the natural persons to which certain official control tasks have been delegated in accordance with this Regulation"*, in order to verify compliance with feed and food law. This Regulation has been recently implemented by the Commission Implementing Regulation (EU) 2019/1793 on the temporary increase of official controls and emergency measures governing the entry into the Union of certain goods from certain third countries [72].

Setting maximum limits for contaminants and procedures for official controls asks for analysis. Sampling is the critical step to obtain reliable and accurate results on contaminant content, to decide the compliance of a product according to a limit and what to do with lots that may be contaminated [73]. A sampling plan and a test procedure must be defined, and an effective sampling procedure for cereal contamination detection and quantification represents a complex challenge for operators [74]. The main topic is the distribution of the contaminated particles. According to the distribution within a lot, food contaminants can be divided into two groups: uniformly distributed (pesticides, additives, heavy metals, PCBs, dioxins, etc.) and non-uniformly distributed (natural toxins, GMOs, etc.) [75].

There are several Regulations covering the topic of food sampling and analysis for official controls (Table 12.3).

Each document gives precise details regarding the methods of sampling for each food, acceptance parameters, the criteria for sample preparation, the analytical performance criteria of the methods of analysis used for the official controls, and the criteria for reporting and interpretation of the results.

Table 12.3 Regulation for food sampling and analysis for mycotoxins, heavy metals and pesticides

Document	Topics	Consolidated version (EurLex link)
Commission Regulation (EC) No 401/2006 [76].	Laying down the methods of sampling and analysis for the official control of the levels of mycotoxins in foodstuffs	https://eur-lex.europa.eu/legal-content/EN/TXT/?uri=CELEX:02006R0401-20140701
Commission Regulation (EC) No333/2007 [77].	Laying down the methods of sampling and analysis for the control of the levels of trace elements and processing contaminants in foodstuffs lead, cadmium, mercury, inorganic tin, inorganic arsenic are included in this regulation	https://eur-lex.europa.eu/legal-content/EN/TXT/?uri=CELEX:02007R0333-20191214
Commission Directive 2002/63/EC [78].	Establishing Community methods of sampling for the official control of pesticide residues in and on products of plant and animal origin	https://eur-lex.europa.eu/eli/dir/2002/63/oj

8 Conclusions

Cereals and highly processed final products from cereals are the world's most important source of food and are used in cuisine for human nutrition. In the Mediterranean area, cereals, vegetables, and citrus fruits account for over 85% of the total agricultural production. From a safety perspective, cereals must receive considerable attention because of their potential role as vehicles of contaminants in the food chain. Cereal contamination with mycotoxins, heavy metals and pesticide residues are the main potential health threat for humans. The European Union has established the most comprehensive regulations for the safety of cereal and cereal derived products to protect consumer health and facilitate world trade. Globalization and the increased global trade associated with cereal production pose the need for EU legislation to face with the different legislative framework of other countries. A lack of a legislative harmonization is an important point to consider in a worldwide discussion regarding the managing risk and regulations in cereal and food security and safety governance.

The legislation in the field of cereal safety is continuously evolving prompted by different factors, such as the availability of new scientific information becoming available in the years, the activity of the EFSA's scientific committees, the results from the monitoring activity. Regarding mycotoxins, future attention should be paid not only to the regulated mycotoxins, but also to the presence of mycotoxin co-contamination, modified and emerging mycotoxins, and secondary metabolites [79–81]. Science-based information concerning the occurrence and the additive/synergistic effects of mycotoxins on human health still need to be evaluated. The availability of this data may support risk management and regulatory bodies to reduce human exposure to dangerous amounts of mycotoxins and to revise legislative limits.

The high variability in contaminant occurrence in cereals and cereal products begs increased awareness and ongoing surveillance. In this scenario, regular, economical and straightforward cereal sampling and testing with regard to a rapid and accurate diagnosis of food quality are needed. The development of rapid methods for use in the field represents a future challenge.

References

1. Thielecke F, Nugent AP. Contaminants in grain—a major risk for whole grain safety? Nutrients. 2018;10:1213–36.
2. FAO. The state of food and agriculture, vol. 2014. Rome: FAO - Food and Agriculture Organization of the United Nations; 2014.
3. FAO. The state of food and agriculture, vol. 2019. Rome: FAO - Food and Agriculture Organization of the United Nations; 2019.
4. FAO 2020 FAO cereal supply and demand brief. http://www.fao.org/worldfoodsituation/csdb/en/
5. Tang Y, Lu L, Zhao W, Wang J. Rapid detection techniques for biological and chemical contamination in food: a review. Int J Food Eng. 2009;5. https://doi.org/10.2202/1556-3758.1744.
6. Pinotti L, Ottoboni M, Giromini C, Dell'Orto V, Cheli F. Mycotoxin contamination in the EU feed supply chain: a focus on cereal byproducts. Toxins (Basel). 2016;8:45–68.
7. UNEP/MAP/BP/RAC. Mediterranean Action Plan (MAP) - Introduction to the Mediterranean Basin; 2009.
8. COCERAL. European Association of cereal, rice, feedstuffs, oilseeds, olive oil, oils and fats and agrosupply trade. 2020. Crop forecasts. http://www.coceral.com/web/june%20 2020/1011306087/list1187970814/f1.html.
9. Eurostat 2009 Eurostat, Statistic in focus. 2009. https://ec.europa.eu/eurostat/web/products-statistics-in-focus/-/KS-SF-13-024.
10. Wollenweber B, Porter JR, Lubberstedt T. Need for multidisciplinary research towards a second green revolution. Curr Opin Plant Biol. 2005;8:337–41.
11. Jacobsen D, Milner AM, Brown LE. Biodiversity under threat in glacier-fed river systems. Nat Clim Chang. 2012;2:361–4.
12. Wang J, Vanga SK, Saxena R, Orsat V, Raghavan V. Effect of climate change on the yield of cereal crops: a review. J Clim. 2018;6:41–60.
13. Paterson RRM, Lima N. Further mycotoxin effects from climate change: a review. Food Res Int. 2011;44:2555–66.
14. Moretti A, Pascale M, Logrieco AF. Mycotoxin risks under a climate change scenario in Europe. Trends Food Sci Technol. 2019;84:38–40.
15. European Commission 1993 Regulation (EEC) No 315/93 of 8 February 1993 laying down Community procedures for contaminants in food. Off J Eur Union 1993 L 37:1–3. Consolidated version at: https://eur-lex.europa.eu/legal-content/EN/TXT/?uri=CELEX:01993R0315-20090807 Accessed 28/07/20.
16. European Commission 2002a Regulation (EC) No 178/2002 of the European Parliament and of the Council of 28 January 2002 laying down the general principles and requirements of food law, establishing the European Food Safety Authority and laying down procedures in matters of food safety. Off J Eur Union L 31/1–24. Consolidated version at http://data.europa.eu/eli/reg/2002/178/2019-07-26 Accessed 28/07/20.
17. RASFF 2018 RASFF - rapid alert system for food and feed. Annu Rep. 2018. https://ec.europa.eu/food/sites/food/files/safety/docs/cs_contaminants_factsheet_en.pdf.

18. European Commission. Health & Consumer Protection. Directorate General. 2008. Food contaminants. https://ec.europa.eu/food/sites/food/files/safety/docs/cs_contaminants_factsheet_en.pdf.
19. Cheli F, Campagnoli A, Ventura V, Brera C, Berdini C, Palmaccio E, Dell'Orto V. Effects of industrial processing on the distributions of deoxynivalenol, cadmium and lead in durum wheat milling fractions. LWT-Food Sci Technol. 2010;43:1050–7.
20. Schaarschmidt S, Fauhl-Hassek C. The fate of mycotoxins during the processing of wheat for human consumption. Compr Rev Food Sci. 2018;17:556–93.
21. Aureli G, D'Egidio MG. Efficacy of debranning on lowering of deoxynivalenol (DON) level in manufacturing processes of durum wheat. Tecnica Molitoria. 2007;58:729–33.
22. Bullerman LB, Bianchini A. Stability of mycotoxins during food processing. Int J Food Microbiol. 2007;119:140–6.
23. Visconti A, Pascale M. An overview on Fusarium mycotoxins in the durum wheat pasta production chain. Cereal Chem. 2010;87:21–7.
24. Cheli F, Pinotti L, Rossi L, Dell'Orto V. Effect of milling procedures on mycotoxin distribution in wheat fractions: a review. LWT-Food Sci Technol. 2013;54:307–14.
25. Karlovsky P, Suman M, Berthiller F, De Meester J, Eisenbrand J, Perrin I, Oswald IP, Speijers G, Chiodini A, Recker T, Dussort P. Impact of food processing and detoxification treatments on mycotoxin contamination. Mycotoxin Res. 2016;32:179–205.
26. D'Mello JPF, Placinta CM, Macdonald AMC. Fusarium mycotoxins: a review of global implications for animal health, welfare and productivity. Anim Feed Sci Technol. 1999;80:183–205.
27. Wild CP, Gong YY. Mycotoxins and human disease: a largely ignored global health issue. Carcinogenesis. 2010;31:71–82.
28. Fink-Gremmels J. Mycotoxins: their implications for human and animal health. Vet Q. 1999;21:115–20.
29. Hussein HS, Brasel JM. Toxicity, metabolism, and impact of mycotoxins on humans and animals. Toxicology. 2001;167:101–34.
30. Streit E, Schatzmayr G, Tassis P, Tzika E, Marin D, Taranu I, Tabuc C, Nicolau A, Aprodu I, Puel O. Current situation of mycotoxin contamination and co-occurrence in animal feed—focus on Europe. Toxins. 2012;4:788–809.
31. Streit E, Naehrer K, Rodrigues I, Schatzmayr G. Mycotoxin occurrence in feed and feed raw materials worldwide: long-term analysis with special focus on Europe and Asia. J Sci Food Agric. 2013;93:2892–9.
32. Lee HJ, Ryu D. Worldwide occurrence of mycotoxins in cereals and cereal-derived food products: public health perspectives of their co-occurrence. J Agric Food Chem. 2017;65:7034–51.
33. Marin S, Ramos AJ, Cano-Sancho G, Sanchis V. Mycotoxins: occurrence, toxicology, and exposure assessment. Food Chem Toxicol. 2013;60:218–37.
34. Schatzmayr G, Streit E. Global occurrence of mycotoxins in the food and feed chain: facts and figures. World Mycotoxin J. 2013;6:213–22.
35. Agriopoulou S, Stamatelopoulou E, Varzakas T. Advances in occurrence, importance, and mycotoxin control strategies: prevention and detoxification in foods. Foods. 2020;9:137–84.
36. Cotty PJ, Jaime-Garcia R. Effect of climate on aflatoxin producing fungi and aflatoxin contamination. Int J Food Microbiol. 2007;119:109–15.
37. Munkvold GP. Cultural and genetic approaches to managing mycotoxins in maize. Annu Rev Phytopathol. 2003;41:99–116.
38. Reyneri A. The role of climatic condition on micotoxin production in cereal. Vet Res Commun. 2006;30:87–92.
39. Wang J.H, Wieser H, Pawelzik E, Weinert J, Keutgen AJ, Wolf GA 2005 Impact of the fungal protease produced by Fusarium culmorum on the protein quality and breadmaking properties of winter wheat. Eur Food Res Technol 220:552–559.
40. Lancova K, Hajslova J, Kostelanska M, Kohoutkova J, Nedelnik J, Moravcova H, Vanova M. Fate of trichothecene mycotoxins during the processing: milling and baking. Food Addit Contam A. 2008;25:650–9.

41. Siuda R, Grabowski A, Lenc L, Ralcewicz M, Spychaj-Fabisiak E. Influence of the degree of fusariosis on technological traits of wheat grain. Int J Food Sci Technol. 2010;45:2596–604.
42. Medina A, Rodriguez A, Magan N. Effect of climate change on Aspergillus flavus and aflatoxin B1 production. Front Microbiol. 2014;5(article 348):1–7.
43. Hejna M, Gottardo D, Baldi A, Dell'Orto V, Cheli F, Zaninelli M, Rossi L. Review: nutritional ecology of heavy metals. Animal. 2018;12:2156–70.
44. Bhargava A, Carmona FF, Bhargava M, Srivastava S. Approaches for enhanced phytoextraction of heavy metals. J Environ Manag. 2012;105:103–20.
45. Govind P, Madhuri S. Heavy metals causing toxicity in animals and fishes. Res J Anim Vet Fish Sci. 2014;2:17–23.
46. Gupta N, Yadav KK, Kumar V, Kumar S, Chadd RP, Kumar A. Trace elements in soil-vegetables interface: translocation, bioaccumulation, toxicity and amelioration – a review. Sci Total Environ. 2019;651:2927–42.
47. Shahid M, Khalid S, Abbas G, Shahid N, Nadeem M, Sabir M, Aslam M, Dumat C. Heavy metal stress and crop productivity. In: Hakeem KR, editor. Crop production and global environmental issues. 1st ed. Berlin: Springer International Publishing; 2015. p. 1–25.
48. Baker AJM, Walker PL. Ecophysiology of metal uptake by tolerant plants. In: Shaw AJ, editor. Heavy metal tolerance in plants: evolutionary aspects. Florida: Boca Raton; 1990. p. 155–77.
49. Gall JE, Boyd RS, Rajakaruma N. Transfer of heavy metals through terrestrial food webs: a review. Environ Monit Assess. 2015;187:201.
50. Tóth G, Hermann T, Da Silva MR, Montanarella L. Heavy metals in agricultural soils of the European Union with implications for food safety. Environ Int. 2016a;88:299–309.
51. Tóth G, Hermann T, Szatmári G, Pásztor L. Maps of heavy metals in the soils of the European Union and proposed priority areas for detailed assessment. Sci Total Environ. 2016b;565:1054–62.
52. Aladesanmi OT, Oroboade JG, Osisiogu CP, Osewole AO. Bioaccumulation factor of selected heavy metal in *Zea mais*. J Health Pollut. 2019;9:191–207.
53. Boussen S, Soubrand M, Bril H, Ouerfelli K, Abdeljaouad S. Transfer of lead, zinc and cadmium from mine tailings to wheat (Triticum aestivum) in carbonated Mediterranean (northern Tunisia) soils. Geoderma. 2013;192:227–36.
54. Singh R, Gautam N, Mishra A, Gupta R. Heavy metals and living systems: an overview. Indian J Pharmacol. 2011;43:246–53.
55. Rajkumar M, Prasad MNV, Swaminathan S, Freitas H. Climate change driven plant–metal–microbe interactions. Environ Int. 2013;53:74–86.
56. European Commission 2005 Regulation (EC) No 396/2005 of the European Parliament and of the Council of 23 February 2005 on maximum residue levels of pesticides in or on food and feed of plant and animal origin and amending Council Directive 91/414/EEC. Off J Eur Union 2005 L 70:1–16. Consolidated version at https://eur-lex.europa.eu/LexUriServ/LexUriServ.do?uri=CONSLEG:2005R0396:20121026:EN:PDF Accessed 28/07/20.
57. European Commission, Food Safety. https://ec.europa.eu/food/plant/pesticides_en
58. Ecobichon DJ. Pesticide use in developing countries. Toxicology. 2001;160:27–33.
59. Bundschuh R, Bundschuh M, Otto M, Schulz RM. Food-related exposure to systemic pesticides and pesticides from transgenic plants: evaluation of aquatic test strategies. Environ Sci Eur. 2019;31:87–99.
60. EFSA. The 2017 European Union report on pesticide residues in food. EFSA J. 2019;17:5743–894.
61. https://eur-lex.europa.eu/legal-content/EN/TXT/?uri=CELEX:01993R0315-20090807.
62. Cheli F, Battaglia D, Gallo R, Dell'Orto V. Review EU legislation on cereal safety: an update with a focus on mycotoxins. Food Control. 2014;37:315–25.
63. Arvanitoyannis IS, Choreftaki S, Tserkezou P. An update of EU legislation (directives and regulations) on food-related issues (safety, hygiene, packaging, technology, GMOs, additives, radiation, labelling): presentation and comments. Int J Food Sci Technol. 2005;40:1021–112.

64. European Commission 2006a Commission Regulation (EC) No 1881/2006 of 19 December 2006 setting maximum levels for certain contaminants in foodstuffs. Off J Eur Union, 2006, L 364:5–24. Consolidated version at http://data.europa.eu/eli/reg/2006/1881/2020-04-01 Accessed 28/07/20.
65. European Commission 2009 Regulation (EC) No 1107/2009 of the European Parliament and of the Council of 21 October 2009 concerning the placing of plant protection products on the market and repealing Council Directives 79/117/EEC and 91/414/EEC. Off J Eur Union 2009 L 309:1–50. Consolidated version at http://data.europa.eu/eli/reg/2009/1107/2019-12-14 Accessed 28/07/20.
66. European Commission 2013 Commission Recommendation of 27 March 2013 on the presence of T-2 and HT-2 toxin in cereals and cereal products. Off J Eur Union 2013 L 91:12–15. Consolidated version at http://data.europa.eu/eli/reco/2013/165/oj Accessed 28/07/20.
67. European Commission 2002b Directive 2002/32/EC of the European Parliament and of the Council of 7 May 2002 on undesirable substances in animal feed. Off J Eur Union 2002 L 140:10–22. Consolidated version at http://data.europa.eu/eli/dir/2002/32/2019-11-28 Accessed 28/07/20.
68. European Commission 2006b Commission Recommendation of 17 August 2006 on the presence of deoxynivalenol, zearalenone, ochratoxin A, T-2 and HT-2 and fumonisins in products intended for animal feeding (2006/576/EC). Off J Eur Union 2006 L 229:7–9. Consolidated version at http://data.europa.eu/eli/reco/2006/576/2016-08-02 Accessed 28/07/20.
69. European Commission 2006c Commission Recommendation of 17 August 2006 on the prevention and reduction of Fusarium toxins in cereals and cereal products. Off J Eur Union 2006 L 234:35–40. Consolidated version at http://data.europa.eu/eli/reco/2006/583/oj Accessed 28/07/20.
70. European Commission 2004 Directive 2004/107/EC of the European Parliament and of the Council of 15 December 2004 relating to arsenic, cadmium, mercury, nickel and polycyclic aromatic hydrocarbons in ambient air. Off J Eur Union 2005 L 23:3–48. Consolidated version at http://data.europa.eu/eli/dir/2004/107/2015-09-18 Accessed 28/07/20.
71. European Commission 2017 Regulation (EU) 2017/625 of the European Parliament and of the Council of 15 March 2017 on official controls and other official activities performed to ensure the application of food and feed law, rules on animal health and welfare, plant health and plant protection products, amending Regulations (EC) No 999/2001, (EC) No 396/2005, (EC) No 1069/2009, (EC) No 1107/2009, (EU) No 1151/2012, (EU) No 652/2014, (EU) 2016/429 and (EU) 2016/2031 of the European Parliament and of the Council, Council Regulations (EC) No 1/2005 and (EC) No 1099/2009 and Council Directives 98/58/EC, 1999/74/EC, 2007/43/EC, 2008/119/EC and 2008/120/EC, and repealing Regulations (EC) No 854/2004 and (EC) No 882/2004 of the European Parliament and of the Council, Council Directives 89/608/EEC, 89/662/EEC, 90/425/EEC, 91/496/EEC, 96/23/EC, 96/93/EC and 97/78/EC and Council Decision 92/438/EEC (Official Controls Regulation). Off J Eur Union 2017 L 95:1–142. Consolidated version at http://data.europa.eu/eli/reg/2017/625/2019-12-14 Accessed 28/07/20.
72. European Commission 2019 Commission Implementing Regulation (EU) 2019/1793 of 22 October 2019 on the temporary increase of official controls and emergency measures governing the entry into the Union of certain goods from certain third countries implementing Regulations (EU) 2017/625 and (EC) No 178/2002 of the European Parliament and of the Council and repealing Commission Regulations (EC) No 669/2009, (EU) No 884/2014, (EU) 2015/175, (EU) 2017/186 and (EU) 2018/1660. Off J Eur Union 2019 L 277:89–129. Consolidated version at http://data.europa.eu/eli/reg_impl/2019/1793/2020-05-27 Accessed 28/07/20.
73. van Egmond HP, Schothorst RC, Jonker MA. Regulations relating to mycotoxins in food. Perspectives in a global and European context. Rev Anal Bioanal Chem. 2007;389:147–57.
74. Cheli F, Campagnoli A, Pinotti L, Fusi E, Dell'Orto V. Sampling feed for mycotoxins: acquiring knowledge from food. Ital J Anim Sci. 2009;8:5–22.

75. Cheli F, Campagnoli A, Pinotti L, Dell'Orto V. Rapid methods as analytical tools for food and feed contaminant evaluation: methodological implications for mycotoxin analysis in cereals. In: Aladjadjiya A, editor. Food production - approaches. Rijeka: Challenges and Tasks Intech; 2012. p. 185–204.
76. European Commission 2006d Commission Regulation (EC) No 401/2006 of 23 February 2006 laying down the methods of sampling and analysis for the official control of the levels of mycotoxins in foodstuffs. Off J Eur Union 2006 L 70:12–34. Consolidated version at http://data.europa.eu/eli/reg/2006/401/2014-07-01 Accessed 28/07/20.
77. European Commission 2007 Commission Regulation (EC) No 333/2007 of 28 March 2007 laying down the methods of sampling and analysis for the control of the levels of trace elements and processing contaminants in foodstuffs. Off J Eur Union 2007 L 88:29–38. Consolidated version at http://data.europa.eu/eli/reg/2007/333/2019-12-14 Accessed 28/07/20.
78. European Commission 2002c Commission Directive 2002/63/EC of 11 July 2002 establishing Community methods of sampling for the official control of pesticide residues in and on products of plant and animal origin and repealing Directive 79/700/EEC. Off J Eur Union 2002 L 187:30–43. Consolidated version at http://data.europa.eu/eli/dir/2002/63/oj Accessed 28/07/20.
79. De Boevre M, Graniczkowska K, De Saeger S. Metabolism of modified mycotoxins studied through *in vitro* and *in vivo* models: an overview. Toxicol Lett. 2015;233:24–8.
80. Gruber-Dorninger C, Novak B, Nagl V, Berthiller F. Emerging mycotoxins: beyond traditionally determined food contaminants. J Agric Food Chem. 2017;65:7052–70.
81. Suman M. Fate of free and modified forms of mycotoxins during food processing. Toxins. 2020;12:448–50.

Index

F. Boukid (ed.), *Cereal-Based Foodstuffs: The Backbone of Mediterranean Cuisine*, https://doi.org/10.1007/978-3-030-69228-5

D

H

X

Y

Z